TRAUMA

TRAUMA
A GENEALOGY

Ruth Leys

THE UNIVERSITY OF CHICAGO PRESS
CHICAGO AND LONDON

The University of Chicago Press, Chicago 60637
The University of Chicago Press, Ltd., London

Printed in the United States of America

09 08 07 06 05 04 03   2 3 4 5

ISBN: 0-226-47765-7 (cloth)
ISBN: 0-226-47766-5 (paper)

Library of Congress Cataloging-in-Publication Data

Leys, Ruth.
    Trauma : a genealogy / Ruth Leys
       p.   cm.
    Includes bibliographical references and index.
    ISBN 0-226-47765-7 (cloth : alk. paper) — ISBN 0-226-47766-5 (paper : alk. paper)
    1. Psychic trauma. 2. Traumatic neuroses. I. Title.
    RC552.T7 L49 2000
    616.85'21—dc21   99-052681

To Michael and Anna

CONTENTS

CONTENTS

ACKNOWLEDGMENTS

For their support, criticism, and help I want especially to thank Mikkel Borch-Jacobsen, Donald Fleming, Ian Hacking, Neil Hertz, Guy McKhann, Toril Moi, Jacqueline Rose, Dorothy Ross, François Roustang, Joan Scott, and Sonu Shamdasani.

I owe a special debt of thanks to two friends, Frances Ferguson and Walter Benn Michaels, both of whom gave me valuable advice at several crucial stages, critical comments on various drafts, and unstinting encouragement from start to finish.

Portions of this book were previously published in somewhat different forms. Chapter 2 appeared as "The Real Miss Beauchamp: Imitation and the Subject of Gender," in *Feminists Theorize the Political*, ed. Judith Butler and Joan Scott, pp. 167–214 (copyright 1992; reproduced by permission of Routledge, Inc.). Chapter 3 appeared as "Traumatic Cures: Shell-Shock, Janet, and the Question of Memory," in *Critical Inquiry* 20 (Summer 1994): 623–62. Parts of "Death Masks: Kardiner and Ferenczi on Psychic Trauma," which appeared in *Representations*, no. 53 (Winter 1996): 44–73 (copyright 1996 by the Regents of the University of California), were used in chapters 1 and 4. My thanks to Routledge Press, the University of Chicago Press, and the University of California Press for permission to use revised versions of these essays here.

Grants from the following sources greatly facilitated my research: The American Council of Learned Societies, Ford Fellowship, 1988–89; National Institutes of Health, National Library of Medicine Grant 1995–97; Wellcome Research Travel Grant, Borroughs Wellcome Fund, May 1997.

I am grateful to the staffs of the libraries of the Milton S. Eisenhower

Library, Johns Hopkins University; the Wellcome Institute for the History of Medicine, London; and the William Alanson White Institute, New York, for their assistance. My thanks also to the Wellcome Institute for the History of Medicine for permission to examine William Sargant's case records and papers, and to the William Alanson White Institute for permission to examine Clara Thompson's papers.

Finally, I wish to acknowledge two others: Michael Fried, first reader, creative critic, and generous companion; and my daughter Anna, whose presence in my life has been a source of such joy that it is barely imaginable how I ever lived without her. I dedicate this book to them.

Introduction

In the spring of 1998 Elizabeth Rubin, in an article in the *New Yorker*, described what had recently happened to a group of young Ugandan girls who had been abducted by a guerilla group, the Lord's Resistance Army (L.R.A.), in order to force them to serve as "wives" and fighters in its war against the Ugandan army. The girls were put through a "hideous initiation" to frighten them away from deserting and revealing L.R.A. hideouts, and to inure them to violence. About a week after the abduction, the girls were smeared with holy oils and white ash "in a kind of L.R.A. baptism"; then the guerilla leader ordered them to finish off a job begun by veteran rebels: "hacking to death, with hoes, axes, and branches, a recently kidnapped girl who had been caught escaping. The students protested, were beaten, and then did as they were told." Rubin reported that since 1988 the L.R.A. had stolen about twelve thousand Ugandan girls and boys for such purposes. Between three and five thousand of those numbed and brutalized children had managed to escape, through their own efforts or by being captured in battle. Once returned to safety, they were sent to two trauma centers for counseling and treatment. "The goals of the trauma centers are modest," Rubin observed, "and therefore fairly realizable. Group therapy, game playing, reenactments of life in the bush, traditional dancing, drawing are all designed to teach the kids to forget. It is a challenging concept: remembering to forget."[1]

In the same spring of 1998 Americans anticipated the trial date of what

---

1. Elizabeth Rubin, "Our Children Are Killing Us," *New Yorker*, 23 March 1998, 58, 63.

1

promised to be the most notorious sexual harassment case in the history of the United States. The lawyers for the claimant, Paula Jones, asserted on the basis of expert testimony that, as a result of the trauma of her alleged sexual harassment by President Clinton, Jones now suffered from post-traumatic stress with long-term symptoms of anxiety, intrusive thoughts and memories, and sexual aversion.[2]

Between them these examples illustrate the spectrum of issues raised by the concept of psychic trauma in our time. On the one hand, there is the absolute indispensability of the concept for understanding the psychic harms associated with certain central experiences of the twentieth century, crucially the Holocaust but also including other appalling outrages of the kind experienced by the kidnapped children of Uganda. On the other hand, it is hard not to feel that the concept of trauma has become debased currency when it is applied both to truly horrible events *and* to something as dubious as the long-term harm to Paula Jones.

Jones's claims for damages appealed to a model of trauma institutionalized in the concept of post-traumatic stress disorder (PTSD), an ailment first officially recognized by the American Psychiatric Association in 1980. Post-traumatic stress disorder is fundamentally a disorder of memory. The idea is that, owing to the emotions of terror and surprise caused by certain events, the mind is split or dissociated: it is unable to register the wound to the psyche because the ordinary mechanisms of awareness and cognition are destroyed. As a result, the victim is unable to recollect and integrate the hurtful experience in normal consciousness; instead, she is haunted or possessed by intrusive traumatic memories. The experience of the trauma, fixed or frozen in time, refuses to be represented *as* past, but is perpetually reexperienced in a painful, dissociated, traumatic present. All the symptoms characteristic of PTSD—flashbacks, nightmares and other reexperiences, emotional numbing, depression, guilt, autonomic arousal, explosive violence or tendency to hypervigilance—are thought to be the result of this fundamental mental dissociation. (On this model, if Paula Jones did suffer from post-traumatic stress, as she claimed, she ought to have been incapable of consciously testifying to her traumatic experience; rather, she should only have been able to repeat it in the mode of a compulsive and repetitive acting out.) Accordingly, the restoration of memory through technologies designed to get the patient to remember by restoring the "pathogenic se-

2. *New York Times*, Saturday, 21 March 1998.

cret"[3] to awareness is one of the major goals of therapy. From the perspective of such an approach to trauma, the more modest goal of forgetting embraced by the therapists of the traumatized children in Uganda may seem irrelevant if not actually harmful. But like the aim of remembering, that of <u>forgetting has a history it is my purpose to explore</u>.

Experts in PTSD often emphasize the vicissitudes of the trauma concept by underscoring the waxing and waning of interest in trauma in the course of more than a century. But they also present PTSD as a timeless diagnosis, the culmination of a lineage that is seen to run from the past to the present in an interrupted yet ultimately continuous way. On that assumption, the history of knowledge about trauma goes, briefly and schematically, like this. Although "people have always known that exposure to overwhelming terror can lead to troubling memories, arousal, and avoidance,"[4] modern understanding of trauma began with the work of the British physician John Erichsen, who during the 1860s identified the trauma syndrome in victims suffering from the fright of railway accidents and attributed the distress to shock or concussion of the spine. Claiming that the traumatic syndrome constituted a distinct disease entity, the Berlin neurologist Paul Oppenheim subsequently gave it the name "traumatic neurosis" and ascribed the symptoms to undetectable organic changes in the brain.[5]

The physiology of shock continued to be a topic of investigation for the next fifty years, as exemplified by the work of the Americans George W. Crile and Walter B. Cannon, and the influential researches of Ivan Pavlov.[6] But the term trauma acquired a more psychological meaning when it was employed by J. M. Charcot, Pierre Janet, Alfred Binet,

---

3. Henri F. Ellenberger, "The Pathogenic Secret and Its Therapeutics" (1966), in *Beyond the Unconscious: Essays of Henri F. Ellenberger in the History of Psychiatry*, ed. Mark Micale (Princeton, New Jersey, 1993), 341–59.

4. Bessel A. van der Kolk, Alexander C. McFarlane, and Lars Weisaeth, eds., *Traumatic Stress: The Effects of Overwhelming Experience on Mind, Body, and Society* (New York, 1996), 47; hereafter abbreviated *TS*.

5. For Oppenheim's work see especially Paul Lerner, "Rationalizing the Therapeutic Arsenal: German Neuropsychiatry in World War I," in *Medicine and Modernity: Public Health and Medical Care in Nineteenth- and Twentieth-Century Germany*, ed. Manfred Berg and Geoffrey Cocks (Washington, D.C., 1996), 121–48; and Claude Barrois, *Les névroses traumatiques: La psychothérapeute face aux détresses des chocs psychiques* (Paris, 1988).

6. Allan Young, *The Harmony of Illusions: Inventing Post-Traumatic Stress Disorder* (Princeton, New Jersey, 1995), 21–26; hereafter abbreviated *HI*.

Morton Prince, Josef Breuer, Sigmund Freud, and other turn-of-the-century figures to describe the wounding of the *mind* brought about by sudden, unexpected, emotional shock. The emphasis began to fall on the hysterical shattering of the personality consequent on a situation of extreme terror or fright.[7] [The traumatized psyche was conceptualized as an apparatus for registering the blows to the psyche outside the domain of ordinary awareness, and hypnotism was used as a psychotherapeutic method for retrieving the forgotten, dissociated, or repressed recollections by bringing them into consciousness and language.] Hypnotic catharsis thus emerged as a technique for solving a "memory crisis" that disturbed the integrity of the individual under the stresses of modernity.[8] The hysterical female epitomized the shattering effects of trauma on the mind. In the 1890s Freud suggested that sexual exploitation was at the core of hysteria by positing that the condition was caused by unconscious, repressed memories of sexual trauma, specifically sexual seduction or assault. But in 1897 he abandoned his belief in the seduction theory and reoriented his work to the study of the effects of repressed erotic infantile wishes and fantasies, thereby denying the significance of actual trauma on the individual psyche.

Joseph Babinski's subsequent dismantling of Charcot's hysteria diagnosis, a massive reaction against hypnosis, and the rise of psychoanalysis, combined to reduce interest in trauma in the years after 1900. But the virtual epidemic of war neuroses during World War I made it impossible to deny the existence in the male of traumatic symptoms which, although gathered together under the rubric of "shell shock," were recognized as not different in kind from those observed in the hysterical female. The majority of physicians remained indifferent to the psychic suffering of the shell-shocked soldier, who was widely accused of malingering. But a small and increasingly influential minority recognized the psychogenic nature of the war neuroses; naturally enough, they turned to Freud's earliest ideas about dissociation and the unconscious for help in understand-

7. For Charcot on trauma see especially Mark S. Micale, "Charcot and *les névroses traumatiques:* Scientific and Historical Reflections," *Revue neurologique* 150 (1994): 498–505; idem, "Charcot and the Idea of Hysteria in the Male: Gender, Mental Science, and Medical Diagnosis in Late Nineteenth-Century France," *Medical History* 34 (1990): 363–411; idem, *Approaching Hysteria: Disease and Its Interpretations* (Princeton, New Jersey, 1995).

8. Michael S. Roth, "Remembering Forgetting: *Maladies de la mémoire* in Nineteenth-Century France," *Representations* 26 (1989): 49–68; idem, "Hysterical Remembering," *Modernism/Modernity* 3 (May 1996): 1–30; Richard Terdiman, *Present Past: Modernity and the Memory Crisis* (Ithaca, New York, 1993); Ian Hacking, *Rewriting the Soul: Multiple Personality and the Sciences of Memory* (Princeton, New Jersey, 1995); hereafter abbreviated *RS*.

ing shell shock, and to the Breuer-Freud cathartic method, which they revived as the treatment method of choice.

After the war, interest in trauma again declined. The Hungarian psychoanalyst Sándor Ferenczi deserves credit for resuscitating Freud's original emphasis on the significance of childhood sexual trauma during the interwar years, but his work remained controversial and he was unable to redirect attention to the problem of trauma among psychoanalysts, who remained largely indifferent to the effects of real traumatic events on the child and adult. Nor did Abram Kardiner's outstanding efforts to codify the nature of the traumatic syndrome, based on his extensive experience with chronic cases from the Great War, or the widespread use of drug catharsis as a therapy for treating "combat fatigue" by William Sargant, Roy Grinker, John Spiegel, and others during the Second World War, prevent an "astounding" loss of memory of war trauma in the years after 1945 (*TS*, 59). Not even the independent psychoanalytic studies of the long-term effects of trauma on survivors of the Holocaust—studies that identified the existence of a chronic "concentration camp syndrome" or "survivor syndrome"—succeeded in arousing widespread interest in trauma. Rather, it was largely as the result of an essentially political struggle by psychiatrists, social workers, activists and others to acknowledge the post-war sufferings of the Vietnam War veteran that the third edition of the American Psychiatric Association's *Diagnostic and Statistical Manual of Mental Disorders* (1980) accorded the traumatic syndrome, or PTSD, official recognition for the first time.[9] Women's advocates, such as the physician Judith Herman, who in the 1970s became concerned with sexual abuse in children, also played a major role in establishing an integrated, post-Vietnam approach to trauma.[10]

9. See especially, Wilbur J. Scott, "PTSD in DSM-III: A Case in the Politics of Diagnosis and Disease," *Social Problems* 37 (1990): 294–310. Cf. Wilbur J. Scott, *The Politics of Readjustment: Vietnam Veterans since the War* (New York, 1993). The concept of stress has played an important role in the prehistory of PTSD. In a large literature on the topic see especially M. J. Horowitz, *Stress Response Syndromes* (New York, 1976). For a quirky, provocative book on the role of the concept of stress in the formulation of the traumatic neuroses, including PTSD, see Robert Kugelman's *Stress: The Nature and History of Engineered Grief* (Westport, Connecticut, 1992). For an attempt, prior to the introduction of PTSD, to pull together the diffuse and heterogeneous literature on the psychological consequences of disaster, including the literature on civilian disasters, Hiroshima, the combat neuroses, and the Holocaust, see Warren Kinston and Rachel Rosser, "Disaster: Effects on Mental and Physical State," *Journal of Psychosomatic Medicine* 18 (1974): 437–56.

10. Judith Herman, *Trauma and Recovery* (New York, 1992).

In an important recent book, the anthropologist Allan Young has co-gently argued that the account of the evolution of the concept of PTSD that I have just sketched is premised on an error: far from being a timeless entity with an intrinsic unity, as its proponents suggest, PTSD is a historical construct that has been "glued together by the practices, technologies, and narratives with which it is diagnosed, studied, treated, and represented and by the various interests, institutions, and moral arguments that mobilized these efforts and resources" (*HI*, 5). And in fact a scrutiny of the pre-1980 psychiatric literature on survivors of the concentration camps and victims of military combat, civilian disasters, and other traumas reveals a wide diversity of opinion about the nature of trauma, a diversity that has been obscured by the post-Vietnam effort to integrate the field. Young rightly observes that PTSD is no less "real" on that account, in the sense suggested by Ian Hacking: in Hacking's phrase, PTSD is a way of "making up" a certain type of person that individuals can conceive themselves as being and on the basis of which they can become eligible for insurance-reimbursed therapy, or compensation, or can plead diminished responsibility in courts of law.[11]

Nevertheless, the field of trauma studies today not only continues to lack cohesion, but the very terms in which PTSD is described tend to produce controversy.[12] For example, in a series of publications that serves as a blueprint for much current research in the field, Bessel A. van der Kolk and his associates have recently shifted the focus of research from the mind back to the *body* by explaining traumatic memory in neurobiological terms. Basing their hypotheses on the model of an animal's response to inescapable shock or stress, van der Kolk and his associates

11. Ian Hacking, "Making Up People," in *Reconstructing Individualism: Autonomy, Individuality, and the Self in Western Thought*, ed. Thomas Heller, Morton Sosna, and David Wellbery (Stanford, California, 1986), 226–36.

12. In spite of the rapid growth in the treatment and study of PTSD since its official recognition in 1980, researchers in the field continue to acknowledge the existence of confusion, even "chaos," a situation that leads to further appeals for order. See G. Boulanger, "A State of Anarchy and a Call to Arms: The Research and Treatment of Post-Traumatic Stress Disorder," *Journal of Contemporary Psychotherapy* 20 (1990): 5–15; Berthold P. R. Gerson and Ingrid V. E. Carlier, "Post-Traumatic Stress Disorder: The History of a Recent Concept," *British Journal of Psychiatry* 161 (1992): 742–48; and Rachel Yehuda and Alexander C. McFarlane, "Conflict Between Current Knowledge About Posttraumatic Stress Disorder and Its Original Conceptual Basis," *American Journal of Psychiatry* 152 (1995): 1705–13. The protean nature of the stress reaction, its tendency, like that of hysteria, to "mime" other disorders, makes it an imitative or mimetic disorder *par excellence*, a point central to my argument.

argue that trauma is preserved in the memory with a timeless accuracy that accounts for the long-term and often delayed effects of PTSD. Thus they hold that the traumatic event is encoded in the brain in a different way from ordinary memory. Specifically, van der Kolk suggests that traumatic memory may be less like what some theorists have called "declarative" or "narrative" memory, involving the ability to be consciously aware of and verbally narrate events that have happened to the individual, than like "implicit" or "nondeclarative" memory, involving bodily memories of skills, habits, reflex actions, and classically conditioned responses that lie outside verbal-semantic-linguistic representation. The hypothesis is that each implicit memory system is associated with a particular area of the brain, and the hunt is now on for the neurohormonal basis of such memories and for techniques of treatment designed to reduce or silence the activities of the central nervous system thought to be the locus of traumatic memory.⌡

At stake in the notion of implicit or non-narrative traumatic memory—a notion van der Kolk traces back to the long-neglected Janet, whose work he has helped resurrect—⌐is the idea that precisely because the victim is unable to process the traumatic experience in a normal way, the event leaves a "reality imprint" (TS, 52) in the brain that, in its insistent literality, testifies to the existence of a pristine and timeless historical truth undistorted or uncontaminated by subjective meaning, personal cognitive schemes, psychosocial factors, or unconscious symbolic elaboration.⌡There is no consensus in the field of memory research regarding such a claim, which I believe to be of dubious validity (see chapter 7). Nevertheless, such an approach to PTSD, which enjoys considerable support both within the scientific research community and among certain postmodernist literary critics, constitutes an original contribution to what Hacking has termed the "memoro-politics" of our time (RS, 210–22),[13] because, by eliminating the question of autobiographical-symbolic meaning, it makes manifest the mechanical-causal basis of much recent theorizing about trauma.⌐When applied to the Vietnam veteran, the model implies that all participants in that war—whether victims of combat who now suffer from repetitive nightmares or perpetrators of atrocities who now feel guilty about what they once did (the two groups of course are not mutually exclusive)—are alike casualties of an external trauma that causes objective changes in the brain in ways that tend to eliminate the issue of moral meaning and ethical assessment.⌡The expan-

13. Cf. Ian Hacking, "Memory Sciences, Memory Politics," in Tense Past: Cultural Essays in Trauma and Memory, ed. Paul Antze and Michael Lambek (New York, 1996), 67–87.

siveness of such a causal approach to trauma—its tendency to collapse distinctions between victims and perpetrators, or simply between victims and others—inevitably leads to skepticism about the reality of trauma. The charge of suggested memory or "false memory" in cases of alleged sexual abuse is one by now familiar expression of that skepticism. Skepticism also surfaces when, as in the Paula Jones matter, the attribution of trauma appears to be made by lawyerly fiat.

In this book I intend to situate the dilemmas, impasses, and controversies that characterize the current field of trauma studies by taking a different approach to the history of trauma from that commonly adopted in the modern literature. I do not proceed as if trauma has a linear, if interrupted, historical development. Rather, I shall take a genealogical approach to the study of trauma, in an effort to understand what Michel Foucault has called "the singularity of events outside of any monotonous finality" and in order to register their recurrence, as he has put it, not for the purpose of tracing "the gradual curve of their evolution, but to isolate the different scenes where they engaged in different roles."[14]

This is not to deny the existence of certain continuities, or rather structural repetitions, in the history of trauma. On the contrary, a major purpose of my book is to demonstrate the centrality to that history of a set of perpetually resurfacing theoretical and practical difficulties all of which may be seen to revolve around *the problem of imitation, defined as a problem of hypnotic imitation.* It is well known that the rise of trauma theory was associated from the start with hypnosis. Hypnosis, or hypnotic suggestion, was the means by which Charcot legitimated the concept of trauma by proposing that the hysterical crises that he suggestively induced in his patients were reproductions of traumatic scenes. What is less well understood is that hypnosis was not just an instrument of research and treatment but played a major theoretical role in the conceptualization of trauma. This is because the tendency of hypnotized persons to imitate or repeat whatever they were told to say or do provided a basic model for the traumatic experience. Trauma was defined as a situation of dissociation or "absence" from the self in which the victim unconsciously imitated, or identified with, the aggressor or traumatic scene in a condition that was likened to a state of heightened suggestibility or hypnotic trance. Trauma was therefore understood as an experience of hypnotic imitation or identification—what I call *mimesis*—an experience that, be-

14. Michel Foucault, "Nietzsche, Genealogy, History," in *Language, Counter-Memory, Practice: Selected Essays and Interviews,* ed. Donald F. Bouchard (Ithaca, New York, 1977), 139–40.

cause it appeared to shatter the victim's cognitive-perceptual capacities, made the traumatic scene unavailable for a certain kind of recollection.

In short, from the beginning trauma was understood as an experience that immersed the victim in the traumatic scene so profoundly that it precluded the kind of specular distance necessary for cognitive knowledge of what had happened. The subject was fundamentally "altered," to use Roustang's formulation, because it was "other."[15] He or she was "nothing more than a series of heterogeneous and dissociated roles," which meant that trauma was defined by "multiple borrowing"[16] (multiplicity was one of the marks of that traumatic-mimetic borrowing; the notion of the identification with the aggressor was another). This meant, too, that the amnesia held to be typical of psychical shock was explained as a kind of post-hypnotic forgetting that risked being irreversible since, according to the hypothesis, the traumatic scene was never present to the hypnotized subject and hence was constitutively unavailable for subsequent representation and recall. (I note for later elaboration that hypnosis itself as a theory and practice needs to be historicized: it cannot be treated as a phenomenon with a timeless essence but rather must be understood as constituted by specific theoretical-technological conditions. The definition of hypnosis as an altered state of consciousness characterized by post-hypnotic amnesia has a genealogy which needs to be unearthed.) All this would seem to suggest that the effort to cure patients by getting them, through the use of hypnotic catharsis or by other means, to recollect and narrate the dissociated traumatic origin was destined to fail.

But it is also my aim to show that a tendency towards the repudiation of mimesis was from the start also at work in the field. It is as though early theorists of trauma were simultaneously attracted to and repelled by the mimetic-suggestive theory, as though the basis of the latter's appeal—its ability to explain the victim's suggestibility and abjection—was also its chief defect—its threat to an ideal of individual autonomy and responsibility. The result was an inclination not only to relegate hypnosis and hypnotic suggestion to a secondary position but to do so in ways that suppressed the mimetic-suggestive paradigm in order to reestablish a strict dichotomy between the autonomous subject and the external trauma. The containment of the mimetic theory was all the more necessary in that, owing to the possibility of confabulation associated with the hyp-

15. François Roustang, foreword to *The Freudian Subject* by Mikkel Borch-Jacobsen (Stanford, California, 1988), viii.

16. Philippe Lacoue-Labarthe, *Typography: Mimesis, Philosophy, Politics* (Cambridge, Massachusetts, 1989), 129.

notic *rapport*, the notion of mimesis tended to call into question the veracity of the victim's testimony as to the veridical or literal truth of the traumatic origin and hence to make traumatic neurosis and traumatic memory a matter of suggested fabrication or simulation. As a consequence, there existed a competing, antimimetic tendency to regard trauma as if it were a purely external event coming to a sovereign if passive victim. The antimimetic model has lent itself to positivist or scientistic interpretations of trauma epitomized by the several neurobiological theories so widely accepted today. Yet, as I also show, mimesis could not simply be willed away—made to disappear—but instead continually reemerged, in the work of Freud, Prince, Ferenczi, and others as the very ground and origin of the traumatic experience. There has thus been a continuous tension or oscillation between the two paradigms, so that even the most resolutely antimimetic theory of trauma has tended to resurrect the mimetic theory itself. The paradoxes and contradictions that have resulted from that tension or oscillation are the subject of this study.

In this book I focus on a series of crucial episodes in the ongoing if interrupted attempt by various physicians and others to define the nature of trauma, all of which episodes evince a struggle in one way or another with the conundrums of imitation-suggestion. As I have said, I do not believe that those episodes are best seen as part of a continuously unfolding historical development. Rather, what is striking about them is their irruptive character; when the same or similar issues recur, they do so as if for the first time and almost with the same quality of shock or disruption that has been attributed to trauma itself, though of course it is also true that each episode bears the distinctive imprint of its historical moment. Throughout this study I try to do justice to the historical specificity of the cruxes I discuss. But my approach, by avoiding the assumptions implicit in a continuous narrative, enables us to see what is recurrent, and in an important sense structural, in the difficulties and contradictions that have tormented conceptualizations of trauma throughout the century. Although I have made use of archives when that seemed desirable, this book is not a social history of the various psychiatric services that have been developed at different times for victims of trauma, or an account of the development of military psychiatry, or an exploration of the invention of the dissociative disorders, although it embraces aspects of all those. Rather, it is a work of intellectual history in which I attempt to elucidate the genealogy of what has come to be seen as one of the signal concepts of our time.

In chapter 1, I focus on the work of Sigmund Freud, an inescapable figure in the genealogy of trauma. Not that Freud was the first to produce a coherent theory of trauma: thanks to the work of Henri Ellenberger and many others we now have a much clearer idea than formerly of the extent to which Freud's early ideas about hysteria belonged to a turn-of-the-century discourse on trauma and dissociation.[17] Nor is his importance for us solely a matter of his subsequent influence on psychiatric thought. Rather, Freud is unavoidable precisely because his aporetic and contradictory writings about the neuroses, including the traumatic neuroses, exhibit a simultaneous preoccupation with and evasion of the question of mimesis in a manner that exemplifies the tensions and paradoxes that have continued to trouble the field to the present day. Many researchers in the field of trauma studies now like to think that Freud's work has been completely superseded. It will become clear that they themselves are heir to difficulties that can only be understood through confronting, not denying, Freud's legacy.

In chapter 2, I provide a detailed discussion of Morton Prince's *The Dissociation of a Personality: A Biographical Study in Abnormal Psychology* (1905), one of the paradigm cases of trauma and dissociation at the turn of the century. From the perspective of recent theorists of multiple personality and trauma, Prince's once-famous study of his patient, Miss Beauchamp, is a founding text in the field, rivaling in importance Freud's contemporaneous Dora case. Prince's text will be discussed in terms of the concepts of trauma, dissociation, hypnosis, and memory that informed his theorizing and practice. I try to demonstrate that there are tensions between mimesis and antimimesis in Prince's analysis of the case such that his efforts to confirm his patient's trauma are thwarted by the very practices of hypnosis that are designed to unearth it. Moreover, in Prince's text the hunt for the patient's "real" or "original" identity is, I also argue, intimately if covertly entangled not only with notions of suggestion or mimesis but also with questions of sexual difference. In the course of my discussion, I ask what relation, if any, obtains between the early history of multiple personality as exemplified by Prince's text and the use of the multiple personality concept in both medical and feminist discourse today. Does the recently revived diagnosis of the multiple self function, as it did in Prince's text, to "repress" the hypnotic-mimetic paradigm? What are the implicit or strategic sexual politics of such a development?

After a precipitous decline in the practice of hypnosis at the beginning of the twentieth century—a decline that was accompanied by a simultaneous abandonment of the diagnosis of multiple personality—Breuer and Freud's hypnotic-cathartic therapy was revived during World War I

17. Henri Ellenberger, *The Discovery of the Unconscious: The History and Evolution of Dynamic Psychiatry* (New York, 1970).

as a medical technology for curing the traumatic neuroses of combat, or shell shock. As a result, the issues that had confronted Prince in his hypnotic treatment of the traumatized female hysteric resurfaced in connection with the therapy of the battle neuroses in the male. In chapter 3, I focus on what proved to be a crucial post-war debate among British physicians William Brown, William McDougall, Charles S. Myers, and others over the nature of catharsis—specifically, whether the therapeutic success of catharsis, if any, depended on the cognitive recovery and integration of traumatic memories or on the emotional intensity of the cathartic discharge, or abreaction. Uncertainty as to this point, a crucial one both theoretically and practically, has been remarkably persistent in the history of conceptualizations of trauma, arising regularly in different historical conjunctures and always remaining unresolved. I seek to demonstrate that the dispute is structured by the same oscillation between mimesis and antimimesis already tracked in the work of Freud and Prince. Moreover, the issues confronted in the context of World War I were of direct relevance to the work of the contemporary French psychologist Pierre Janet, a major figure in the history of trauma and hypnosis. I show that Janet's approach to trauma has been fundamentally misunderstood by recent commentators, who have failed to recognize the significance of the fact that his cures often depended on getting the patient not to remember but to *forget* the traumatic origin.

In recent years, Sándor Ferenczi, Freud's most talented disciple, has come to be seen by psychoanalytically informed theorists of trauma as a central figure because, unlike his colleagues who followed the master by emphasizing the role of infantile libido in the etiology of the neuroses, he attempted to revive Freud's earliest ideas about sexual trauma, dissociation, and catharsis. Although Ferenczi's work earned him the enmity of Freud and the psychoanalytic movement, it is now widely acclaimed as an important anticipation of current theories about the causative role of childhood abuse in the dissociative disorders. In chapter 4 I focus on Ferenczi's *Clinical Diary*, a detailed record of the development of his theoretical ideas and therapeutic experiments written in 1932, one year before his untimely death (in part because of its revisionary content, it was only recently published). In particular, I concentrate on the conflict between the mimetic and antimimetic paradigms in his work in order to show that, in spite of his ostensible commitment to recovering the traumatic origin, according to the terms of his own theorizing the mimetic-identificatory nature of the traumatic experience was such that the origin could not be recuperated. In the course of my analysis I discuss not only Ferenczi's work on female patients, but also his important contributions to the study of the traumatic neuroses of World War I. I also discuss the views of Abram Kardiner, whose early efforts to theorize the war neuroses drew inspiration from Ferenczi's ideas.

In chapter 5, I focus on another aspect of Ferenczi's thought, the theme of hysterical lying, in order to address a topic of major importance in the genealogy of trauma, that of malingering or *simulation*. The problem of malingering has never been far from the problem of hysteria, especially in times of war. Freud's concept of the unconscious served to protect the female hysteric or war neurotic from the charge of simulation by positing the existence of a "psychic reality" that dissolved the traditional opposition between truth and lies. It is therefore all the more surprising that Ferenczi subscribed to a notion of lying, or simulation, in his approach to the problem of trauma. Moreover, Ferenczi's hypnotic methods could be seen—and were seen by one of his patients, the American psychoanalyst Clara Thompson—as exacerbating the problem of lying they were designed to cure by encouraging the patient to *feign* or *simulate* the traumatic scene or origin.

Thompson's objections have something in common with the recent critique of one of the most brilliant modern commentators on psychoanalysis, Mikkel Borch-Jacobsen, who, repudiating an earlier commitment to the mimetic hypothesis, has proposed that the relivings characteristic of hypnotic-cathartic methods are in principle incapable of producing evidence of the traumatic origin because they belong to the order of simulation, that is, to the order of fictive "games" carried out voluntarily between patient and hypnotist. Accordingly, I look closely at Borch-Jacobsen's arguments about simulation in the light not only of Ferenczi's suggestive-therapeutic practices but also of the post–World War II shifts in theorizing about hypnosis that inform Borch-Jacobsen's skeptical conclusions.

Some further preliminary remarks are in order about the importance I attach to Borch-Jacobsen, whose work occupies a special position in this study. From my perspective his writings fall into two phases, which is not to deny that there are points of consistency between his earlier and later views or that one might claim to discern a trajectory connecting them. In *The Freudian Subject* (first published in French in 1982) and related essays, Borch-Jacobsen conducts a close and sustained reading of a wide range of Freudian texts in the general mode of Derridean deconstruction in order to show that the mimetic paradigm serves as a key structuring principle in Freud's thought even as the latter continually and often self-contradictorily struggled against it.[18] This work by Borch-Jacobsen has been central to my approach in the present study, undergirding my claim, explicit in the previous pages, that mimesis functions as one of two fundamental and unstable poles in all major theorizations of trauma. In the second, relatively recent phase of his career, Borch-Jacobsen has emerged as one of Freud's most formidable critics, one whose skepticism about the entire psychoanalytic project, indeed about the very concept of trauma, in turn demands to be scrutinized. His new position, based on a reconceptualization of the nature of hypnosis, involves a repudiation of his previous understanding of hypnosis as an absorptive or "blind" mimesis or identification with the other (or hypnotist) in favor of an antimimetic emphasis on the autonomy of the subject. As will emerge, I regard his new position as an attempt to *resolve* the oscillation between mimesis and antimimesis that has structured the history of trauma all along, which also means that for all its sophistication it exemplifies the perhaps insoluble difficulties that have attended all attempts to effect such a resolution. In chapter 5, therefore, I pay considerable attention to the second phase of Borch-Jacobsen's work, particularly as it bears on the issue of suggestion and simulation.

During World War II, British and American physicians reintroduced a version of Freud's catharsis in the treatment of the traumatic neuroses by using barbiturates and other drugs to induce abreaction in soldiers traumatized by combat, thereby reviving the World War I debate over the nature of trauma, the cathartic-abreactive cure, and the possibilities of remembering. In chapter 6, I use the work of the British psychiatrist William Sargant as a vehicle for evaluating the vicissitudes of drug catharsis during and after the war. I focus on the tensions between the therapeutic requirement to remember the trauma, central to the insight-based approach of much cathartic or abreactive treatment during the

---

18. The use of the term "mimesis" to mean hypnotic identification seems to have originated in the Strasbourg circle of Borch-Jacobsen, Philippe Lacoue-Labarthe, and Jean-Luc Nancy. In *The Freudian Subject* Borch-Jacobsen employs the term "mimesis" rather than "imitation" on the grounds that mimesis as he understands it in *The Freudian Subject* does not refer to the simple imitation of a *model* or to fictive simulation, both of which presume the existence of the very spectatorial or spectating subject that is in question here (Borch-Jacobsen, "Dispute," in Leon Chertok, Mikkel Borch-Jacobsen, et al., *Hypnose en psychanalyse* [Paris, 1987], 203–6). The texts to which I refer use the terms "imitation," "suggestion," "mimicry," and "mimesis" (as in "mimetic" identification) interchangeably with regard to both mimetic and antimimetic processes, a fact I understand as expressive of the tension or oscillation between the mimetic and antimimetic tendencies at work in them.

war, and the contrary requirement to forget or erase the past, implicit in Sargant's work. I also ask the larger question as to whether any of the methods used actually succeeded in their aim. This leads, in turn, to a brief review of the impact of preventive approaches to the management of trauma on the practice of cathartic abreaction, as well as to a discussion of the recent revival of abreaction as a means of treating dissociative disorders and PTSD.

The history of trauma itself is marked by an alternation between episodes of forgetting and remembering, as the experiences of one generation of psychiatrists have been neglected only to be revived at a later time. Just as it took World War II to "remember" the lessons of World War I, so it took the experience of Vietnam to "remember" the lessons of World War II, including the psychiatric lessons of the Holocaust. The delay in the full appreciation of literature of the Holocaust helps explain why, in a book on the genealogy of trauma, I have not devoted a chapter to the psychiatric literature on the Holocaust itself. But that omission calls for further explanation.

For all sorts of reasons the psychiatric response to the Holocaust was belated, in that it was not until some years after the war that survivors and psychiatrists alike began to be aware of the devastating long-term psychic and medical costs of the experience of the concentration camps. In fact, the notion or observation that the symptoms of stress may make their first appearance after a considerable lapse of time, even years after the traumatic event, is among the important contributions that the literature on the Holocaust survivor is now seen to have made to the definition of PTSD. Yet that same body of work on the camp survivors remained somewhat isolated from the literature on the combat neuroses, and to some extent also from the literature on civilian trauma, until it was assimilated into the post-Vietnam literature on PTSD. (Subsequently, of course, it has acquired considerable status on its own.) But the assimilation has occurred at a price: much (although not all) of the concentration camp literature, especially the American literature, is psychoanalytically inspired, which the later approach to PTSD generally is not, with the result that many of the distinctions and qualifications characteristic of earlier work on the Holocaust have been lost in the translation. At the same time, the process of viewing the literature of the Holocaust through the lens of Vietnam and PTSD has produced a simplification that works to the benefit of the former, in the sense that it can now be seen to have contributed directly to the development of current research on PTSD. For these reasons, the Holocaust now appears, retroactively so to speak, not only to have been *the* crucial trauma of the century, but also the one that can be fully understood only in the light of our knowledge of PTSD. By the same token, the diagnosis of PTSD represents the culmination of an attempt to do justice to the earlier psychiatric literature on the Holocaust survivor by integrating it into a unified theory that applies to the victim of natural disaster, the combat victim, the Holocaust survivor, the victim of sexual abuse, and the Vietnam veteran alike. In the spirit of genealogy, therefore, I approach the literature on the Holocaust indirectly, as a contribution to the modern definition of PTSD.

In chapter 7 I examine in detail the neurobiological theories of Bessel A. van der Kolk, who has recently emerged as a leading theorist of PTSD. I have chosen to focus on his ideas because they represent an important approach to trauma in our present culture, especially in the United States, where biological paradigms are in the ascendant in psychiatry. In particular, I critically examine van der Kolk's central claim that traumatic memory involves a literal imprint of an external trauma that, lodged in the brain in a special traumatic memory system, defies all possibility of representation. Consequently, his ideas have helped solidify a powerful trend in the humanities to recognize in the experience of trauma, especially the trauma of the Holocaust, a fundamental crisis for historical representation (and at the limit, for representation as such). I argue that such a literalist view of trauma is not only theoretically incoherent but also poorly supported by the scientific evidence. I also demonstrate that although van der Kolk's work gains prestige by being associated with paradigms, technologies, and practices that conform to the dominant model of what constitutes good psychiatric science today, it offers a causal analysis of trauma as fundamentally external to the subject that is not only poorly formulated but is haunted by the same problem of mimetic suggestibility that the theory is designed to forestall.

In chapter 8, my final chapter, I undertake a detailed criticism of the work of literary theorist Cathy Caruth, whose ideas about trauma are today much in vogue in the humanities (especially in the United States). Her work epitomizes the contemporary literary-critical fascination with the allegedly unrepresentable and unspeakable nature of trauma, especially the trauma of the Holocaust, which in effect stands in for trauma generally. But what gives her work a certain distinction and seeming authority is that she combines a postmodernist literary-theoretical approach to trauma, of the kind associated with the work of the post-structuralist critic and theorist Paul de Man, with an appeal to post-Vietnam research on PTSD, making use in particular of the neurobiological claims and findings of van der Kolk and others. Her work thus represents a surprising, but, as I show, entirely coherent alliance of a certain version of deconstructive criticism with empirical neuroscience.

In my discussion of Caruth's book *Unclaimed Experience: Trauma, Narrative, and History* (1996), I concentrate chiefly on her interpretation of Freud, a major focus of her study. In the course of my analysis I try to show that tensions between mimesis and antimimesis silently control Caruth's commentaries. Thus Caruth posits an absolute opposition between external trauma and victim in ways that have been associated historically with the repudiation of mimesis. But by imagining that trauma stands outside representation altogether, she also embraces a version—in fact it is more like an inadvertent parody—of the mimetic theory. Indeed, I demonstrate that although Caruth does not discuss the problem of hypnosis, historically central to the genealogy of trauma, the question of mimesis surfaces in her text in her insistence on the contagious effects of trauma. That is, she links a notion of the dichotomy between the external trauma and the victim with a de Man-inspired version of the idea of the mimetic-contagious transmission of psychic suffering to others, even to later generations, with the result that trauma becomes unlocatable in any particular individual. Caruth thereby contributes to that collapse of distinctions to which I point at the outset of this introduction as a general problem in current approaches to trauma."[19]

My book ends with a brief conclusion, in which I summarize my findings and say one or two things about their implications for current debates over trauma.

19. During the final stages of preparing this book for publication, I became aware of an impressive practical and theoretical critique of the emphasis on Western-inspired notions of trauma and PTSD by policy makers providing relief in recent genocidal and other wars, *Rethinking the Trauma of War*, ed. Patrick J. Bracken and Celia Petty (London, 1998). My thanks to Martin Wilkinson for alerting me to this work.

Freud and Trauma

In the introduction, I invoked the Paula Jones case as an example of the way in which the concept of trauma can lend itself to trivialization. "I would say it is a big joke," President Clinton's lawyer, Robert S. Bennett, remarked in discounting the claim made by Paula Jones's lawyers that as a result of Jones's alleged sexual harassment by Clinton she now suffered from PTSD. "All of a sudden," Bennett went on, "to fill a gap in their pleading . . . there pops up a Ph.D. in education who gives them an affidavit that there is now a damage claim of sexual aversion. It is a joke. . . . I mean no more than that. No more than that. As Freud said, sometimes a good cigar is just a good cigar."[1] Bennett's misreference to Freud,[2] here enlisted on the side of common sense in a case that would seem to have nothing to do with psychoanalysis, may be taken to suggest that Freud is somehow an ineluctable figure in the genealogy of trauma. But why, or in what sense, is Freud ineluctable?

Freud is ineluctable if for no other reason than that, as Hacking has put it, he "cemented" the idea of psychic trauma—specifically, the trauma of sexual assault, Freud's famous seduction theory. In other words, Freud is a founding figure in the history of the conceptualization of trauma. At the same time, as Hacking also observes, no figure is more reviled by present-day theorists of childhood trauma, precisely because in 1897 Freud famously abandoned the very theory of sexual seduction that is crucial to today's recovered memory movement.[3] Yet if we are to evaluate Freud's

1. *New York Times*, Saturday, 21 March 1998.

2. Freud is supposed to have said: "Sometimes a cigar is just a cigar."

3. Ian Hacking, "Memory Sciences, Memory Politics," in *Tense Past: Cultural Essays in Trauma and Memory*, ed. Paul Antze and Michael Lambek (New York, 1996), 74–75.

role in the genealogy of trauma, as we must, it is essential to understand that the terms in which modern trauma theorists tend to describe Freud's "betrayal" reveal a fundamental misunderstanding of his thought. In a word, Freud's theory of seduction was never the simple causal theory of trauma that contemporary critics, such as van der Kolk, Herman, Jeffrey Moussaieff Masson, and others have portrayed it to be.[4] But it is not a simple matter to characterize Freud's contribution to the topic of trauma.

Trauma was originally the term for a surgical wound, conceived on the model of a rupture of the skin or protective envelope of the body resulting in a catastrophic global reaction in the entire organism. Yet as Laplanche has emphasized, it is not easy to retrace the "transposition" of this medicosurgical notion into psychology and psychiatry. Indeed, for so long has the notion of a shock with a physical "break in" and that of danger to life been the model for an allegedly psychical symptom that to this day psychical trauma is still bound to the concept of surgical shock. "There would be a series of gradations linking major impairments of tissue to decreasingly perceptible degrees of damage, but that would nevertheless be of the same nature," Laplanche has observed: "histological damage and, ultimately, intracellular damage. The trauma would proceed, as it were, to a kind of self-extenuation, but without losing its nature, until it reached a certain limit, that limit being precisely what we call 'psychical trauma.'"[5]

Laplanche's description holds good for the modern neurobiological definition of PTSD, which is explicitly modeled on a physiological-causal theory of shock. But as he has rightly pointed out, an entirely different direction was taken by Freud. This is not just because Freud followed Charcot and others in attributing traumatic hysteria to psychological rather than anatomicophysiological changes. More significantly, Freud stressed the role of a post-traumatic "incubation," or latency period of psychic elaboration, in ways that made the traumatic experience irreducible to the idea of a purely physiological causal sequence. In *Studies on Hysteria* (1895), coauthored by Josef Breuer, and even more explic-

4. In the large, skeptical literature on the seduction theory see especially Jeffrey Moussaieff Masson, *The Assault on Truth: Freud's Suppression of the Seduction Theory* (New York, 1984); Frederick Crews (and his critics), *The Memory Wars: Freud's Legacy in Dispute* (New York, 1995); and Mikkel Borch-Jacobsen, "Neurotica: Freud and the Seduction Theory," *October* 76 (Spring 1996): 15–43. For a Freudian-feminist refutation of the simple schema of internal and external that governs Masson's criticisms of Freud see Jacqueline Rose, "Feminism and the Psychic," in *Sexuality in the Field of Vision* (London, 1986), 1–23.

5. Jean Laplanche, *Life and Death in Psychoanalysis* (Baltimore, 1976), 129–30; hereafter abbreviated *LDP.*

itly in "The Aetiology of Hysteria" (1896), Freud argued that the symptoms of hysteria could only be understood if they were traced back to experiences that had a traumatic effect, specifically early experiences of sexual "seduction" or assault. But what many critics of Freud fail to grasp is that, even at the height of his commitment to the seduction theory, Freud problematized the originary status of the traumatic event by arguing that it was not the experience itself which acted traumatically, but its delayed revival as a *memory* after the individual had entered sexual maturity and could grasp its sexual meaning. More specifically, according to the temporal logic of what Freud called *Nachträglichkeit*, or "deferred action," trauma was constituted by a relationship between two events or experiences—a first event that was not necessarily traumatic because it came too early in the child's development to be understood and assimilated, and a second event that also was not inherently traumatic but that triggered a memory of the first event that only then was given traumatic meaning and hence repressed. For Freud, trauma was thus constituted by a dialectic between two events, neither of which was intrinsically traumatic, and a temporal delay or latency through which the past was available only by a deferred act of understanding and interpretation. Increasingly, Freud emphasized that owing to the peculiar unevenness of its temporal development, human sexuality provided an eminently suitable field for the phenomenon of deferred action. Thus from the outset, even when he was committed to the seduction theory, Freud rejected a straightforward causal analysis of trauma according to which the traumatic event assaults the subject from the outside (according to which, in other words, inside and outside are absolutely distinct from one another).[6]

In sum, for Freud traumatic memory is inherently unstable or mutable owing to the role of unconscious motives that confer meaning on it. That premise underlies Freud's studies of parapraxes in *The Psychopathology of Everyday Life* (1901). It is also the theme of his paper, "Screen Memories" (1899), in which he speaks of the "tendentious nature of our remembering and forgetting" and, because of the role of *Nachträglichkeit*, con-

6. Sigmund Freud, "The Aetiology of Hysteria" (1896), in *The Standard Edition of the Complete Psychological Works of Sigmund Freud*, trans. and ed. James Strachey (London, 1953–74), 3: 191–221; and Jean Laplanche and J.-B. Pontalis, *The Language of Psycho-Analysis* (New York, 1973), s.v. "deferred action." Freud's concept of *Nachträglichkeit* developed out of the idea, formulated by Charcot among others, that the traumatic memory goes through a process of elaboration or incubation after the event, a process that gives it its subsequent force and fixity (Michael Roth, "Hysterical Remembering," *Modernism/Modernity* 3 [May 1996]: 4–5).

cludes by questioning "whether we have any memories at all *from* our childhood: memories *relating to* our childhood may be all that we possess."[7] Nor does Freud's new emphasis on the role of fantasy after the so-called abandonment of the seduction theory in 1897 invalidate the concept of deferred action, as Freud's reliance on it in the Wolfman case of 1916–18 clearly shows.[8]

Nevertheless, there is something about trauma that troubles the Freudian project. The concept of *Nachträglichkeit* calls into question all the binary oppositions—inside versus outside, private versus public, fantasy versus reality, etc.—which largely govern contemporary understandings of trauma. However, Freud's rejection of the notion of trauma as direct cause and his emphasis on psychosexual meaning involved a tendency within psychoanalysis to interiorize trauma, as if the external trauma derived its force and efficacity entirely from internal psychical processes of elaboration, processes that were understood to be fundamentally shaped by earlier psychosexual desires, fantasies, and conflicts. The infantile internal drives thus became the properly etiological ground. In Laplanche's formulation: "What defines psychical trauma is not any general quality of the psyche, but the fact that the psychical trauma comes from within. . . . Everything comes from without in Freudian theory, it might be maintained, but at the same time every effect—in its efficacity—comes from within, from an isolated and encysted interior" (*LDP*, 42–43). Logically, therefore, Laplanche has questioned the value of the idea of traumatic neurosis: "the traumatic neurosis appears as nothing more than an initial, purely descriptive approximation which cannot survive any deeper analysis of the factors in question."[9]

But it is not at all obvious that the concept of the traumatic neuroses can be relegated to insignificance in this way—certainly not for Freud himself, for whom the traumatic neuroses of war in World War I helped precipitate a major reconsideration of his position on the primordial importance of the infantile psychosexual drives. Were not the thousands of cases of combat hysteria observed in apparently healthy adult men the direct result of the external trauma of trench warfare? This was the view of the majority of physicians who, unlike Freud, had firsthand experience of the war neuroses. Thus on the one hand, the Freudian movement bene-

---

7. Sigmund Freud, "Screen Memories" (1899), *Standard Edition*, 3: 322.

8. See Jean Laplanche and J.-B. Pontalis, "Fantasy and the Origin of Sexuality," *International Journal of Psychoanalysis* 49 (1968): 1–17.

9. Laplanche and Pontalis, *Language of Psycho-Analysis*, 472. Hence what "disarms" the ego in trauma, for Laplanche, is always the psychosexual drive (*LDP*, 47).

fited from the war because, after it became clear to some physicians that victims of shell shock fell ill not from organic lesions but from psychical causes, psychoanalysis seemed to be the only theoretical-therapeutic approach capable of interpreting and treating the functional disorders associated with the massive traumas of modern warfare. A small group of doctors in Britain and Germany turned to Freud's ideas about psychogenesis for guidance in the analysis and treatment of the war neuroses, with the result that catharsis was reinstated as a therapeutic method. In that regard, psychoanalysis emerged from the war with its reputation considerably enhanced. On the other hand, most of those same physicians remained dubious about Freud's specific emphasis on the role of the sexual drives in the origin of the neuroses. The challenge Freud thus faced was how to assimilate the experience of shell shock into his already well-established theoretical system, especially the libido theory and the theory of the psychosexual origins of the neuroses.[10]

Freud and the Traumatic Neuroses

Freud's initial response to that challenge was to suggest that the war neuroses were the consequence of a conflict, not between the ego and the sexual drives, but between different parts of the ego itself, that is, between the soldier's old peace-loving ego, or instinct for self-preservation, and his new war-loving ego, or instinct for aggression. Those egos were now defined, according to Freud's new theory of narcissism, as themselves sexually or libidinally charged. Such an explanation had the merit of recuperating the traumatic neuroses of the war for the libido theory and of assimilating them to the category of the ordinary transference neuroses.[11] At the International Congress of Psychoanalysis held in Budapest in 1918, Freud's disciples faithfully echoed Freud's proposition by treating the symptoms of the war neuroses as regressions to an earlier, narcissistic, stage of libidinal development.[12]

10. A point made by Robert Jay Lifton, *The Broken Connection: On Death and the Continuity of Life* (New York, 1983), 165.

11. Sigmund Freud, introduction to *Psycho-Analysis and the War Neuroses* (1919), *Standard Edition*, 17: 207–10. Freud thus suggested a motive for the soldier's "flight into illness," while distancing himself from the prevalent moralism and suspicion of cowardice or malingering by emphasizing the unconscious nature of the conflicts involved.

12. Sándor Ferenczi, Karl Abraham, Ernst Simmel, and Ernest Jones, "Symposium Held at the Fifth International Psycho-Analytical Congress at Budapest, September 1918," in *Psycho-Analysis and the War Neuroses* (London, 1921).

From the start, however, Freud responded to the problem posed by the war neuroses somewhat differently by emphasizing, or reemphasizing, the importance of those economic considerations that had always been central to his metapsychology. "The term 'traumatic,'" he wrote in 1916 in an early reflection on the war neuroses, "has no other sense than an economic one. We apply it to an experience which within a short period of time presents the mind with an increase of stimulus too powerful to be dealt with or worked off in the normal way, and this may result in permanent disturbances of the manner in which energy operates."[13] Precisely the same economic definition informed Freud's new theory of the death drive. As is well known, the general problem of repetition, especially the tendency of traumatized people to repeat painful experiences in their dreams—a tendency difficult to account for as an attempt to achieve libidinal satisfaction—compelled Freud in *Beyond the Pleasure Principle* (1920) to acknowledge the existence of a "beyond" of pleasure, or death drive, acting independently of and often in opposition to the pleasure principle. In that work, Freud posited the existence of a protective shield or "stimulus barrier" designed to defend the organism against the upsurge of large quantities of stimuli from the external world that threatened to destroy the psychic organization. Trauma was thus defined in quasi-military terms as a widespread rupture or breach in the ego's protective shield, one that set in motion every possible attempt at defense even as the pleasure principle itself was put out of action. "There is no longer any possibility of preventing the mental apparatus from being flooded with large amounts of stimulus," Freud wrote, "and another problem arises instead—the problem of mastering the amounts of stimulus which have broken in and of binding them, in the psychical sense, so that they can be disposed of."[14] (I draw attention to Freud's use of "binding" here, a key term in his lexicon and one that implies its opposite, "unbinding"; the pairing of the two notions plays a crucial role in his thought, as will become clear in a moment.)

13. Sigmund Freud, *Introductory Lectures on Psycho-Analysis* (1915–17), *Standard Edition*, 16: 275.

14. Sigmund Freud, *Beyond the Pleasure Principle* (1920), *Standard Edition*, 18: 29–30; hereafter abbreviated *BPP*. Such a theory, which as Freud himself observed seemed to reinstate the "old, naive theory of shock," was subsequently made use of by Walter Benjamin to define a specifically modern structure of perception. See Walter Benjamin, "On Some Motifs in Baudelaire," in *Illuminations: Essays and Reflections*, ed. Hannah Arendt, trans. Harry Zohn (New York, 1969), 155–200; and Wolfgang Schivelbusch, *The Railway Journey: Trains and Travel in the Nineteenth Century* (New York, 1977), 152–60.

According to Freud, the failure of such attempts at mastery and binding, a failure due to the role of fright and the ego's lack of preparedness, produced the general disorganization and other symptoms characteristic of psychic trauma. In sum, according to Freud the traumatic neuroses represented a radical "unbinding" of the death drive. Admitting for the first time an exception to his proposition that dreams represented the fulfillment of infantile-erotic wishes, Freud observed that

> it is impossible to classify as wish-fulfilments the dreams we have been discussing which occur in traumatic neuroses, or the dreams during psychoanalyses which bring to memory the psychical traumas of childhood. They arise, rather, in obedience to the compulsion to repeat, though it is true that in analysis that compulsion is supported by the wish (which is encouraged by "suggestion") to conjure up what has been forgotten and repressed. . . . If there is a "beyond the pleasure principle," it is only consistent to grant that there was also a time before the purpose of dreams was the fulfilment of wishes. (*BPP*, 32–33)

Freud's hypothesis of the death drive presaged a subtle shift in his theorizing from the analysis of desire to what he came to call "the analysis of the ego," a shift that was accompanied by a general revision and widening of the concept of defense. Many of Freud's texts of the 1920s can be seen as attempts to define the various mechanisms of defense the ego was held capable of deploying against stimulation, as well as the consequences for the psyche when those defenses failed. It is as though Freud during those years came to realize that the concept of repression, which, after the publication of *Studies on Hysteria* in 1895, had emerged as the psyche's fundamental response to excitation, needed to be supplemented by a variety of other modes of defense, the relations between which remained obscure and unresolved.[15] Among the mechanisms of defense increasingly invoked by Freud in the 1920s were "disavowal" (*Verleugnung*), linked by him not only to the fear of castration but also of death and the problem of mourning, "rejection" or "repudiation" (*Verwerfung*, Lacan's "foreclosure"), "negation" (*Verneinung*), "splitting of the ego" (*Ichspaltung*), and "primal repression" (*Urverdrängung*), some of which went back to Freud's earliest, prepsychoanalytic speculations on the operations of the psychic apparatus. In *Inhibitions, Symptoms and Anxiety* (1926)—a text in which the

---

15. This has been noted by André Green, *Le travail du négatif* (Paris, 1993), 163, who comments on the increasing importance Freud attached to mechanisms of defense other than repression after 1920 and on the difficulty Freud experienced in generalizing his findings and stabilizing his ideas.

traumatic neuroses of war illustrated the problem of anxiety in one of its most characteristic forms—Freud for the first time distinguished the more general notion of defense from that of repression by treating the latter as the defense specific to hysteria, stating in this regard that "repression is only one of the mechanisms which defence makes use of."[16]

Yet it cannot be emphasized too strongly that, in spite of the developments I have summarized, Freud's writings of the 1920s and 1930s remained fraught with doubt and vacillation. In particular, everything he wrote about the ego's defenses in the traumatic neuroses of war was marked by hesitation and contradiction. This is especially evident in *Inhibitions, Symptoms, and Anxiety*, the last of his metapsychological essays, where the danger to which the ego responds in traumatic situations was constantly redefined in libidinal terms as the danger or threat of castration or loss of the mother, with the result that even as the traumatic neuroses of war were systematically linked to the economics of unbinding and the death drive, they were simultaneously construed in terms of the theory of childhood psychosexual desire and the mechanism of repression from which they had been ostensibly released. In short, Freud's writings in the 1920s raised questions about the role of repression and sexuality that those same writings were unable fully to resolve.

It is against the background of these conceptual difficulties that the problem of psychic trauma and psychic violence has come back to haunt the theory and practice of psychoanalysis. In recent years, in texts by various authors who frequently bear no explicit relation to one another yet are linked by a set of common apprehensions, the idea of trauma has come to the fore in ways that express serious metapsychological and therapeutic dissatisfactions. It is as though psychic trauma represents an obstacle to psychoanalysis, one that constantly threatens to overturn its most basic assumptions. Thus Henry Krystal, in a series of articles based on his clinical experience with concentration camp survivors, has deplored the vagueness of psychoanalytic uses of the term trauma, and has urged a return to Freud's work on anxiety in order to reconceptualize both infantile and adult post-traumatic phenomena.[17] In a related devel-

---

16. Sigmund Freud, *Inhibitions, Symptoms and Anxiety* (1926), *Standard Edition*, 20: 114 (cf. 115, 163); hereafter abbreviated *ISA*. For the evolution of Freud's ideas of defense see Laplanche and Pontalis, *Language of Psycho-Analysis*, s.v. "repression."

17. See especially Henry Krystal, "Trauma and Affects," *Psychoanalytic Study of the Child* 33 (1978): 81–116; idem, "Alexithymia and Psychotherapy," *American Journal of Psychotherapy* 33 (1947): 17–31; idem, "Trauma and the Stimulus Barrier," *Psychoanalytic Inquiry* 5 (1985): 131–61.

opment, Jonathan Cohen and Warren Kinston have rejected the value of the concept of psychosexual repression in explaining the severely narcissistic, "borderline," or other states of extreme mental disorganization they associate with the trauma of the concentration camps and other disasters, and have revived instead Freud's earliest ideas about defense in order to account for the "immemorial" yet "unforgettable" residues of trauma.[18] Similarly, in her studies of survivors of the Holocaust—studies made without reference to the literature on combat neuroses or other traumas—Ilse Grubrich-Simitis has argued that the quasi-psychotic anxieties, peculiarly concrete or demetaphorized modes of thinking, traumatic fixations, dissociative doublings or splittings, actings-out, and memory disturbances observed especially in the children of Holocaust victims need to be reconceptualized as the consequence of a profound impairment of the ego's most basic symbolizing and other functions.[19] Grubrich-Simitis has drawn attention in this connection to the work of Marion Oliner who, on related grounds, has likewise suggested that the defensive depersonalizations, altered states of consciousness, and mental "absences" found in children of Holocaust survivors be understood as transitory "hysterical psychoses" or dissociations of the ego of the kind discussed by Freud and Breuer in their "pre-psychoanalytic" work on hysteria.[20]

In spite of differences of approach and conceptualization among these psychoanalytic critics, they all share a concern with the role of external reality (or the "environment") in the etiology of trauma. This is also the dominant theme of researchers in the field of PTSD today, one that therefore tends to unite psychoanalysts, cognitive psychologists, and neurobiologists alike. Characteristically, within psychoanalysis the desire to do justice to the real or "objective" danger of trauma finds expres-

18. Jonathan Cohen, "Structural Consequences of Psychic Trauma: A New Look at 'Beyond the Pleasure Principle,'" *International Journal of Psychoanalysis* 61 (1980): 421–32; idem, "Trauma and Repression," *Psychoanalytic Inquiry* 5 (1985): 163–89; Jonathan Cohen and Warren Kinston, "Repression Theory: A New Look at the Cornerstone," *International Journal of Psychoanalysis* 65 (1983): 411–22; Warren Kinston and Jonathan Cohen, "Primal Repression: Clinical and Theoretical Aspects," *International Journal of Psychoanalysis* 67 (1986): 337–55; Warren Kinston and Rachel Rosser, "Disaster: Effects on Mental and Physical State," *Journal of Psychosomatic Research* 18 (1974): 437–56.

19. Ilse Grubrich-Simitis, "From Concretism to Metaphor: Thoughts on Some Theoretical and Technical Aspects of the Psychoanalytic Work with Children of Holocaust Survivors," *Psychoanalytic Study of the Child* 39 (1984): 301–29.

20. Marion M. Oliner, "Hysterical Features Among Children of Survivors," in *Generations of the Holocaust*, ed. Martin S. Bergmann and Milton E. Jucovy (New York, 1990), 267–86.

sion in the call for a return to what has been forgotten or neglected in the development of Freud's ideas. In particular, Krystal and Cohen have reinstated Freud's concepts of automatic anxiety and primal repression, respectively, in an effort to establish a theory of trauma that resolves the aporias in Freud's arguments while incorporating recent clinical findings (PTSD is today defined as an anxiety disorder). By attempting to reevaluate two of Freud's most fundamental concepts these authors have made important efforts at clarification. But in their desire to disentangle the contradictions in Freud's thought, especially in their aspiration to replace Freud's economic theories of anxiety and primal repression by "operational" or "structural" approaches respectively, Krystal and Cohen fail to address the formidable issues raised by Freud's economic concepts.

Anxiety, Primal Repression, Mimesis

Inhibitions, Symptoms and Anxiety is Freud's key text on anxiety as well as on the obscure notion of primal repression—as we shall see, the two concepts are closely linked. In *Inhibitions,* Freud appears to privilege a "signal" theory of anxiety, according to which the ego signals the approach of a recognizable danger, over an economic or "automatic" theory of anxiety, involving the breaching or breaking through of the protective shield against stimuli (of which the traumatic neuroses of war are the paradigm). Anxiety "should not be explained from an economic point of view," Freud states at the outset. "Anxiety is not newly created in repression; it is reproduced as an affective state in accordance with an already existing mnemic image. . . . Affective states have become incorporated in the mind as precipitates of primaeval traumatic experiences, and when a similar situation occurs they are revived like mnemic symbols" (*ISA*, 93). Anxiety is "only an affective signal" in the production of which "no alteration has taken place in the economic situation" (*ISA*, 126). He thus subordinates the economic dimension of anxiety in favor of an account that historicizes and narrativizes it, by taking the danger that threatens the ego to be the reproduction of a prior situation that the ego can in principle signal, indicate, and represent: the threat of the father (castration) or, more primordially, the danger of the loss of the mother, or her breast (*ISA*, 128–29). On this model, anxiety serves the purpose of protecting the psyche's coherence by allowing the ego to represent and master a danger situation that it recognizes as the reproduction of an earlier situation involving the threatened loss of an identifiable libidinal object.

Freud would like to assimilate the traumatic neuroses to the same libidinal model. He acknowledges in this connection that, as a result of

World War I, many physicians had been tempted to regard the war neuroses as a direct result of the fear of death and hence to dismiss the question of castration. Against them, he argues that the introduction of the concept of narcissism, which libidinizes the ego and the instinct for self-preservation, rules that dismissal out of court. Moreover, he thinks it is "highly improbable" (*ISA*, 129) that a neurosis could come into being merely because of the objective presence of danger, without any participation of the deeper levels of mental functioning. Since according to Freud the unconscious knows nothing of death or negation,[21] he suggests that the fear of death should be regarded as analogous to the fear of castration) and that the "situation to which the ego is reacting is one of being abandoned by the protecting super-ego—the powers of destiny—so that it no longer has any safeguard against all the dangers that surround it" (*ISA*, 130).

But these assertions leave a remainder or supplement, in the form of a reinstatement of the very economic approach to anxiety and trauma that has ostensibly been rejected. "In addition," Freud immediately goes on, "it must be remembered that in the experiences which lead to a traumatic neurosis the protective shield against external stimuli is broken through and excessive amounts of excitation impinge upon the mental apparatus; so that we have here a second possibility—that anxiety is not only being signalled as an affect but is also being freshly created out of the economic conditions of the situation" (*ISA*, 130). But Freud's second possibility is also the first possibility, because the breaking through, or breaching of, the protective shield defines the mechanism that Freud calls "primal repression"—that archaic or primal form of repression that comes before repression proper, and on which the latter depends. And as Freud also observes, primal repression can only be described in economic or quantitative terms: "It is highly probable that the immediate precipitating causes of primal repressions are quantitative factors such as an excessive degree of excitation and the breaking through of the protective shield against stimuli" (*ISA*, 94). So Freud characterizes anxiety simultaneously as the ego's guard against future shocks *and* as what plunges it into disarray owing to a breaching of the protective shield: <u>anxiety is both *cure* and *cause* of psychic trauma</u>. The result is that the opposition between the signal theory of anxiety and the automatic or economic theory of anxiety cannot be sustained. For the historical situation of threatened loss (of the phallus or the mother) is itself defined as a situation of helplessness or

21. Sigmund Freud, "Thoughts for the Times on War and Death" (1915), *Standard Edition*, 14: 289.

"unbinding" (primal repression) due to an excess of stimulation that by traumatically breaching the boundary between inside and outside shatters the unity and identity of the ego.

There is more. In the last sentence I have deliberately followed Freud in using the term "unbinding" to denote the piercing or breaching of the ego's protective shield and the consequent release of energy or affect associated with the traumatic situation. I have done so in order to bring out the importance of that term to his conceptualization of psychic trauma. The term "unbinding" belongs to the pair "*Bindung-Entbindung*," terms that are associated from the start with Freud's economic hypotheses. For Freud, as Laplanche, Pontalis, and others have shown, binding furnishes the general concept of union, that is, the formation of coherent, homogeneous, and massive unities.[22] In Freud's earliest writings, binding is the process that binds "free" or "unbonded" energy in order to establish stable forms—for example, the ego, which requires a mass of neurones whose energy is in a bound state. In *Beyond the Pleasure Principle*, binding is the most important function of the psychical apparatus, which binds the destructive external quantities of excitation in order to master them, even before the intervention of the pleasure principle. *Binding* is thus the mechanism that serves to protect the organism against the unpleasurable *unbinding* of the ego caused by excessive stimulation, or trauma. It is only when the ego is caught unprepared and insufficiently "cathected" to bind additional amounts of inflowing energy that its protective shield is breached and a massive release of unbound or unpleasurable energy occurs. By binding excitations, the organism defers its own death drive. Binding also carries an explicitly political meaning: by binding or bonding the individual with the other or outside in an emotional bond of identification that constitutes the homogeneous group or mass, individuals neutralize their lethal tendency to disband into a disorderly panic of all against all (*ET*, 5).

According to Freud, it is Eros or libido that binds subjects to the objects of their desires, including the Father, Führer, or Chief. But as Mikkel Borch-Jacobsen has shown in *The Freudian Subject* and related essays, Freud's texts are quietly disorganized by a gesture that threatens the libidinal economy. This is because alongside the theory of love or libido he simultaneously postulates the existence of a principle that binds the

---

22. Laplanche and Pontalis, *Language of Psycho-Analysis*, s.v. "binding"; and Mikkel Borch-Jacobsen, who has characterized Freud's concept of binding as "one of the most decisive (and problematic) notions in the Freudian apparatus," *The Emotional Tie: Psychoanalysis, Mimesis, and Affect* (Stanford, California, 1992), 4; hereafter abbreviated *ET*.

individual to the other based not on desire for an object but on an emotional bond of identification that is "*anterior* and even *interior* to any libidinal bond" (*ET,* 8). Freud also calls that emotional bond of identification "feeling," a term that overlaps with a whole group of psychological concepts, such as sympathy and mental contagion, and implies an entire theory of *imitation,* or "mimesis." "Identification is known to psychoanalysis as the earliest expression of an emotional tie with another person," one that is "already possible before any sexual-object choice has been made," Freud observes in *Group Psychology and the Analysis of the Ego,* a text that precedes *Inhibitions, Symptoms, and Anxiety* and sets the stage for it. Moreover, according to Freud violence is inherent in the imitative-identificatory process, which he describes as a cannibalistic, devouring, incorporative identification that readily turns into the hostile desire to rid oneself of the other, or enemy, with whom one has just merged. "Identification, in fact, is ambivalent from the very first," he states; "it can turn into an expression of tenderness as easily as into a wish for someone's removal. It behaves like a derivative of the first, *oral* phase of the organization of the libido, in which the object that we long for and prize is assimilated by eating and is in that way annihilated as such. The cannibal, as we know, has remained at this standpoint; he has a devouring affection for his enemies and only devours people of whom he is fond."[23] A related text is Freud's "Mourning and Melancholia" (1917 [1915]), in which he also emphasizes the emotional ambivalence of identification; the terms in which he does so have been used to explain the characteristic depression and guilt of the survivor, as the symptoms become an expression of both the repressed hostility toward and love for the lost object.[24]

The primordiality Freud ascribes to the process of mimetic identification upsets the logic of desire and repressed libidinal representations that ostensibly governs his analysis of the history of the individual subject by proposing that, prior to the history of the repressed representations of

23. Sigmund Freud, "Identification," in *Group Psychology and the Analysis of the Ego* (1921), *Standard Edition,* 18: 105.

24. Sigmund Freud, "Mourning and Melancholia," *Standard Edition,* 14: 243–58; cf. M. Straker, "The Survivor Syndrome: Theoretical and Therapeutic Dilemmas," *Laval Medical* 42 (1971): 37–41. Although mourning is treated by Freud as normal and melancholia as its pathological version, a close reading of the text shows that the mechanism of ambivalent incorporation and identification held to be characteristic of melancholia is also the very ground of possibility of any relation to an object, including the child's first "object," the mother, an important theme in the conceptualization of traumatic identification, as I show.

the Oedipus complex, lies a *pre*history of unconscious emotional identi-fications with or incorporative bindings to the other, identifications that precede the distinction between subject and object on which the analysis of desire, even unconscious desire, depends. For Freud, the paradigm for those unconscious emotional identifications is crucially the hypnotic relationship or *rapport*, understood by him, as by the majority of his contemporaries, as an altered state of consciousness, or condition of *un-consciousness*, that involves an absorption in or identification with the hyp-notist or role so profound that the other, or role, is not perceived as other, or object. In short, Freud places a hypnotic-suggestive tie or bond at the center of the traumatic paradigm.

The primacy Freud accords to such nonlibidinal, hypnotic bonds of emotional identification silently undermines the Oedipal (erotic-repressive) logic of his approach to trauma. Thus on the one hand, Freud attempts to establish the singularity of psychoanalysis by breaking with hypnosis and grounding the patient's neurosis instead in (real or fanta-sized) repressed libidinal representations whose recovery through recol-lection or construction is the task of analysis. The unconscious, for Freud, is the repository of those repressed infantile representations, and it is the latter that, transferred secondarily to the person of the analyst, are held to become accessible to consciousness and recollection in the form of the patient's self-narration, or *diegesis*. For Freud in this mode, the patient's speech during the hypnotic trance does not constitute such a *diegesis* for that speech is a hypnotic-mimetic performance that occurs precisely in the absence of consciousness and self-representation.

On the other hand, as Freud soon discovers, the transference, far from facilitating recollection, proves rather to be its major stumbling block. Instead of remembering, patients *repeat* the earlier scenes or memories in the present, in a "positive" transference onto the analyst that, for all the absence of overt suggestion, or rather precisely because of the analyst's deliberate self-effacement, manifests all the more clearly that affective bond of identification with the "other" that for Freud is emotional iden-tification—or mimesis. In other words, following Borch-Jacobsen we can say that if Freud continues to believe that the transference constitutes a resistance to recollection by disguising or dissimulating a prior Oedipal affective tie, his own writings of the 1920s strongly suggest that no such dissimulation is involved. This is because the patient's transferential re-sistance rests on an affective bond or tie that, as Freud observes, cannot be repressed but can only be felt or experienced in the immediacy of an act-ing out or repetition in the present that is unrepresentable to the subject and that—like the unconscious itself—knows no delay, no time, no doubt, and no negation.

In short, where the notion of recollection becomes problematic in Freud is where he states, in his speculations in *Inhibitions, Symptoms, and Anxiety* and related texts, that the Oedipal tie which is supposed to be recalled in transference is itself a derivative of an even more archaic "affective tie" or "primary identification"—an identification that can never be remembered by the subject precisely because it precedes the very distinction between self and other on which the possibility of self-representation and hence recollection depends. It follows that the origin is not present to the subject but is on the contrary the condition of the latter's "birth."

If this is true of the origin, is it also true of trauma? Ever since the work of Sándor Ferenczi and Anna Freud we have become accustomed to think of the identification with the aggressor as one of the subject's characteristic responses to, or defenses against, psychic trauma.[25] But what if—as Freud suggests—trauma is understood to consist in *imitative or mimetic identification itself*, which is to say in "the subject's *originary* 'invasion'" or alteration?[26] This would be to attribute the patient's lack of memory of the trauma not to the repression of a representation of the traumatic event, but to the vacancy of the traumatized subject or ego in a hypnotic openness to impressions or identifications occurring prior to all self-representation and hence to all rememoration. So if the victim of a trauma identifies with the aggressor, she does so not as a defense of the ego that represses the violent event into the unconscious, but on the basis of an unconscious imitation or mimesis that connotes an abyssal openness to all identification. This would explain why the traumatic event cannot be remembered, indeed why it is "relived" in the transferential relationship not in the form of a recounting of a past event but of a hypnotic identification with another in the present—in the timelessness of the unconscious—that is characterized by a profound amnesia or absence from the self. It would also suggest an explanation, grounded in Freud's conception of trauma as the archetrauma of identification, of why the vic-

25. Although the concept of identification with the aggressor is usually attributed to Anna Freud's *The Ego and Mechanisms of Defense* (1936), it was first formulated by Sándor Ferenczi in "Confusion of Tongues Between Adults and the Child" (1933), in *Final Contributions to the Problems and Methods of Psychoanalysis*, ed. Michael Balint (New York, 1955), 162.

26. Mikkel Borch-Jacobsen, "Dispute," in *Hypnose et psychanalyse: réponses à Mikkel Borch-Jacobsen*, ed. Leon Chertok (Paris, 1987), 203.

tim's memory of the traumatic event is so often difficult if not impossible to recover (something Freud appears to recognize when, in discussing the interminableness of analysis, he acknowledges the implacable nature of the death drive, or compulsion to repeat).[27]

From this perspective, the traumatic "event" is redefined as that which, precisely because it triggers the "trauma" of emotional identification, strictly speaking cannot be described as an event since it does not occur on the basis of a subject-object distinction. (Hence the ambiguity of the term "trauma," which is often used to describe an *event* that assaults the subject from outside, but which according to the theme of mimetic identification is an experience or "situation" of identification that strictly speaking does not occur to an autonomous or fully coherent subject.) The archetrauma of birth, defined on this model as a primary identification or hypnotic repetition that occurs prior to any conscious perception or any repression, is unrepresentable to the subject, which is why in *Inhibitions* Freud opposes Rank's account of trauma as the repetition of a birth *event* (*ISA*, 135). Trauma is thus imagined as involving not the shattering of a pregiven ego by the loss of an identifiable object or event but a dislocation or dissociation of the "subject" prior to any identity and any perceptual object.[28] It is no accident, I think, that whenever Freud in *Inhibitions, Symptoms, and Anxiety* is in danger of forgetting this proposal and hence of treating the ego's reaction to trauma as a reaction to a specific, determinate event or object that it can in principle indicate, signal, and confront, the war neuroses symptomatically reappear in his text as the paradigm of trauma defined in economic terms as involving the breaching of the protective shield, which is to say, of trauma defined as mimetic or imitative identification.

There is one more point to be made here. Throughout my discussion I have aligned trauma both with the breaching of the protective shield, or unbinding, and with mimetic identification, or binding. This is because

27. Sigmund Freud, "Analysis Terminable and Interminable" (1937), *Standard Edition*, 23: 242–44.

28. Freud's criticisms of Rank suggest that for Freud, the traumatic situation does not designate a determinate objective reality but something vague and indeterminate that overwhelms it: a situation of helplessness due to fright that he will also interpret in hypnotic-identificatory terms. For Ferenczi's similar criticisms of Rank see "Zur Critique der Rankschen 'Technik der Psychanalyse'" (1927), in *Bausteine zur Psychanalyse* (Leipzig, 1927), 2: 116–28. Cf. Philippe Lacoue-Labarthe and Jean-Luc Nancy, "The Unconscious is Destructured Like an Affect: (Part I of 'The Jewish People Do Not Dream')," *Stanford Literature Review* 6 (Fall 1989): 200; and Samuel Weber, *The Legend of Freud* (Minneapolis, Minnesota, 1982), 48–60.

in the economic terms associated with Freud's ideas, the traumatic experience involves a fragmentation or loss of unity of the ego resulting from the radical unbinding of the death drive, but it also entails a simultaneous binding (or rebinding) of cathexes: both unbinding and binding—hate and love—are constitutive of the traumatic reaction. This thesis is implicit in the work of Freud and related figures, even as they fail to thematize it or pursue its implications. Thus for Freud and his followers the overwhelming of the ego's protective shield and mimetic disappearance of the ego represent a defusion or decoupling of the life and death drives (love and hate) and a consequent unbinding of the cathexes of the death drive. "In traumatic neurosis we have . . . a splitting apart of the instinctual components of cathexes," Abram Kardiner observes. "In the traumatic moment . . . all cathexes are abruptly cut through and the defused destructiveness which is turned against the ego is manifest in the form of loss of consciousness."[29] Ferenczi will give the process of unbinding a political reading when, on the basis of Freud's interpretation of the politics of identification and the crowd or mass, he compares the psychical disorganization and multiple identifications consequent on the loss of the ego's leadership in severe traumatic states to the panic reaction of the crowd when it loses its political leader or fuhrer.[30]

Kardiner appears to imagine that unbinding (the death drive, or Thanatos) can be contrasted with binding (the life drive, or Eros) as two terms in an oppositional process, such that an unbinding or trauma can be succeeded by its opposite, a rebinding and hence an attempt at cure. He thus follows Freud in interpreting the traumatic nightmare as a retroactive attempt at binding, or mastery, of this kind.[31] But a close reading of Freud's texts on the economics and politics of identification shows that the opposition between unbinding and binding is constitutive. Panic—in individual terms the *unbinding* or splitting of the subject into mimetic identifications, in political terms the unbinding of the ties between individuals in the unruly crowd of all against all—is simultaneously and irreducibly a *binding* consequent on the very mimetic identifications that suggestively or contagiously bind the individual to the other, or mass.

29. Abraham Kardiner, "The Bio-Analysis of the Epileptic Reaction," *Psychoanalytic Quarterly* 1 (1932): 461. Sometime later, Abraham Kardiner changed his name to Abram.

30. Stefan Hollos and Sándor Ferenczi, *Psychoanalysis and the Psychic Disorder of General Paresis*, trans. Gertrude M. Barnes and Gunther Keil (New York, 1925), 47–48.

31. Kardiner, "Epileptic Reaction," 461.

Mikkel Borch-Jacobsen, who has written the most penetrating analysis of this aspect of Freud's thought, puts it as follows:

> The paradox is this: since "sympathy" (i.e., co-feeling or suffering-with) truly constitutes the most immediate possible bond with others, the disappearance of the bond of love with the chief does not, as Freud wished it to, liberate the pure and simple disbanding of Narcissi [independent egos or subjects]. In a sense it does not liberate anything at all, and certainly not autarchic subjects (individuals), since panic is precisely uncontrollable breaching by the ego by (the affects of) others, or . . . a mimetic, contagious, suggested narcissism. What comes out in panic phenomena . . . is everything that Freud had violently rejected under the rubric of "suggestion" or "suggestibility," understood as the relation of immediate, "hypnotic" fusion with another. . . . The acme of the "sympathetic" relationship with others is simultaneously the ultimate nonrelationship with others: each imitates the "every man for himself" of the others; here assimilation is strictly equivalent to a disassimilating dissimilation. The panic bond goes beyond the alternatives of bonding and disbanding . . . A disbanding band must be qualified as both narcissistic and nonnarcissistic, egoistic and altruistic, asocial and social. (*ET,* 9)

"*Asocial and social.*" Doesn't that characterize the behavior of the traumatized soldier as described by Kardiner and others? Paradoxically, the asocial traumatized soldier who is so antimimetically withdrawn from the world that he is completely numb to it is simultaneously so socially identified with it that the boundaries between himself and others are completely effaced. Thus Kardiner depicts the victim's behavior as so rigid, his face so lacking all mimicry and expressions of feeling that he gives the impression of being utterly detached from others; but he also portrays the same victim as so mimetically identified with the world's dangers as to be completely impressionable or suggestible. The response of the traumatized soldier thus at one and the same time represents the achievement of defense and the failure of defense, the success of protection and the breaching of the protective shield—antimimesis and mimesis.[32]

Mimesis and Antimimesis

To summarize: in his discussion of the concepts of automatic anxiety and primal repression to which the problem of trauma seems to warrant a re-

---

32. Cf. Ruth Leys, "Death Masks: Kardiner and Ferenczi on Psychic Trauma," *Representations* 53 (Winter 1996): 44–73.

turn, Freud placed a hypnotic-mimetic process of binding and unbinding at the center of the traumatic situation—a process that is rendered invisible by Krystal's, Cohen's, and Winston's rejection of Freud's economic concepts. Trauma was thus defined by Freud as a situation of unconscious identification with, or "primary repression" of, the traumatic scene or person that occurs in a state akin to the trance state and that is independent of a libidinal relation to the object. Hypnotic suggestibility was the key to the traumatic experience defined in this way—a claim that situates Freud among his contemporaries, such as Charcot, Janet, Prince, and others, for whom the conceptualization of trauma was inevitably connected with the rise of hypnosis as a legitimate field of inquiry and research. Hypnosis provided Freud with a model for unconscious identification because, according to the interpretation of hypnosis dominant at the turn of the century, hypnosis seemed to involve a subjection or immersion in the other or scene that is "blind" in the sense that the subject of hypnosis is unconscious or unaware of the hypnotist's commands, which she or he enacts or repeats without seeing that she or he is submitting to them—commands that are not remembered afterwards because they were not present to the subject in the form of a self-representation in the first place. Similar notions can be found in the writings of a whole range of authors who are crucial for the genealogy of trauma, including Prince, Ferenczi, Kardiner, and others whose work I examine. In short, Freud's work crystallizes and makes manifest a problematic of hypnotic-mimetic identification that was central to the origin of theorizing about trauma at the turn of the century.

Nevertheless, as Borch-Jacobsen has demonstrated, Freud constantly repudiated the hypnotic-dissociative indistinction between subject and other that was held to characterize the traumatic situation. He attempted to evade the uncanny loss of individuality or dedifferentiation between self and other that was held to take place in hypnosis by reinterpreting the effects of suggestion as the product not of the relationship between hypnotist and subject, but of the subject's sexual desire. What Freud found disturbing about hypnosis-suggestion, and what he therefore struggled to suppress, was the idea that in suggestion my thoughts do not come from my own mind or self but are produced by the imitation or suggestion of another—the hypnotist or, in psychoanalytic practice, the analyst. Freud's theory of the unconscious may thus be seen as an attempt to solve the problem of the hypnotic rapport by transforming suggestion into desire. In the work of other authors, the repudiation of mimetic identification takes place within the theorization of suggestion itself. Thus Morton Prince, lauded today as a pioneer in the study of the dissociative disorders, attributed the effects of trauma simultaneously to a hypnotic dissociation or splitting that was immemorial *and* to the spontaneity of a subject who could *see* the scene of trauma and so represent it to herself. The same double structure—of mimesis and antimimesis—is found in the writings of Ferenczi, Kardiner, and many others.

The antimimetic turn within the mimetic theory has several important consequences that are apparent in conceptualizations of trauma even today. According to the mimetic hypothesis, the traumatic repetition which the victim is encouraged to dramatize in the hypnotic-cathartic treatment takes the form of an acting out of the real or fantasized scene of trauma (for in the trance state the scene in question may well contain fictive elements, as Freud and others were aware)—an acting out that, because it takes place in the mode of a emotional identification that constitutes the hypnotic rapport, is unavailable for subsequent recollection. At the same time, the reliving of the real or fictive traumatic situation under hypnosis is also understood differently, or antimimetically, by Freud and others: not as a dramatic *mimesis* but as a verbalization or *diegesis,* in which the patient recounts and recollects the traumatic scene in full consciousness—even if, as my discussion in subsequent chapters of the cathartic treatment of the traumatic neuroses of war in both World Wars shows, within the understanding of hypnosis at the time the demand for the patient's recollection and self-knowledge cannot easily be met.

Equally crucial, the antimimetic turn within the mimetic paradigm—the therapist's demand that the patient *be a subject* capable of distancing herself from the traumatic scene—is simultaneously the moment when emphasis tends to shift from the notion of trauma as involving a mimetic yielding of identity to identification to a notion of trauma as a purely external cause or event that comes to an already constituted ego to shatter its autonomy and integrity. Passionate identifications are thereby transformed into claims of identity, and the negativity and violence that according to hypothesis inhere in the mimetic breaching of the boundaries between the internal and the external are violently expelled into the external world, from where they return to the fully constituted, autonomous subject in the form of an absolute exteriority. The result is a rigid dichotomy between internal and external such that violence is imagined as coming to the subject entirely from the outside.[33] The value to

33. These insights have been applied to the phenomenon of serial killing and the culture of violence in the United States by Mark Seltzer, *Serial Killers: Death and Life in America's Wound Culture* (New York, 1998).

proponents of the view that violence is utterly external to the subject is that it serves to forestall the possibility of scapegoating by denying that the victim participates, or in any way colludes with, the scene of abjection and humiliation. But it is a view of the location of violence that also has its costs:

1. It makes unthinkable, or renders incoherent, the mimetic-suggestive dimension of the traumatic experience, a dimension that, I have tried to show, calls into question any simple determination of the subject from within or without and that is present in the tendency to suggestibility that is still recognized as symptomatic of patients suffering from trauma. The hypnotic suggestibility of the victim of dissociation or PTSD makes the patient's testimony about the historical truth of the traumatic origin inherently suspect owing to the potential for hypnotic confabulation and "false memories," yet (as my discussion of the work of van der Kolk in chapter 7 shows) that suggestibility is untheorizable within the terms of an analysis of trauma that rejects any acknowledgment of the mimetic dynamic.

2. Indeed, such an analysis tends to produce a conceptualization of the dissociated or traumatic memory as completely literal in nature, as if an account of the traumatic experience as absolutely true to external reality, uncontaminated by any subjective, unconscious-symbolic or fictive-suggestive dimension, is necessary in order to reinforce a rigid polarization between inside and outside that is otherwise threatened by the mimetic dynamic. Yet the theory of the literal nature of traumatic memory has been, and continues to be, challenged by evidence proposing the presence of a subjective-suggestive component in the constitution of the traumatic experience.

3. The same dichotomy between internal and external reinforces an opposition between absolute aggressor and absolute victim in such a way as to render untheorizable the violence and ambivalence that, according to the mimetic hypothesis, necessarily inhere in the victim of the traumatic scenario. The mimetic theory makes it possible to sympathetically acknowledge the hideous ways in which the victim can come to psychically collude in the scene of violence through fantasmatic identifications with the scene of aggression. Whereas the complete rejection of any idea of the mimetic renders the source of such identifications mysterious.

4. The rigid dichotomy held to exist between the external and internal inevitably reinforces gender stereotypes by conceptualizing the already-constituted female subject as a completely passive and helpless victim. The irony is that, when an unexamined notion of contagion or infection is added to such a stark opposition between the outside and the inside, it

turns out that the aggressor too can become a victim, as I show in my discussion of the recent work on trauma by Cathy Caruth.

5. In a radically different direction, the same antimimetic turn within the mimetic paradigm can call into question the entire validity of the concept of trauma. Thus an alternative version of the same demand that there be a subject capable of distancing herself from the traumatic-mimetic scene revalorizes the notion of *simulation* as a kind of voluntary game between subject and hypnotist and does so in ways that make the very concept of trauma suspect. When Borch-Jacobsen wrote *The Freudian Subject* he accepted the common turn-of-the-century definition of hypnosis as a non-specular or "blind" identificatory mimesis that preceded the subject-object divide.[34] At the time, he seems to have believed he was defining the essence of hypnosis, rather than a historical conceptualization that happened to be dominant at the turn of the century. But in his more recent writings he has rejected his earlier emphasis on the blindness of the hypnotic relationship in order to describe hypnosis as a *specular* game carried out with the subject's complete awareness of her performance. In short, Borch-Jacobsen now wants to resolve the tension between imitation as blind mimesis and imitation as spectatorial distantiation by deciding in favor of lucid simulation. His new position goes hand in hand with a rejection of any concept of the unconscious and any notion of (libidinally repressed or mimetically dissociated) traumatic memories.

34. Hypnosis is "properly speaking, a subjection, in the very strong sense of this word," he wrote: "the subject is rendered subject, assigned as subject. The hypnotic commandment does not present itself to a consciousness that is already there to hear it; rather, it takes hold of it (and posits it) prior to itself—in such a way that it never presents itself to consciousness. It falls into a radical 'forgetting' that is not the forgetting of any memory, of any (re)presentation. It gives no order *to* a subject; it orders the subject. . . . Far from replying, then, to the discourse of the other, the hypnotized person quotes it in the first person, acts it out or repeats it, without knowing that he is repeating [as Borch-Jacobsen observes, this is precisely Freud's definition of the 'repetition compulsion']. He does not submit himself *to* the other, he *becomes the other,* comes to be like the other—who is thus no longer an other, but 'himself.' No property, no identity, and in particular no subjective liberty precede the commandment, here. . . . In short, hypnosis involves the birth of the subject—perhaps not a repetition of the birth event, but birth as repetition, or as primal identification: in it the subject comes into being (always anew: this birth is constantly repeated) as an echo or duplicate of the other, in a sort of lag with respect to its own origin and identity. An insurmountable lag, then, since it is a constitutive one, and one that without any doubt constitutes the entire 'unconscious' of the subject, prior to any memory and any repression. The (constraint to) repetition, as Freud has indeed said, is the unconscious itself" (Mikkel Borch-Jacobsen, *The Freudian Subject* [Stanford, California, 1988], 229–31).

Consequently, for him there is no genuine forgetting of the mimetic performance, which is merely a suggested scenario undertaken with the patient's voluntary compliance. All cathartic "reenactments" or "relivings" are characterized by him in the same terms, with the result that he goes so far as to intimate that the traumatic neuroses of war belong to the same category of simulated, hypnotic inventions—an argument that comes disturbingly close to the traditional view of traumatic neurosis as a form of malingering.[35]

Yet Borch-Jacobsen's skeptical ideas generate several contradictions, demonstrating yet again that mimesis cannot be simply made to disappear. Rather, according to the discourse that has shaped the conceptualization of trauma from the start, both mimesis and antimimesis are internal to the traumatic experience. We might put it that the concept of trauma has been structured historically in such a way as simultaneously to invite resolution along the lines of an antimimetic repudiation of the mimetic dimension and to resist it, or at any rate to suggest that the desire to resolve the oscillations internal to that paradigm is a response to the anxieties that are constitutive of it.

Freud's texts exhibit that structure in a particularly exemplary way, as does Morton Prince's book on the Beauchamp case, to which I now turn.

35. Mikkel Borch-Jacobsen, *Remembering Anna O.: A Century of Mystification* (New York, 1996); idem, "Neurotica: Freud and the Seduction Theory"; idem, "L'effet Bernheim (fragments d'une théorie de l'artefact généralisé)," *Corpus* 32 (1997): 147–73.

II

The Real Miss Beauchamp:
An Early Case of Traumatic Dissociation

In the spring of 1898 a young woman of modest Irish-American background and education came to Morton Prince, a well-known New England psychotherapist, for the treatment of a condition he diagnosed as a typical if extreme example of neurasthenia or hysteria. Almost immediately, the case took an unexpected turn when, failing to improve his patient's health by "conventional" methods of treatment, Prince began to use hypnotic suggestion.[1] The patient, when awake and when hypnotized, was usually weary, depressed, and extremely passive. One day, however, under deep hypnosis, she suddenly changed: she became lively, bold, saucy—and difficult to control. In this new state, she insisted that she was an entirely different person from the morbidly conscientious, reticent, anxious, self-sacrificing, and pious individual whose body she shared and from whom she differentiated herself by referring to the latter contemptuously in the third person as "She." Soon convinced that this second vivacious personality's claim to independence was genuine, Prince named her "BIII," "Chris," or "The Devil," in order to distinguish her from her other self, whom he called "BI," "Christine," or "The Saint" (BII appeared to be BI in the hypnotic state). Subsequently, "of her own accord," Chris in a "spirit of fun" adopted the name "Sally Beauchamp" after "a character in some book" (DP, 29–30). Nor was this the end of the multiplication of the patient's selves, for about a year later still another personality emerged, "BIV," whose independence, "frailties of temper, self-concentration, am-

1. Morton Prince, *The Dissociation of a Personality. A Biographical Study in Abnormal Psychology*, 2d ed. (London, 1910), 20; hereafter abbreviated *DP*. By conventional methods of treatment, Prince meant hydrotherapy and electrotherapy.

41

bition, and self-interest" (*DP,* 17) led Prince to call her "The Woman" or "The Realist," and whom Sally scornfully dubbed "The Idiot."

The confrontations, struggles, and embarrassments that resulted from the dramatic appearances and disappearances of the three major personalities in the case furnished the plot of Morton Prince's pioneering study of multiple personality, *The Dissociation of a Personality: A Biographical Study in Abnormal Psychology* (1905). The case only came to an end six years later when Prince succeeded through his hypnotic powers in identifying the trauma at the origin of his patient's condition and in "resurrecting" Miss Beauchamp's "real, original or normal self" (*DP,* 1) by annihilating Sally and synthesizing or "integrating" the other personalities into a single, stable identity. At last restored to the self "that was born and which she was intended by nature to be" (*DP,* 1), Miss Beauchamp— "like the traditional princess in the fairy story"—was awakened by her Prince and "soon married and 'lived happily ever afterward.'"[2]

Prince's text invokes the genres of the detective and the adventure story. But as the last quotation shows, the narrative also alludes to the genre of the fairy tale. As in all fairy tales, *The Dissociation of a Personality* is preoccupied with change, with the magical—i.e., hypnotic—transformation of the self.[3] And as in a fairy tale, the narrative's resolution is at once conventional, enforcing a social moral—the destiny of women is marriage—and arbitrary. One feels the arbitrariness above all in the way the text comes to an abrupt stop. It would be tempting to say that the abruptness of the ending amounts to a kind of violence, epitomized by the narrative's representation of Prince's relentless, Svengali-like struggle for Miss Beauchamp's soul as a war to the death and the suppression of his patient's various alter egos as "psychical murder" (*DP,* 248). Yet it might be truer to say that the abruptness of the ending is a way of *avoiding* violence. Sally must be made to "go back to where she came from" (*DP,* 414, 138–39, 405, 524), an ambiguous expression that could be a eu-

2. Morton Prince, "Miss Beauchamp: The Theory of the Psychogenesis of Multiple Personality" (1920), in *Clinical and Experimental Studies in Personality* (Cambridge, Massachusetts, 1929), 208. Soon after the end of her treatment, the upwardly mobile Miss Beauchamp (actually Clara Norton Fowler) managed to obtain the rudiments of a college education by enrolling as a special student for three semesters at Radcliffe; in 1912, aged thirty-nine, she married Dr. George A. Waterman, a leading Boston psychotherapist who was a close associate of Prince.

3. It seems obvious that Olive Higgins Prouty's novel, *Now, Voyager* (1941) (made into a film with the same title starring Bette Davis), with its alternate personality of Renée Beauchamp and its emphasis on the transformation of the self, is based in part on Prince's Beauchamp case.

phemism for her murder or suicide but that could also be taken to suggest that she isn't exactly dead—as if Prince had contrived at once to eliminate and to spare the playful, flirtatious, heartless, rebellious, dangerous Sally (one reviewer compared her to an unruly city mob)[4] without whom the case would lose much of its narrative drive and interest and whom Prince finds "delightfully attractive" (*DP*, 53), indeed "irresistible" (*DP*, 110). (It was typical of such cases that the most prominent of the selves to emerge during treatment was more athletic, outgoing, spontaneous, reckless, and irresponsible—in short, more *juvenile*—than the patient's primary self.)[5] The great, even excessive length of Prince's narrative—it goes on for more than five hundred pages, while another American case-history closely modeled on the Beauchamp case is a numbing fifteen hundred pages long[6]—likewise suggests the author's reluctance if not inability to bring the narrative to a close. Prince's friend and colleague, William James, was also fascinated by Sally: "But *who* and *what* is the lovely Sally? That is a very dark point."[7] The reader too would like to know what the enthralling but threatening Sally represents in the case.

Prince's Beauchamp case is arguably the most impressive of that group of multiple personalities cases that, as Hacking has argued, became an object of knowledge for the first time at the end of the nineteenth century.[8] Deliberately aimed at a general audience, Prince's book "created a

4. Cited by Otto Marx, "Morton Prince and the Dissociation of a Personality," *Journal of the History of the Behavioral Sciences* 6 (1970): 124.

5. A point stressed by Hillel Schwarz, "The Three-Body Problem and the End of the World," in *Fragments for a History of the Human Body*, ed. Michael Feher (New York, 1989), 4. 420–24.

6. Walter Franklin Prince (and James Hervey Hyslop), *The Doris Case of Multiple Personality; A Biography of Five Personalities in Connection with One Body* (New York: Proceedings of the American Society for Psychical Research, 1915–17).

7. William James to Morton Prince, September 28, 1906, cited by Saul Rosenzweig, "Sally Beauchamp's Career: A Psychoarchaeological Key to Morton Prince's Classic Case of Multiple Personality," *Genetic, Social, and General Psychology Monographs* 113 (1987): 8. Two days later Prince wrote James: "And to think that Sally has 'gone back to where she came from' when she might have told me so much that I wished I knew!"—a remark that testifies to Prince's sense that the case might be over but that it was not *resolved*. Prince's letter to James is cited by Michael G. Kenny, "Multiple Personality and Spirit Possession," *Psychiatry* 44 (1981): 347, n. 19.

8. Ian Hacking, "The Invention of Split Personalities," in *Human Nature and Natural Knowledge*, ed. Alan Donagan, Anthony N. Perovitch, Jr., and Michael V. Wedin, *Boston Studies in the Philosophy of Science* 89 (1989), 63–85. I have also profited from the following sources: Henri Ellenberger, *The Discovery of the Unconscious: The History and Evolution of*

sensation wherever English was read,"[9] quickly went through two edi-
tions and many more printings, and provided the theme for more than
five hundred plays, one of which drew capacity audiences on Broadway.[10]
Yet by 1905, the by then familiar terms of its analysis, derived chiefly from
Janet—emotional shock or trauma, mental dissociation, and hypnotic
suggestion—were already being challenged by Freud (the Dora case appeared the same year). That is one reason why hypnosis and with it the
multiple personality concept were progressively abandoned in the years
that followed.[11] And in fact it is difficult to read Prince's text without
tending more or less automatically to place it in a psychoanalytic frame—
without thinking, for example, of Sally's investment in Prince in libidi-
nal-conflictual terms,[12] or of Prince's desire to hunt down the "real" Miss
Beauchamp in Lacanian terms as a wish to recover a presence or origin
that in principle cannot be recuperated but only fantasized or displaced.
But to yield to that tendency would, I think, be a mistake for the simple
reason that it would remove us all too quickly and efficiently from the

---

Dynamic Psychiatry (New York, 1970); E. T. Carlson, "The History of Multiple Personality
in the United States: Mary Reynolds and Her Subsequent Reputation," *Bulletin of the History of Medicine* 58 (1974): 72–82; idem, "The History of Multiple Personality in the
United States: 1. The Beginnings," *American Journal of Psychiatry* 138 (1981): 666–68;
Jacques Quen, ed., *Split Minds/Split Brains* (New York, 1984); Michael G. Kenny, *The Passion of Anselm Bourne: Multiple Personality in American Culture* (Washington, D.C., 1986);
Ian Hacking, "The Making and Molding of Child Abuse," *Critical Inquiry* 17 (1991): 253–
88; idem, "Two Souls in One Body," *Critical Inquiry* 17 (1991): 838–67; idem, *Rewriting the
Soul: Multiple Personality and the Sciences of Memory* (Princeton, New Jersey, 1995). For a
valuable discussion of a case of multiple personality often cited by Prince see Sonu Shamdasani, introduction to *From India to the Planet Mars: A Case of Multiple Personality with
Imaginary Languages* by Theodore Flournoy (1900; reprint Princeton, New Jersey, 1994),
xi–li.

9. Henry A. Murray, "Dr. Morton Prince: A Founder of Psychology," *Harvard Alumni
Bulletin* 32 (1930): 491.

10. See Nathan G. Hale, Jr., *James Jackson Putnam and Psychoanalysis: Letters between
Putnam and Sigmund Freud, Ernest Jones, William James, Sándor Ferenczi, and Morton Prince,
1877–1917* (Cambridge, Massachusetts, 1971), 17.

11. For a discussion of the reasons for the decline of hypnotism, a complex development that awaits its historian, see Alan Gauld, *A History of Hypnotism* (Cambridge, 1992),
559–67; and Hacking, *Rewriting the Soul*, 129–32.

12. For an attempt at interpretation along these lines see Harold Grier McCurdy, "A
Note on the Dissociation of a Personality," *Character and Personality* 10 (1941): 35–41.

contradiction-riddled but also intensely interesting discursive field of Prince's famous work.

Now a fundamental assumption of Prince's narrative is that the different personalities are autonomous subjects whom he understands as having emerged as a consequence of an emotional trauma and the patient's pathological condition. But his text is also haunted by the fear or possibility that Sally and the others have been created by the physician's own hypnotic or suggestive powers. What complicates matters further is that Prince's book is written under the sign of, and in relation to, a paradigm of *mimetic-hypnotic identification* in which he himself is necessarily implicated.

But what is identification? In chapter 1 I argued that if we want to understand the genealogy of trauma we cannot uncritically follow Freud's tendency to regard identification as the result of the subject's unconscious desire for a loved object, as commentators in various fields have tended to do. Rather, we must understand the extent to which identification was also understood by him as involving the hypnotic or mimetic imitation by one "self" of an "other" that to all intents and purposes was indistinguishable from the first and as a result of which desire was provoked or *induced*. "In the beginning is mimesis: as far back as one goes in anamnesis (in self-analysis, we might say, if the *self* were not precisely what is in question here), one always finds the identification from which the 'subject' dates (the 'primary identification,' as Freud later puts it)," Mikkel Borch-Jacobsen has written in reopening the problem of suggestion and imitation in Freud:

> This is why the chronology Freud most frequently indicates has to be inverted. Desire (the desiring subject) does not come first, to be *followed* by an identification that would allow the desire to be fulfilled. What comes first is a tendency toward identification, a primordial tendency which then gives rise to a desire; and this desire is, from the outset, a (mimetic, rivalrous) desire to oust the incommodious other from the place the pseudo-subject already occupies in fantasy.

"If desire is satisfied in and through identification," Borch-Jacobsen adds in commenting on this aspect of Freud's thought, "it is not in the sense in which a desire somehow precedes its 'gratification,' since no desiring subject (no 'I,' no ego) precedes the mimetic identification: identification brings the desiring subject into being, and not the other way around." The hypnotic relationship between the subject and the hypnotist precisely exemplified the workings of that "primordial," unconscious, identification.[13]

Hypnosis threatened to dissolve the distinction between self and other to such a degree that the hypnotized subject came to occupy the place of the "other" in an unconscious identification so profound that the other was not apprehended *as other.* We might put it that in hypnosis the imitation or mimesis was unrepresentable to the subject. Yet as Borch-Jacobsen has argued, this lack of distinction between self and other had to be "acted out" (*FS*, 40). This meant that mimesis was continually relegated to a secondary position: the hypnotized person was conceived not as imitating the "other" in a scene of unconscious, non-specular identification unavailable to subsequent recall, but rather as occupying the vantage point of a spectator who, distanced from the scene, nevertheless could see herself in the scene, could represent herself to herself *as other,* and hence could distinguish herself from the model. In short, mimesis was continually being converted into the specular order of the hated double or rival. (I draw attention again to the idea that mimesis is always ambivalent, it always produces a sadistic, paranoid desire to annihilate the "other" who is also "myself": "I hate her, I just hate her," [*DP,* 130] says Sally Beauchamp of her alter ego, Miss Beauchamp. "I wish she were dead!" [*DP,* 169]. The same rivalry and animosity between selves, or what today are called "alters," is often reported in more recent cases of multiple personality.) The identity between self and other that was constitutive of mimesis constantly succumbed to the antimimetic requirement that there be a subject who preceded mimesis and out of which identification was produced.

We might put it that Freud sought to evade the radical dedifferentiation implicit in hypnotic suggestion by reinterpreting the effects of hypnosis as the product, not of the relationship between the hypnotist and "subject," but of the subject's sexual desires. For Borch-Jacobsen, therein lies precisely Freud's originality and historical specificity. "Always [Freud] imputed to an improbable hypnotised *subject* the responsibility and the initiative for this unappropriable, unimputable 'rapport,'" Borch-Jacobsen states:

13. Mikkel Borch-Jacobsen, *The Freudian Subject* (Stanford, California, 1988), 47; hereafter abbreviated *FS*. My analysis is also indebted to Philippe Lacoue-Labarthe, *Typography: Mimesis, Philosophy, and Politics,* ed. Christopher Fynsk (Cambridge, Mass., 1989); Lacoue-Labarthe and Jean-Luc Nancy, "La panique politique," *Cahiers confrontation* 2 (1980): 33–54; François Roustang, *Dire Mastery: Discipleship from Freud to Lacan* (Baltimore, 1976); and idem, *Psychoanalysis Never Lets Go* (Baltimore and London, 1983).

Of suggestion, of the "transmission" or "transference of thoughts," . . . of magic and hypnotic demonism in general, he simply wanted to know nothing: "She, the hysteric, is not me." Or again: "It is not I who influence her, it is she who influences herself—her or her unconscious." And above all: "'It is not I who speaks with her mouth, not I the demon who possesses her—it is her, or Another in her.'"[14]

In relation to Prince, the implications of such a re-interpretation of Freud are no less charged, for it suggests that an analogous tendency towards the suppression of the mimetic paradigm might powerfully if silently be at work in the writings of Prince and the American school of imitation-suggestion.[15] Thus Prince and his colleagues marvelled both at the multiple roles or "impersonations" that their patients were capable of performing while under deep hypnosis, and at the fact that those patients afterwards seemed to remember nothing of what they had said or done. Such a radical self-estrangement or psychic splitting was the hallmark of posthypnotic suggestion, in which the awake patient carried out, as if in a trance, actions that had been suggested in a previous hypnotic state and that were characteristically attributed to another person, "another me," who was present to the hypnotic order in a way the patient never was. "They speak of this person as of a stranger," Binet wrote;[16] or as Prince remarked of Miss Beauchamp: "I was startled to hear her, when hypnotized, speak of herself in her waking state as 'She'"(*DP*, 26). From this standpoint the notion of hypnotic suggestion as involving a somnambulistic, unconscious mimicry so deep that the patient "blindly" or unconsciously, without the possibility of subsequent recollection or narration, takes the place of or incarnates the other, can scarcely if at all be distinguished from the phenomenon of multiple personality as it was viewed and discussed in the early twentieth century.

But this is not how Prince and his fellow hypnotists tended to deploy the concept of multiple personality. Instead, they interpreted the scandalous estrangement or splitting of the "self" in hypnosis not as the effect of mimetic identification but as the sign of *another subject or personality*—

14. Mikkel Borch-Jacobsen, "In statu nascendi," in *Hypnoses*, ed. Mikkel Borch-Jacobsen, Eric Michaud, and Jean-Luc Nancy (Paris, 1984), 71.

15. For a discussion of the centrality of theories of imitation and hypnosis to an emerging twentieth-century American account of the origin of the self see Ruth Leys, "Mead's Voices: Imitation as Foundation, or, The Struggle Against Mimesis," *Critical Inquiry* 19 (Winter 1993): 277–307, reprinted in *Modernist Impulses in the Human Sciences, 1870–1930*, ed. Dorothy Ross (Baltimore, 1994), 210–35.

16. Alfred Binet, *Les altérations de la personnalité* (Paris, 1890), 75.

as the manifestation of a part of the self that had hitherto been latent or concealed (William James's "subliminal" or "hidden self") but that had now been revealed or "dissociated" by a trauma or by suggestion.[17] The hypnotized person was conceived not as a sleepwalker given over body and soul to the hypnotist's suggestions such that she *was* the other, but as a subject who could distance herself from the scene and hence could observe the other *as* other, that is, could distinguish the model from the copy. That is why the act of naming assumed such importance in cases of this kind, for conferring a name on the patient's second self helped establish and reinforce the idea that the patient's actions were not the effect of hypnotic or mimetic identification but were produced by component parts of the patient's consciousness each of which under normal (James) or abnormal (Prince) circumstances might thus constitute a distinct identity or "personality."

"If the concept of an unconscious is to be retained," Borch-Jacobsen has observed, "it needs to be definitively liberated from the phantom of the *other subject* and the *other consciousness* inherited from late nineteenth-century psychology and psychiatry (Azam's 'somnambulic' or 'hypnotic consciousness,' Binet's 'personality alterations,' Breuer's 'hypnoid state,' the early Freud's 'dissociation of consciousness,' and so on)" (*FS*, 41). But this is to suggest that *both* Prince and Freud may be viewed as seeking to provide a solution to the historical-theoretical problem of imitation-suggestion: the later Freud through a theory of unconscious sexual desire, and both the earlier Freud and Prince through a theory of the "dissociation of consciousness"—that is, multiple personality. But perhaps it would be more accurate to say that Prince's *The Dissociation of a Personality* stages a struggle in which the multiple personality concept functions as a switch-point between two competing models, the first mimetic and the second antimimetic, of personal identity and, I argue, gender-formation—for Prince's Beauchamp case exemplifies the important fact that most cases of multiple personality have been female. Therein, I want to claim, lies *its* originality and historical specificity.

I will only add by way of further introduction that the tension between mimesis and antimimesis that structures Prince's investigation reflects a founding tension in the conceptualization of hypnosis itself. Like so many others, Prince was struck by the automatism of the subject when under the influence of hypnotic suggestion. At the same time he rejected the idea, voiced especially by P. Déspine and R. Heidenhain, that hyp-

17. William James, "The Hidden Self," *Essays in Psychology* (Cambridge, Massachusetts, 1983), 263.

notic behavior was a purely reflex action. Instead, influenced by a general late nineteenth-century reaction against mechanism in psychology, Prince tended to conceptualize the hypnotic performance as the work of a hidden intelligence or consciousness presumed to have many of the same capacities found in ordinary awareness. There would soon come a moment when hypnotists Hippolyte Bernheim and Joseph Delboeuf would deny that hypnosis involved any kind of altered mental state. Yet not even Bernheim or Delboeuf reduced hypnosis to a voluntary play of fictive roles, as many modern theorists of hypnosis have done. Prince, especially, remained close to the views of Edmund Gurney, Janet, and others who saw in hypnotic behavior the work of a submerged intelligence or consciousness whose performances were marked by the experience of involuntariness and posthypnotic forgetting.[18]

Borch-Jacobsen has recently described the "paradox" of hypnosis as follows: How can one induce a person to become passive (suggestible) if passivity requires prior consent? If the person consents, it is because he or she must really have wanted to do so. But if he or she wanted to consent, can one still say that she passively executes the suggestion? He notes that another way of putting the paradox is to ask: Is suggestion an effect of the operator or of the subject?[19] Gurney's description of the hypnotic state as a "psychical reflex action" captures that paradox of intelligence and spontaneity combined with automaticity.[20] Neither Bernheim, Del-

---

18. For Prince's views on hypnotism and his debt to the theories of Gurney, Janet, and others see especially his "Some of the Revelations of Hypnotism. Post-Hypnotic Suggestion, Automatic Writing, and Double Personality," in *Psychotherapy and Multiple Personality: Selected Essays*, ed. Nathan J. Hale, Jr. (Cambridge, Massachusetts, 1975), 37–60. For Gurney's views see "The Stages of Hypnotism," *Proceedings of the Society for Psychical Research* 2 (1884): 61–72; idem, "The Problems of Hypnotism," ibid., 2 (1884): 265–92; idem, "Peculiarities of Certain Post-Hypnotic States," ibid., 4 (1887): 268–323; idem, "Stages of Hypnotic Memory," ibid., 4 (1887): 515–31; idem, "Hypnotism and Telepathy," ibid., 5 (1888): 215–59. For the general background to the concept of unconscious psychical action see Ruth Leys, "Background to the Reflex Controversy: William Alison and the Doctrine of Sympathy Before Hall," *Studies in History of Biology* 4 (1980): 1–66; Ruth Leys, *From Sympathy to Reflex: Marshall Hall and His Critics* (New York, 1991); and Marcel Gauchet, *L'Inconscient cérébral* (Paris, 1992).

19. Mikkel Borch-Jacobsen, "L'Effet Bernheim (fragments d'une théorie de l'artefact généralisé)," *Corpus* 32 (1997): 147.

20. "The hypnotised 'subject' who carries out complicated orders is a conscious, and often even a reckoning and planning, automaton" (Gurney, "The Problems of Hypnotism," 268).

boeuf, nor Prince was able to resolve the paradox. Instead, Prince and his colleagues tended to vacillate between two alternatives—between imagining that in hypnotic acting the subject was so compulsively immersed in her role that she was incapable of representing it to herself and others (and accordingly was unable afterwards to remember anything of the experience), and imagining the same acting as a spectatorial performance that could in principle be represented and recollected. To restate that vacillation in the terms I have been developing, the immersive, or (following Borch-Jacobsen's *The Freudian Subject* and related texts), the "mimetic" version of hypnosis imagines the subject to be so sincerely and completely caught up in her role that she is completely unaware of herself and the fact that she is in effect on the stage. In contrast, the antimimetic or "spectatorial" version of hypnosis holds that the subject always remains aware of her role, a theory which leads readily to the charge that the hypnotized subject is a clever actress who cooly feigns the emotions she represents in the hypnotic performance.

It is hardly surprising that the relationship between hypnosis and stage acting was a focus of investigation at this time. In 1888 the influential British theater critic William Archer relaunched a long-standing debate over the nature of stage acting by rejecting Denis Diderot's notorious portrayal in *Le paradoxe sur le comédien* of the ideal actor as a cold-blooded simulator who at every moment remains highly conscious of his words and gestures. Instead Archer maintained, on the basis of answers to questionnaires sent to some of the leading performers in the English theater, that the best actors identify with their parts to the extent of actually feeling the emotions they imitate on the stage. In *Principles of Psychology* (1890), William James approvingly cited Archer's work in support of his own discussion of the relationship between the hysterical performance, hypnosis, and imitation. In the 1890s, too, Alfred Binet, basing himself on a survey similar to Archer's, likewise adopted the anti-Diderotian position; according to Binet, the hypnotic performance exemplified that mode of sincere or immersive imitation. Archer, however, also rejected the idea that the actor was completely or somnambulistically possessed by his part; owing to the "duplex" or "multiplex" nature of the mind, he argued, even at the height of his passion the actor remains aware of his role, which is to say that Archer's analysis remained suspended between a mimetic and an antimimetic account of the nature of acting. The same vacillation between mimesis and antimimesis can be detected in James and Binet. It follows that the equation commonly made between the hypnotic performance and stage acting risks begging numerous questions if it fails to acknowledge the different models of acting that have been proposed historically for both hypnotic and dramatic imitation.[21]

The Birth of the Subject

It is against this background that I approach Prince's text. My argument is that Prince's analysis of the Beauchamp case simultaneously instantiates and suppresses the mimetic theory. It follows that we might expect to find scenes of mimetic identification and of their virtually immediate denial by way of effects of distance and "originary" difference at strategic junctures in his text. Put the other way round, we might expect the narrative to contain scenes of specular representation that, even as they appear to prove the spontaneity and independence of the subject in question, tacitly but unmistakably depend on the very structure of mimetic identification they appear to deny.

One such scene may be regarded as exemplary. One day early on, Chris, the patient's recently named second self, who until then had always had her eyes closed in conformity with Prince's hypnotic command, *opens her eyes*. For some time, Chris—who, unlike the passive, "very suggestible" (*DP*, 15) Miss Beauchamp, "from the outset showed a will and individuality of her own, which was in no way subject to anybody else's influence" (*DP*, 56–57)—had been rubbing her closed eyelids with her hands, asserting that it was "her deliberate purpose to get her eyes open," and even going so far as to "threaten insubordination, insisting that she *would* see, and that she 'had a right to see.' She had complained rather piteously that it was not fair Miss Beauchamp should be allowed to see, while she was forbidden" (*DP*, 91). Prince was determined to prevent Chris from opening her eyes, on the theory that if she did so she "might become educated into an independent personality" (*DP*, 92). But in this contest of wills Prince's attempts to "limit the mental experiences of

21. See William Archer, *Masks or Faces?* (1888; reprint, New York, 1957), cited by William James, *The Principles of Psychology* (1890; reprint, Cambridge, Massachusetts, 1983), 1079; Alfred Binet, "Réflexions sur le paradoxe de Diderot," *Année Psychologique* 3 (1897): 279–95. See Jacqueline Carroy, *Les personnalités doubles et multiples: entre science et fiction* (Paris, 1993), 149–70, for a somewhat different interpretation of aspects of these issues. For discussions of Diderot's views on the theater in *Le paradoxe* and earlier texts, see Michael Fried, *Absorption and Theatricality: Painting and Beholding in the Age of Diderot* (Berkeley, California, 1980); and David Marshall, *The Surprising Effects of Sympathy: Marivaux, Diderot, Rousseau, and Mary Shelley* (Chicago, 1988), chap. 4.

Chris" proved "hopeless" for as he says with admiration: "She proved herself made of different stuff" (*DP*, 92–93). Prince emphasizes the willful character of Chris's—or, as she comes to be called, Sally's—disobedience by adding that the mechanical movement of rubbing her eyes was not in itself sufficient for her to achieve her goal:

> Besides rubbing her eyes she was obliged to "will" to come. "Willing," as a part of her conscious processes, plays a very prominent part in the psychological phenomena manifested by this personality, particularly in those which are the effect of her influence upon the others.
>
> "How did you make her do this or that?" I frequently asked.
>
> "I just 'willed,'" was the reply. (*DP*, 93)

Here is how Prince describes the scene in which Chris/Sally succeeds in opening her eyes:

> One day toward the end of June Miss Beauchamp was sitting by the open window reading. She fell into what Chris afterward called a half "mooning" state. She would read a bit, then look out of the window and think; then turn to her book and again read. Thus she would alternately read and dream,—daydreaming, it was.... Here was Chris's opportunity.... So while Miss Beauchamp was dreaming in her chair, Chris took both her hands,—Miss Beauchamp's hands,—rubbed her eyes, and "willed"; then, for the moment, Miss Beauchamp disappeared and "Sally" came, mistress of herself, and, for the first time, able to see. From this time on, we shall call Chris by the name of Sally; for though it was much later that Chris took the name, the complete independent existence of this personality dates from this event.
>
> Sally had gotten her eyes open at last, and with the opening of her eyes she may be said to have been truly born into this world, though she claimed to have really existed before. Sally was delighted with her success, so she must celebrate her birthday by smoking two cigarettes. Her belief in the naughtiness of it all, and a consciousness of the displeasure which it would occasion Miss Beauchamp, added to her enjoyment. (*DP*, 95–96)

Sally's delight was tempered by the fear that she had somehow killed Miss Beauchamp and perhaps would not be able to bring her back. But, ever ingenious, she remembered that Prince sometimes used a "strong Faradic battery" (*DP*, 96) to wake up the hypnotized Miss Beauchamp when she, Chris, herself would not disappear on his command. Accordingly, she took her lighted cigarette and, imitating Prince, (sadomasochistically) burned her (alter-ego's) arm so that Miss Beauchamp "woke up" (*DP*, 96).

The significance of the scene is that it marks Sally's birth as an independent person, which is to say that it marks the birth of the subject *as a subject.* If sleep is the sign of the hypnotic trance, as Prince and others tended to believe,[22] then when Sally opens her eyes she proves (or rather, demonstrates) her claim that she is not an artifact of suggestion, not merely a version of Miss Beauchamp when the latter is asleep or hypnotized, but a genuinely independent person. It's not just the fact that Sally opens her eyes, but *how* she does it that matters: she opens her eyes in an act of volition that not only establishes her difference from the Miss Beauchamp's "aboulia" or lack of will (*DP,* 15) but is so powerful that henceforth she appears to defeat even Prince's most determined efforts to control her. *Willing* is here conjoined with *looking* (more broadly, with *specularity*) as that which defines personal identity, and one is not surprised to learn later in Prince's text that Sally plans to write a "willing book" (*DP,* 340).[23]

Another point that should be stressed is that this ostensible scene of origins really constitutes a *second* birth, for Prince had already devoted a previous chapter to the "Birth of Sally," when he had first conferred a name, Chris, on his patient's hypnotic self. Yet as Prince recognizes, the act of naming alone fell short of securing Sally's autonomy as a subject, if only because the very fact that Chris adopted her name at Prince's suggestion casts doubt on her genuineness as a second personality. "How did you get the name of Chris?" Prince asks. To which Chris replies, "You suggested it to me one day, and I remember everything" (*DP,* 41). Moreover, Prince doesn't think it's sufficient to appeal, as James, Janet, Binet, and others do, to the phenomena of automatic writing and posthypnotic suggestion as evidence for the existence of a second or hidden self, for these are experimental phenomena which may also be artifacts of suggestion (*DP,* 26–27). Similarly, Chris's/Sally's testimony as to her ignorance of the literature of multiple personality, her ability to assert her will by hypnotizing Miss Beauchamp into telling outrageous lies (*DP,*

22. The most common method of hypnotism involved getting the subject or patient to close her eyes, usually by concentrating her attention on the hypnotist's own eyes, telling her that she was feeling sleepy and that she would close her eyes. "Look right at me and think only of falling asleep," ordered Bernheim, whom Prince visited in 1893 (*De la suggestion dans l'état hypnotique et dans la veille* [Paris, 1884], 5). Similarly, when Sally wants to hypnotize "The Idiot" she imitates Prince by stating: "'[A]s you read, slowly, slowly, your lids grow heavy—they droop, droop, droop; you're going, going, gone'" (*DP,* 320); conversely, when she's trying to be cooperative, she promises Prince she will go to sleep.

23. "I will, therefore I am," Maine de Biran declares in a statement that became the starting point of Janet's psychology (cited by Henri Ellenberger, "Pierre Janet and His American Friends," in *Psychoanalysis, Psychotherapy, and the New England Medical Scene,* ed. George E. Gifford, Jr. [New York, 1978], 64).

58–60), drinking wine (*DP*, 59), and doing many other things distasteful to the latter's moral standards, and her personal conviction that she is a completely different personality from Miss Beauchamp (*DP*, 42–49) all might be adduced as evidence in favor of Sally's independence as a subject. Yet these too, Prince argues, might be the product of suggestion or simply delusions. "Spontaneous phenomena were essential for proof" (*DP*, 50).[24] And perhaps the most dramatic demonstration of Sally's "spontaneity"—the event that marks her (re)birth as a stable character and genuinely independent subject—is her ability to successfully will her eyes open.

What makes Sally's act of volition all the more interesting to the historian is that it seems to contradict William James's assertion in his *Principles of Psychology* (1890) that "no creature not endowed with divinatory power can perform an act voluntarily for the first time."[25] Until we have performed an action at least once, James argues, "we can have no idea of what sort of a thing it is like, and do not know in what direction to set our will to bring it about." That is, a randomly occurring or reflex movement must, through the "kinesthetic" impressions associated with it, first leave an image of itself in the memory before the movement can be "desired again, proposed as an end, and deliberately willed." The idea or representation of the movement must always precede its execution. "[W]e need to know at each movement just *where we are in it*," James writes, "if we are to will intelligently what the next link shall be"; thus if a suitably predisposed person is told during the hypnotic trance that he can't feel his limb, "he will be quite unaware of the attitudes into which you may throw it."[26] This is precisely Sally's condition, for, until she wills her eyes open, like many hysterics she is completely anaesthetic and so lacks the guiding sensations James believes are necessary for voluntary movement. When her eyes are shut, Sally can "feel nothing" (*DP*, 147), with the result that Prince can place a limb in any posture he chooses without her being able to recognize the position it's in. "In reality the movement cannot even be *started* correctly in some cases [of hysteria] without the kinesthetic impression," James observes, adding: "M. Binet suggests . . . that in those

24. As Prince states in his book: "I emphasize the word 'spontaneous,' because, by artificial means (hypnotic suggestion) a mental dissociation in some apparently *healthy* people can be experimentally induced which is capable of exhibiting such automatic phenomena (posthypnotic suggestion, etc.)" (*DP*, 284).

25. James, *Principles of Psychology*, 1099.

26. Quotations from James's "What the Will Effects," *Essays in Psychology*, 218–19, and *Principles of Psychology*, 1099, 1102.

who cannot move the hand at all the sensation of light is required as a 'dynamogenic' agent."[27] Prince seems to have the same idea in mind when he observes of Sally, "But let her open her eyes and look at what you are doing, let her join the visual sense with the tactile or other senses, and the lost sensations at once return. The association of visual perceptions with these sensations brings the latter into the field of her personal consciousness" (*DP*, 147–48).

But this leaves unexplained how Sally *first* voluntarily opens her eyes, an action whose origin therefore appears utterly mysterious. It is as if Sally achieves a state of personhood *ex abrupto*, to use James's expression:[28] as she puts it, "I just 'willed.'" The absoluteness of this feat of willing is for Prince the hallmark of Sally's authenticity as a distinct personality, which is to say of the repudiation of mimesis-suggestion as an explanation of her existence. Yet the possibility can't be ignored that in that scene of origins Sally succeeds in authorizing herself as a subject (parthenogenetically giving birth to herself by willing her eyes open) precisely by identifying mimetically with the hypnotist, Prince, who enjoys a privileged position of spectatordom and who is described (by BIV) as seeing everything with his "'eagle eye'" (*DP*, 244). Indeed Sally's subsequent burning of Miss Beauchamp with her cigarette, thereby imitating Prince's use of a Faradic battery for the same purpose, may be taken as spelling out that identification. But this is to say that Sally's action of opening her eyes is simultaneously and irreducibly mimetic and antimimetic, her birth as a subject at once a product of mimetic repetition and an act of self-production. (Sally "dates her whole independent existence from this day, and she always refers to events as being 'before' or 'after she got her eyes opened.' That is the central event in her life, just as mothers date periods before or after the birth of a child," Prince wrote in a preliminary account of the case.[29] The scene as Prince thematizes it thereby announces the problematic of the mother, to which I return.) Understood in these terms, Sally's celebration of her birth by smoking not one but *two* cigarettes is highly significant. For even before she gets her eyes open Sally (still Chris), in a "Bohemian" (*DP*, 55) act designed to provoke the disapproval of Miss Beauchamp, to whom smoking is "absolutely repugnant" (*DP*, 55), imitates Prince by smoking one of his ciga-

27. James, *Principles of Psychology*, 1102–3.

28. James, "What the Will Effects," 219.

29. Prince, "The Development and Genealogy of the Misses Beauchamp: A Preliminary Report of a Case of Multiple Personality" (1900–1901), in *Psychotherapy and Multiple Personality*, 142.

rettes (*DP*, 54). (Thereafter cigarettes mark both Sally's difference from her other selves and her male-identified transgression of conventional standards of female behavior.) In that sense James was right after all—as if in order for Sally successfully to will to smoke she had to have *already* smoked, not as a willed action but rather as one of involuntary or mimetic identification. Indeed for James, hypnotic suggestion and the will turn out to be virtually indistinguishable from one another: to illustrate what he considers the simplest or most fundamental type of voluntary action—that which follows "*unhesitatingly and immediately* the notion of it in the mind"—James cites the example of hypnotized subjects who "repeat whatever they hear you say, and imitate whatever they see you do"[30]—a point of view that Prince, for reasons that are already plain, could hardly afford to make his own.

The shift from mimetic identification to specularity exemplified by Sally's opening her eyes is accompanied in Prince's text by a thematic of surface and depth that also works to establish, if not always the autonomy of the subject in question, at any rate the antimimetic nature of certain subject-*effects*. So for example Sally is a brilliant mimic who repeatedly deploys her mimetic skills in order to get her own way. Sally used to "impersonate Miss Beauchamp," Prince writes, "copying as far as she was able her mannerisms and tone" (*DP*, 117). "When blocked in some design I have seen Sally over and over again attempt to pass herself off as Miss Beauchamp" (*DP*, 118). But if Sally can fool her friends, she can't deceive Prince, for she's incapable of feeling the inner emotions that involuntarily determine Miss Beauchamp's outward expression. As those expressions are "purely automatic," it is impossible for one personality to completely simulate the other. "When Sally tries to impersonate Miss Beauchamp the best she can do is to look serious; but as she does not *feel* serious, or actually have the emotion or mood of Miss Beauchamp, her face does not assume the expression of that personality" (*DP*, 123).

Not that Miss Beauchamp is for Prince a paragon of psychic independence; on the contrary, for the most part she epitomizes passivity, obedience, indeed suggestibility. But Prince's unease with suggestion and mimesis is such that he is at pains to emphasize what might be called Miss Beauchamp's *limited* autonomy in the face of Sally's efforts to mimic her persuasively. At the same time, Sally's very failure to deceive Prince serves to dramatize her own (relatively) *absolute* autonomy or say her sheer psy-

---

30. James, *Principles of Psychology*, 1130, 1132. In James's account the relation of will to automatism is reversible: if volition is essentially a matter of suggestion, consciousness is impulsive and dynamic.

chic distinctness, the specific individuality that sets her apart from the other selves and that Prince finds so attractive. And of course it also further establishes Prince as the master of internal/external relations, the one person in the book in whose eyes Miss Beauchamp and Sally cannot fail to look like themselves, in their irreducible difference from each other.

Further light is thrown on the dynamics of specularity in Prince's text by a scene that occurs much later on. BIV, who has been locked in a "life or death struggle, a fight to a finish" (DP, 476) with her hated rival, Sally, has a "very remarkable experience" (DP, 360). A day or two earlier she had written an "ultimatum" (DP, 359) to Sally, threatening to have her banished to an asylum if she did not comply with BIV's demands for control. Now BIV, in a "depressed, despondent, rather angry frame of mind" (DP, 360) looks at herself in a mirror. She was combing her hair, and at the time thinking deeply, when suddenly she saw, "notwithstanding the seriousness of her thoughts, a curious, laughing expression—a regular diabolical smile—come over her face. It was not her own expression, but one that she had never seen before. It seemed to her devilish, diabolical, and uncanny" (DP, 361). Prince identifies that expression as the peculiar smile of Sally. "IV had a feeling of horror come over her at what she saw. She seemed to recognize it as the expression of the thing that possessed her. She saw herself as another person in the mirror and was frightened by the extraordinary character of the expression" (DP, 361). Realizing that her attempt to question the "thing" was "absurd," BIV hits on the idea of using automatic writing to get Sally to respond. Placing some paper on her bureau, and taking a pencil in her hand, "she addressed herself to the face in the glass. Presently her hand began to write" and an exciting interrogation of Sally took place—"for, of course, the 'thing' was Sally" (DP, 360–61).

On the one hand, the scene in question represents a moment of identification as BIV confronts her own image in a mirror. On the other (to my mind more importantly), it dramatizes the constitutive instability of *all* moments of identification in Prince's strongly antimimetic text, as Sally's image—more precisely, her *expression* (both less and more than an image in its own right)—instantly takes the place of BIV's own. Of course what makes the displacement uncanny, what gives it its horrifying force, is that difference is established only on the basis of a fundamental likeness (the image in the mirror is of BIV but its "expression" is different). To point up the contrast with a Freudian reading of the same scene, we might say that on this account the uncanny double is not the product of the subject's repressed infantile desires, as it is for Freud, but arises directly out of the mimetic relationship itself, as the person with whom BIV identifies is immediately converted into the rival who is *seen* to occupy *her* place.[31]

A closely related aspect of the mirror scene is that it reinforces the idea, thematized though never theorized throughout the book, that from the perspective of mimesis the specular as such is linked to the assertion of difference, of essential otherness, indeed of *violent rivalry* with that other who is, as the scene makes clear, a version of the same. The annexation of automatic writing, elsewhere in the text a means of suspending conscious control and therefore associated with suggestion-imitation, in order to give expression to a rival will is consistent with this idea: BIV's brilliant stratagem allows Sally to make use of her, BIV's, hand as an instrument for expressing her, Sally's, "independent" personality (I imagine this automatic writing as dependent on the framing context of the mirror). The scene thus converts the automatic into the volitional, BIV's loss of control into Sally's intentions. "The fact that the writing was spontaneous . . . deserves to be emphasized," Prince says, "because this spontaneity removes it from the class of artifacts unwittingly manufactured by the observer." Then, as if to reassure himself and the reader on this point, Prince adds: "BIV, as she testifies, was not in an abstracted state while the writing was done, but was alert, conscious of her surroundings, excited, and extremely curious to know what the hand was writing." And finally,

> It is to be regretted that she was not under observation at the time,—though this would have given rise to the suspicion that the doubling of consciousness was an artifact,—but I have seen the same feat performed under substantially similar conditions. (*DP*, 364)

Prince wishes he had something more than BIV's word to go on, but he also insists that the fact that she was alone (with Sally, so to speak) guarantees the antimimetic nature of the events she describes. At the same time, he wants us to know that he has witnessed equivalent scenes—as if the reassurance this offers can somehow escape the taint of mimesis he correctly recognizes is inseparable from his presence.

31. Freud in one passage describes the double in mimetic terms, stating that the relation to the double is "accentuated by mental processes leaping from one of these characters to another—by what we should call telepathy—so that the one possesses knowledge, feelings, and experiences in common with the other. Or it is marked by the fact that the subject identifies himself with someone else, so that he is in doubt as to which his self is, or substitutes the extraneous self for his own" ("The Uncanny," *Standard Edition*, 17: 234, cited by Borch-Jacobsen, "In statu nascendi," 68).

Gender and Imitation

Sally gets her eyes open at last and thereby gains her full independence as a very special kind of subject. For Prince, what makes her special, what for him defines her as psychologically the most interesting of all the personalities in the case—"the one who had to be reckoned with" (*DP*, 266)—is that she is what he calls a "co-consciousness": not only does she alternate with the other selves, appearing in their place when they disappear, but she permanently coexists with them. Unlike Miss Beauchamp, "The Idiot," and other multiples, all of whom when submerged are unaware of what the other personalities are thinking and doing, Sally has a continuous, uninterrupted existence: "It meant the co-existence of two different combinations at one and the same time, each with a self-consciousness. There were two I's then in existence."[32]

Being a coconsciousness, Sally is able to observe the other personalities at all times and hence can see, hear, and know things of which they are entirely unaware. "I am always present," she says (*DP*, 158). "*I am always alive*" (*DP*, 339). As a permanent "spectator of their lives" (*DP*, 147), Sally serves as Prince's surrogate by providing him with crucial information about the case that he is unable to obtain for himself. "I can't conceive of things being done without my knowledge, even in hypnosis," she says. "They never have been, you know, since that very, *very* early time when I used to sleep" (*DP*, 318). Always conscious, Sally claims never to sleep and hence never to dream. But precisely because she's always conscious she knows the dreams of her other selves, even those the latter are unable to remember on waking, which is why she can produce a "unique" account of Miss Beauchamp's dream life.[33] Similarly, Sally's coconsciousness permits her to write out for Prince a remarkable autobiography providing a record of two independent sets of thoughts, memories, and wills, her own and Miss Beauchamp's, going back to earliest childhood, that are known in their entirety only to herself; for unlike Miss Beauchamp and BIV, who have gaps in their memories corresponding to their condition as alternating states, Sally has a remarkably comprehensive and continu-

32. Prince, "Miss Beauchamp: The Theory of the Psychogenesis of Multiple Personality," 133.

33. Though Prince doubts Sally's claim that she never sleeps, observing that she is probably unaware of the passage of time while she sleeps, he nevertheless wants to believe Sally's assertion for it helps establish the case for her permanent spectatorship, that is, for her continuity as an independent personality (*DP*, 153–54, 330–33). Disturbances in the sense of time (and space) are of course held to be characteristic of the hypnotic trance (Milton H. Erickson, *The Nature of Hypnosis and Suggestion* [New York, 1980], 380).

ous memory (*DP,* 238)—along with volition, the mark of personal identity according to Janet and Ribot.[34] "Perhaps it would be better if I divided the page [into parallel columns] and carried it on that way, would it?" Sally suggests in her first attempt to express her duality of thought in a suitable literary form. "That was really the way the thought went, you know, until I got my eyes open" (*DP,* 316).[35]

It follows that Sally is internally divided in ways the other personalities in the case are not. Whereas Miss Beauchamp and "The Idiot" differ from each other (and from all the other selves), Sally differs *from herself* in ways that make her virtually incarnate the notion of double or multiple personality as such.[36] Sally's internal heterogeneity finds expression in a variety of ways, starting with her choice of a last name, "Beauchamp," which is marked by a discrepancy between its (French) spelling and its (English) pronunciation, "*Beecham*" (*DP,* 1). But the aspect of her heterogeneity that I want to focus on here is the fact that she is gendered both female and male.

Prince's commitment to the idea that the various personalities in the case are not artifacts of suggestion but are genuinely (if unequally) autonomous subjects motivates his attempt to draw sharp lines between them by differentiating their tastes, educational attainments, feelings, morals, and other traits (even their handwriting differs). He thereby creates what amounts to a virtual typology of early twentieth-century concepts of the feminine. In Prince's analysis, the morbidly impression-

34. "It is memory that establishes the continuity of psychological life" (Janet, *L'Automatisme psychologique* [Paris, 1894], 323). "It would seem that the identity of the self rests entirely on memory," writes Théodule Ribot in *Maladies de la mémoire* (Paris, 1881), cited by Michael S. Roth, "Remembering Forgetting: *Maladies de la mémoire* in Nineteenth-Century France," *Representations* 26 (1989): 54.

35. In another of case of multiple personality treated by Prince, the patient took the autobiographical approach one step further and actually published an account of her experience of duality: "B.C.A." [Nellie Parsons Bean], "My Life as a Dissociated Personality," *Journal of Abnormal Psychology* 3 (1908): 240–60; "B" [Nellie Parsons Bean], "An Introspective Analysis of Co-Conscious Life," *Journal of Abnormal Psychology* 3 (1908–9): 311–60. Michael Kenny has commented that the patient's letters show that she "reconstructed" her experiences in terms of the concept of multiple personality and suggests that Prince shaped or "educated" the case of "neurasthenia" into a case of multiple personality (Kenny, *The Passion of Anselm Bourne,* 141).

36. In this regard, Sally personifies a particular form of what I have elsewhere described as an emerging, barely theorizable, concept of individuality based on a relation of self-difference, the "type of one" (Ruth Leys, "Types of One: Adolf Meyer's Life Chart and the Representation of Individuality," *Representations* 34 [Spring 1991]: 1–28).

able, passive, modest, and self-sacrificing Miss Beauchamp, who has a history of somnambulism, daydreaming and trancelike states, represents a neurasthenic version of traditional womanhood. She is the antithesis of the self-reliant, realistic, enormously strong-willed, indeed "belligerent" (*DP*, 324) and emphatically unmaternal BIV (*DP*, 293), who imagines that she is "quite capable of running the world" (*DP*, 292) and who in Prince's disapproving, misogynistic description amounts to a parody of the "New Woman." Between these extremes, Sally occupies the position of the untutored, irresponsible young girl whose love of adventure, play, and excitement and dislike of books and the passive, scholarly life declare her unmistakably to be the adolescent of the American psychologist Stanley Hall's recent, influential definition.[37] "[O]ne of the most marked peculiarities of Sally's personality is its *childlike immaturity*. Sally is a child" (*DP*, 152), Prince writes, characterizing her as a young girl of about twelve or thirteen (*DP*, 112). "[S]he looked at everything from a child's point of view. Her general attitude of mind and her actions were those of a very young girl, as were some of her ideas of fun, and particularly her love of mischief" (*DP*,112).

As an adolescent, Sally is highly labile in mood and character, representing that unique moment of pubescent transition from relatively undifferentiated sexuality to gender decision, when according to Hall youth is "plastic and suggestible to an amazing degree" and when girls especially are prone to gender confusion and even complete role reversal.[38] Flirtatious and saucy, Sally acts her heterosexual feminine role to the hilt as she plays (the safely married) Prince off against the other important male in the story, the mysterious "William Jones," whose dramatic intervention in Miss Beauchamp's life I shall examine in a moment. "Know all men by these presents that I, Sally, being of sound mind and in full possession of all my senses, do hereby most solemnly promise to love, honor, and obey Morton Prince, M.D., situate in the city of Boston, state of Massachusetts, from this time forth, *toujours*. Amen, amen, amen. *Toujours* is French, you know," Sally writes Prince in a mock imitation of a marriage vow (*DP*, 138). Yet she is also "faithless" to Prince by repeatedly making forbidden "engagements" to meet her "caro amico" Jones, with whom she hopes to run off and love "for always" (*DP*, 111). At the same time, in conformity with Hall's concept of the pervasiveness of homosexuality in adolescent girls, Sally is also very attached to several older

---

37. Stanley Hall, *Adolescence: Its Psychology and Its Relations to Physiology, Anthropology, Sociology, Sex, Crime, Religion, and Education*, 2 vols. (New York, 1904).

38. Hall, *Adolescence*, 1: 316–17.

women, mother-substitutes as she recognizes, who make shadowy appearances in the text.

For Hall, this otherwise normal homosexual phase of development risks progressing to outright lesbian masculinization if not properly checked.[39] Just so, Sally is insistently phallicized throughout Prince's text, starting with her name "Chris" (the masculine version of "Christine")[40] and her liking for Prince's "masculine" cigarettes. At moments of conflict with Miss Beauchamp, Sally's voice, coming like an explosion "suddenly out of the depths," changes to a "bass note" (*DP*, 157). Resembling her father in looks (*DP*, 12), she is careless of her appearance, threatening to "ape" a man by cutting off Miss Beauchamp's hair. "You will cut off your own hair; it is your hair," Prince points out. "[Laughing.] 'I don't care. She will look like a guy—just like one of those monkeys. I don't care how I look'" (*DP*, 169). (At one point, Prince compares Sally to a monkey [*DP*, 313].) Like a "guy," she is unafraid of the spiders, snakes, toads, and mice with which she torments the terrified, conventionally feminine Miss Beauchamp (*DP*, 71, 161, 208).

More generally, Sally's male-coded refusal to be "one small victim" (*DP*, 98), her enormously energetic and powerful will,[41] and her love of practical jokes, of an "outdoor, breezy life; sports, amusements, physical activity, games, and the theatre," and of the "tales of adventure and of outdoor life, of hunting and riding" which the playful, energetic, and athletic Prince often has to tell her "to satisfy her longing" (*DP*, 129)—all signal Sally's identification with Prince's virile, if similarly "adolescent" or "boyish" modes of behavior.[42] This is especially apparent in Sally's autobiography. Written at Prince's suggestion and designed, like the *Journal* of Marie Bashkirtseff (for Hall the very type of the "exaggerated

39. Hall, *Adolescence*, 2: 107–8.

40. Rosenzweig suggests that the patient—or rather, Sally—took the name "Beauchamp" from the hero of George Meredith's novel, *Beauchamp's Career* (1876), which if true would mean that she identifies with its male protagonist (Rosenzweig, "Sally Beauchamp's Career," 24).

41. A contemporary of Prince treats a strong will in a woman as a "virile" manifestation (Charles Godfrey Leland, *The Alternate Sex* [New York, 1904], 35).

42. A friend described Prince as "boyish" in his enthusiasm (Kenny, *The Passion of Anselm Bourne*, 131); and Ernest Jones characterized him as having a "boyish love of fighting" (Hale, *James Jackson Putnam*, 263). On Prince's love of sports, especially riding and yachting, see Henry Murray, "Dr. Morton Prince: Sketch of His Life and Work," *Journal of Abnormal and Social Psychology* 52 (1956): 294.

adolescent confessionalist")[43] to reveal "every little secret emotion of her soul" (*DP*, 365), Sally's autobiography provides further evidence for Hall's alarming claim that nearly half of American adolescent girls rejected the role of wife and mother and much preferred the freer life of the male sex.[44] "[N]asty squally little thing" is how the unmaternal Sally describes a baby sister who died in her arms (*DP*, 387). Whereas the bookish Miss Beauchamp had enjoyed school, Sally had "wanted to play 'hookey.' I thought it would be awfully exciting because the boys did it" (*DP*, 373). Like the ambitious Bashkirtseff who rebels against the restrictions imposed on women and frequently cries out, "If only I were a man,"[45] Sally longs for the opportunities and mobility of the male sex, even to the extent of "donning boys' clothes on several occasions in search of adventure."[46]

Sally's monkey-like ability to imitate gender roles is in keeping with contemporary notions of the fluidity of identity in the prenubile state. Hall appeals to the "new psychology of imitation" in order to explain what belongs to adolescence as "one of its most intrinsic traits," its disposition to "ape positions, expressions, gait and mien" so that "every peculiarity is mimicked and parodied" and every youth has a "more or less developed stock of phrases, acts, and postures, expressive of mimetic love, anger, and fear." According to Hall, this mental plasticity explains the frequency, especially in young girls, of somnambulism, reverie, "transliminal modes of psychic action," and "multiplex personality." It also explains why, at a time before full sexual differentiation, both girls and boys are "more or less plastic" to the will not only of persons of the opposite sex but also of somewhat older persons of the same sex. Carrying the theory of suggestion, as embodied in contemporary notions of

43. Hall, *Adolescence*, 1: 554. The *Journal* of the aspiring painter (dead from tuberculosis at 25) caused a sensation in Europe and especially in the United States when it appeared, not least because, at a time of intense debate over the "woman question," Bashkirtseff was perceived, in Hall's words, as a "veritable spy upon woman's nature" (*The Journal of Marie Bashkirtseff*, trans. Mathilde Blind [London, 1985], 629).

44. Hall, *Adolescence*, 2: 391–92.

45. Rozsika Parker and Griselda Pollock, introduction to *Journal of Marie Bashkirtseff*, ix.

46. Prince, "Miss Beauchamp: The Theory of the Psychogenesis of Multiple Personality," 180. In Prince's book, Sally is said to daydream about "the things I would like to do if I could. When some men who were quarrelling passed, I thought of them for a long while, and envied them, for it was very late . . . and they had the street to themselves" (*DP*, 340–41).

sexual fetishism, to its logical conclusion, Hall draws the "momentous inference" that in adolescence *the sexual glow may come to be associated with almost any act or object whatever and give it an unique and otherwise inexplicable prominence in the life of the individual,* and that even the Platonic love of the eternally good, beautiful, and true is possible because of this early stage of indetermination and plasticity."[47]

In other words, desire for an object is secondary to mimesis, which comes first. What is particularly interesting in this connection is that the same year he first encountered Miss Beauchamp, Prince also analyzed the origin of sexual identity in mimetic-suggestive terms. In a paper of 1898, Prince agreed with Binet and others that homosexuality was a "vice" and not a disease because it was acquired by corrupt habits and wasn't the expression of a congenitally diseased nervous system, as Krafft-Ebing and the majority of sexologists believed. As a vice it deserved to be condemned; but as a vice it might also be cured or prevented. To give renewed credibility to a position that risked being viewed as moralizing and hence as unscientific, Prince made use of recent studies of hypnotism to argue that the sexual aberrations might originate in accidental, external suggestions that were subsequently forgotten, just as hysterical symptoms might be caused by forgotten subconscious influences.

Within a general framework that always tended to represent homosexuality as involving gender role "inversion" of various degrees of severity, Prince rejected the view that there are fixed, essential male, female, or homosexual identities in nature, as the congenitalists maintained. Instead, he envisaged a continuum of sexual possibilities, ranging from the "strong, vigorous, masculine characters" at one end, through the intermediate types—the "men with female personalities" and the "masculine females"—to the "strongly marked feminine personalities" at the other end. For Prince in 1898, the differentiation between the sexes was a product of the "total environment" or "education" broadly defined, including the effects of "intentional education," "unconscious mimicry," "external suggestion" and "example" in determining the "tastes and habits of thought and manners" of the child. Nor did he shrink from the transgressive implications of this claim, which was that the homosexual option was equally available to all men and women. It is "extremely probable," he writes, "that if a boy were brought up as a girl or a girl as a boy" under conditions that were absolutely free of counter-influences, "each would have

47. These citations are from Hall, *Adolescence,* 1: 239–40, 286–87, 316–37; 2: 105–108 n.; emphasis in original.

the nonsexual [i.e., the characterological-erotic or nonanatomical] tastes and manners of the other sex."[48]

As Prince sees it, the process of education by which sexuality is molded isn't a matter of simple social conditioning, since each person is complicit in the "cultivation" of his or her desires. But nor is it a matter of inherent moral corruption, since healthy people develop normal heterosexuality precisely because they *are* healthy, whereas the homosexual's willful opposition to society's unwritten social laws involves a weakened resistance to external influence that is itself grounded in a congenital nervous taint. This double determination in "culture" and "nature," in "character" and the body, completely obscures the opposition between vice and disease with which Prince begins. It also produces a narrative that works to undermine the text's official condemnation of homosexuality. For in the mimetic scenario of origins that Prince now develops, homosexuality turns out to be the most "natural," heterosexuality the most "unnatural," form of sexual desire.

"It is questionable whether only abnormally the vita sexualis of the male is excited by the *female*, and conversely" ("SP," 94, emphasis added), Prince writes. But doesn't he mean to say that it is questionable whether it is only abnormally the male is sexually aroused by the *male?* That even under normal conditions a male may be excited to desire another *man?* Prince's extraordinary gender slip has the effect of making it appear doubtful whether the male is *ever* naturally excited by the female, as if heterosexuality is always superinduced on an authentic form of homosexual desire. And this is exactly what Prince immediately goes on to imply, remarking that there "is every reason to believe" that in "some perfectly healthy individuals" erotic feelings may be excited by the "sight or touch" of the sexual organs of a person of the "same sex," and that "at any rate, thoughts (pertaining to anatomy) so excited may very naturally awaken secondarily associated sexual feelings." As Prince explains: "[T]he vita sexualis in a boy is first associated with his own sexual organs. Later, the sight of those of another boy awaken the association of ideas by the well-known law, and then, in a degenerate, cultivation does the rest" ("SP," 94). Nor is the origin of homosexual desire in suggestion restricted to the male sex, since Prince claims that Krafft-Ebing's first case of homosexuality in a girl may "readily" be explained in the same way. Thus in Prince's scenario of origins, the initial sexuality of both boys and girls is equally not hetero-

---

48. Morton Prince, "Sexual Perversion or Vice? A Pathological and Therapeutic Inquiry" (1898), reprinted in Prince, *Psychotherapy and Multiple Personality*, 93–94; hereafter abbreviated as "SP."

sexual but homosexual, in the sense that it first arises in the form of an autoeroticism that develops via suggestion into a homosexual love of the same before it is later transformed into heterosexual desire.

From one point of view the scene Prince alludes to is a scene of representation, because in order to imitate the other boy the child must recognize the latter *as other*, that is to say, must *represent the other to himself*. But from another perspective, the scene in question takes place outside all "specular" representation, for—like the hypnotized subject to whom Prince compares him—the child is so in the grip of the "blind," unconscious reflex, so caught up in a mimetic identification with the other boy, that in effect he becomes the other boy without being able to represent him to himself. One might say that the child's hypnotically or mimetically induced homosexual response isn't truly "sexual" at all. Rather, it's an automatic, involuntary response of the body that has no psychical, representational, or cultural meaning in that the child is unaware of the significance of his act or that society considers it wrong. In fact, this is exactly what Prince immediately goes on to argue, quoting his opponent Krafft-Ebing to the effect that at the beginning of the child's sexual development "'the psychical relation to persons of the opposite sex is still absolutely wanting, and the sexual acts during this period partake more or less of a reflex spinal nature'"("SP," 94). Here we can appreciate the ambiguity of Prince's narrative, for although the passage from Krafft-Ebing seems to concern the reflex nature of *heterosexual* love, Prince cites it in order to confirm the reflex character of *homosexual* desire. Thus it's not until adolescence, Prince claims, again quoting Krafft-Ebing, that these purely reflex reactions acquire a psychological dimension. Not until the anatomical and functional maturation of the "'generative organs'" associated with puberty and the "'differentiation of form belonging to each sex'" does the child develop "'rudiments of a mental feeling corresponding with the sex'" and begin to feel the "'powerful effect'" of "education and external influence'" ("SP," 94). Again, Prince's inability to decide between the primacy of biology or social training in the production of gender is striking. Are the rudimentary feelings associated with physical maturation what determine desire, in which case the effects of society are secondary to those of biology, or are external influences and education decisive? Nothing seems more fragile than the distinction between the two causes. "Now, in a person of perfectly healthy mind and body," Prince continues, "all social customs, habits of thought, unwritten laws, and moral precepts tend to suppress any existing homo-sexual feeling and its gratification, and to encourage hetero-sexual feeling" ("SP," 94). On the other hand, a person of "tainted constitution" does everything in his power to "foster, indulge and cultivate the perverse instinct" ("SP," 94) which may, as in the female hysteric, through constant repetition acquire the "monstrous force" of a compulsive psychosis. Hence his conclusion:

Thus may arise a perversity that had its origin in a normal reflex, but the accidental cause of which is forgotten with much else of the psychical life of childhood, or, if not forgotten, considered abnormal because of its future monstrous development.... What is really pathological in this aberration is the extraordinary intensification of the sexual feelings and the unbridled lack of restraint with which the subject indulges his senses and seeks every opportunity for gratification. These, without doubt, depend upon the neuropathic constitution. ("SP," 95)

But the inescapable implication of this scenario is that a violent act of cultural repression is required to force desire away from its natural origins in a mimetically induced homosexual desire. Figuring the order of authority, law, and difference, culture produces a split in the self between a youthful, presexual homosexuality and an adult, reproductive heterosexuality that has to be brutally imposed because it has no simple basis in natural desire. In a text that never stops reinscribing the biological in the cultural, the natural becomes "monstrous" and corrupt only *because* it is socially prohibited, as if only this violent social prohibition gives a natural and normal homosexuality its unnatural, pathological power of growth.

In sum, according to Prince's mimetic hypothesis, sexual identity is a consequence or effect of mimetic identification and not, as Freud will argue, the other way round. But this mimetically induced multiplicity of desire and identity is precisely what must be repudiated. For doesn't imitation-suggestion threaten to dissolve the boundaries between the sexes? Prince's writings, like Hall's, thus join the increasingly abrasive turn-of-the-century debate over the "crisis" of masculinity associated with the New Woman's appropriation of male roles and the breakdown of the traditional separation between male and female spheres.[49] A tacti-

49. In a large body of literature on this topic, see especially Joe L. Dubbert, "Progressivism and the Masculinity Crisis," *Psychoanalytic Review* 61 (1974): 443–55; Carroll Smith-Rosenberg, *Disorderly Conduct: Visions of Gender in Victorian America* (New York, 1986); Harry Brod, ed., *The Making of Masculinities* (Boston, 1987); J. A. Mangan and James Walvin, eds., *Manliness and Morality: Middle-Class Masculinity in Britain and America, 1800–1940* (New York, 1987); Gail Bederman, "'The Women Have Had Charge of the Church Work Long Enough': The Men and the Religion Forward Movement of 1911–1912 and the Masculinization of Middle-Class Protestantism," *American Quarterly* 41 (September 1989): 432–65.

cal retreat from mimesis is therefore required, an antimimetic gesture that will reassert the radical difference between the sexes and will involve an attack on feminism for encouraging the homogenization of sexual identity associated with the mimetic-suggestion hypothesis.

Exorcism

That retreat takes a variety of forms. In *Adolescence* Hall undercuts his own appeal to the mimetic lability of gender by condemning the women's movement for ignoring the natural, innate differences between the sexes. Criticizing coeducation for creating exhausted, effeminate males and dominant, masculine females, Hall characterizes three types of dysfunctional women produced by "the new woman movement," the "witch," the "egoist," and the "saint," in terms that strikingly resemble Prince's portraits of Sally, "The Idiot" (or "Realist"), and Miss Beauchamp respectively.[50] Similarly, if in certain passages sexologists like Havelock Ellis and others appear to accept the mimetic plasticity of sexual identity, in other passages it is precisely because this presumed plasticity makes heterosexual desire as accidental as homosexuality that they (on antifeminist grounds) object to the theory of acquired sexual perversion, emphasizing instead the role of inborn factors in the origin of desire.[51]

But what I want to stress is less the sexologists' appeal to biology or instinct as a means of containing mimesis than the antimimetic tendencies within the structure of mimesis itself. In his 1898 scenario of the mimetic origins of sexual desire and sexual difference, Prince surreptitiously grounds that difference in a scene that presumes the existence of the desiring subject whose origin his theory of imitation-suggestion is designed to explain. The child first autoerotically desires himself, Prince observes: his "vita sexualis is first associated with his own sexual organs." But by a specular logic that is nowhere made explicit, in order to love himself the child must get outside himself—more precisely, *he must imitate or identify with another boy in whom he sees an image of himself*, that is, whom he believes he *resembles*. In Prince's scene of origins, the child can only love himself if he is penetrated by an internal difference that is itself produced by the child's analogical perception of resemblance or identity. "Later,

50. Hall, *Adolescence*, 2: 561–646.

51. Havelock Ellis, "Sexual Inversion: With an Analysis of Thirty-Six New Cases," *Medico-Legal Journal* 13 (1895–96): 261–62. Like many congenitalists, Ellis retains the notion of a "pseudo" homosexuality of the classroom, the prison, etc., conceived as the effect of a "spurious imitation."

the sight of those [sexual organs] of another boy awakens the association of ideas by the well-known law, and then, in a degenerate, cultivation does the rest" ("SP," 94). The child is already object-oriented, he already autoerotically loves his own genitals, and desires only doubles—objects that resemble his own (or himself). (Resemblance so understood is an antimimetic relation.) In the process, Prince makes the desiring subject, which is to say the very phenomenon of gender, not the consequence or effect of identification, as his theory of suggestion-imitation ostensibly maintains, but rather its basis and origin. He thus proposes a version of Freud's thesis that an autoerotic or narcissistic stage followed by a homosexual stage is the normal sequence in the origin of heterosexual desire.[52]

I think we can see in this compulsion to defeat mimesis why homosexuality paradoxically occupies a privileged position in Prince's 1898 account of the mimetic origin of desire. It is as if the *homosexual subject* is present in Prince's text as an answer to—one might say it serves to suppress—the far greater problem of dedifferentiation that structures mimesis itself. And whereas mimesis cannot be cured (if for no other reason than that on the mimetic theory it is what makes subjecthood possible), the homosexually desiring subject, like the female hysteric, can: for what abnormal "cultivation" may achieve in the way of consolidating a perverse desire, hypnotic suggestion may undo, as Prince believes the hypnotic treatment of the perversions demonstrates.[53] The scandal of homosexuality can thus be averted, according to Prince's normalizing narrative, by resort to a hypnotic therapy designed to undo the homogenization implicitly at work in his account of the mimetic origins of desire by mesmerizing—or scaring—the male and female population straight.[54]

Moreover, for Prince the retreat from mimesis is all the more necessary because mimesis has primarily to do with the female, specifically the mother. To return to the Beauchamp case, it may be argued that a prob-

52. Sigmund Freud, *Three Essays on the Theory of Sexuality (1905), Standard Edition,* 7: 145–46.

53. Prince refers to the American translation of Albert Philibert Franz Schrenck von Notzing's book, *Therapeutic Suggestion in Psychopathia Sexualis* (Philadelphia, 1895).

54. Put slightly differently, in my interpretation of Prince's text a notion of the genesis of homosexuality as the result of (specular) *resemblance* displaces the greater "scandal" of *dedifferentiation or sameness* implied by the mimetic-suggestive relation to the other. I put the term "scandal" in quotation marks here to indicate that, although for Hall and others the scandal is indeed the blurring of gender boundaries and hence of identity, the figuration of homosexuality as love of the same can be seen as essentially domesticating in that homosexuality can and does involve highly complex modes of differentiation.

lem of maternal identification or imitation is central to its genesis and unfolding. As Sally reveals in her autobiography over the objections of Miss Beauchamp and BIV (*DP*, 375, 427), the patient as a child was "terrified" of her violent father (*DP*, 12), "worshipping, literally worshipping" instead her unhappily married mother, "who did not however care for her and paid her slight attention" (*DP*, 374). "C.[hristine Beauchamp]'s whole life, all her thought and action and feeling, centred about her mother," Sally reports. "She believed that God wanted her to save mamma from some dreadful fate, and that in order to do this she must, before the day should come, have attained a certain ideal state, mentally, morally and . . . spiritually" (*DP*, 380). "[A]s for mamma, she never wanted C. near her after we grew older. She didn't even want to see her, but was always saying, 'Keep out of my sight'" (*DP*, 387). "*Haunted*" (*DP*, 380) by this impossible ideal day and night, the thirteen-year-old child became "half delirious" or "disintegrated" (*DP*, 12) when her mother died and, consumed with guilt, believed herself to be "the victim of fierce persecution" (*DP*, 389). (Miss Beauchamp's subsequent choice of a career in nursing, which Sally detests, may be linked to the theme of maternal identification, as may her "Madonna worship" and religious conversion [*DP*, 344–55].)

As these citations suggest, according to Prince Miss Beauchamp's identification with the mother is ambivalent from the start: her idealization of her mother, self-reproaches, and paranoia barely conceal a murderous aggressiveness towards the maternal figure. It is as though her melancholic reaction to her mother's death exemplifies the ambivalent process of identificatory incorporation and devouring entombment that Freud will later describe in his account of pathological mourning. According to Prince's text, mimesis is grounded in maternal dependency, which is why it is so threatening, especially for the male, for whom the original maternal "subjection" connotes passivity, feminization—a theme that, as I show in chapter 3, reappears in the description of the shell-shocked soldier of World War I. Moreover, according to Prince's own mimetic hypothesis, Miss Beauchamp's identifications produce in her rivalrous and envious desires of a mimetic type. Thus it is a curious fact, though hardly surprising, that according to Sally an image of herself as an entombed living corpse or "Egyptian mummy"—"for to kill the double is of course to kill oneself" (*FS*, 92)—haunts the paranoid Miss Beauchamp's dreams (*DP*, 338).[55] In short, Miss Beauchamp's imita-

---

55. Sally states that as a child Miss Beauchamp's terrifying dreams were "usually about her mother" (*DP*, 376) and associates one dream to "Egyptian mummy images" (*DP*,

tive incorporation of the mother provokes a highly labile desire that from its inception is infused with rivalry and hatred and—my central point here—can only be disavowed by an emphasis on difference and autonomy.

The concept of multiple personality will be harnessed to that project of disavowal by way of three closely related developments. *First*, the malleability and radical heterogeneity of the "subject" that is internal to the mimetic theory will be countered by an antimimetic concept of the self conceived as the sum or aggregate of more or less fixed component parts (Prince's "traits" or "dispositions," Janet's psychical "elements," McDougall's "instincts," etc.) that can be shuffled together in a variety of different combinations—Prince employs the metaphor of a pack of cards[56]—to produce the more or less unified, functionally adapted personality. And whereas the mimetic paradigm expressly holds that no "real" self exists prior to mimesis, the concept of the self as a multiple of component traits or dispositions lends itself to the commonsense idea that there exists a "real" or "normal" self that can be identified and recuperated. As Prince states: "Common experience shows that, philosophize as you will, there is an empirical self which may be designated the real normal self" (*DP*, 233).

Second, the concept of the self as a multiple of traits or dispositions will be assimilated to a theory of *psychic hermaphroditism* or *bisexuality* that functions antimimetically to reinforce a compulsory heterosexuality. The political and social threat of mimesis is staved off by assuming that, according to Godfrey Leland, men and women are "radically different as regards both body and mind" and that "in proportion to the female organs remaining in man, and the male in woman, there exists also in each just so much of their peculiar mental characteristics. . . . [W]hat of late

---

341). It is also relevant to the theme of identification with the mother that at a very early moment, before her second self acquires the name Chris, Miss Beauchamp tentatively identifies her "other self" with the all-powerful, two-thousand-year-old, murderous matriarch "She" of Rider Haggard's immensely popular novel of the same name (*DP*, 28). "She"—or "She-Who-Must-Be-Obeyed"—rules over a city ("Kor") of murdered, entombed bodies and at the end of the narrative is herself withered to the condition of an Egyptian mummy. For the place of Rider Haggard's text in the *fin de siècle* imagination see Sandra M. Gilbert, "Rider Haggard's Heart of Darkness," in *Coordinates: Placing Science Fiction and Fantasy*, ed. George E. Slusser, Eric S. Rabkin, and Robert Scholes (Carbondale, Illinois, 1983), 124–38.

56. Prince, "Why We Have Traits—Normal and Abnormal: An Introduction to the Study of Personality" (1929), in *Clinical and Experimental Studies in Personality*, 129.

years occupied much thought as the Subliminal Self, the Inner Me, the Hidden Soul, Unconscious Cerebration, and the like, may all be reduced to or fully explained by the Alternate Sex in us."[57] In this process of assimilation, the theory of bisexuality, hitherto associated with a reprobated homosexuality, will be largely stripped of its perverse connotations and put to heterosexual uses. Again according to Leland, the male or female "alternate" in us invariably desires its heterosexual opposite. The female in the male loves the male in the female and *vice versa*, for within the structure of bisexuality there can be no same-sex desire. In short, the concept of bisexuality emerges as a *solution* to the problem of mimesis by virtue of its positing the existence of a *bisexual subject* in whom two heterosexual desires coincide.

Furthermore, this reheterosexualization of desire is accompanied by a denial of the character-inversion previously associated with the theory of bisexuality: since all men and women have second selves or traits of the opposite sex, the existence of a bisexual constitution or alternately gendered self is normal and, at least in the case of the man, does not compromise his virility. Indeed, according to Edward Carpenter, theorist of inversion or the "intermediate sex," even the male homosexual remains completely masculine in his nonsexual behavior.[58] "Among the loosely organized complexes in many individuals, and possibly in all of us," Prince writes in 1914, "there are certain dispositions towards views of life which represent natural inclinations, desires, and modes of activity which, for some reason or another, we suppress or are unable to give full play to. . . . Thus a person is said to have 'many sides to his character,' and exhibits certain alternations of personality which may be regarded as normal prototypes of those which occur in abnormal mental states." A favorite example of Prince's in this regard is the Scottish writer, William Sharp (1855–1905), the creative and imaginative side of whose personality Sharp "distinctly felt to be feminine in type" and whose writings he published under the pen name "Fiona Macleod." Such an example, Prince observes, "brings home to us the recognition of psychological facts which we all, more or less, have in common."[59]

57. Leland, *The Alternate Sex*, v.

58. Edward Carpenter, *The Intermediate Sex: A Study of Some Transitional Types of Men and Women* (London, 1921), 126–27 (first published in 1896 as an essay entitled "The Intermediate Sex").

59. Prince, *The Unconscious: The Fundamentals of Human Personality Normal and Abnormal* (New York, 1914), 294–99. In a gesture that serves to fend off the sexual dangers of identification with Freud, Prince in 1910 in a letter to James Jackson Putnam will express

Third, the concept of the self as a composite of male and female traits will be placed in the service of a misogyny that functions to contain the threat of mimesis by scapegoating the female. On the one hand, the theory of the alternate self contributes to a dissolution of gender boundaries by expanding the definition of masculinity and femininity to include aspects of the self that had previously been excluded. On the other hand, in such an expanded definition of the self, specifically in the "more interesting case of the male," the "alternate" self is described in such stereotypical terms that it—she—comes to lack all the attributes by which subjecthood is defined. The theory of bisexuality that accords a feminine side to the male reconfigures masculinity and femininity in such starkly oppositional terms as not only to reinforce the traditional superiority of the former but to do so by representing the latter as pure plasticity or negativity. Thus if, according to Otto Weininger's well-known theory of bisexuality, only the male can claim absolute difference and originality, if only the male can resist the influence of the model (for he "objects to being thought a mere echo"), this is because only man possesses the will, the superior, continuous memory, and consciousness on which the existence of an incommensurable and permanent ego depends. By contrast woman, for Weininger absolute mother or absolute prostitute, lacks all originality—she is the very essence of the undifferentiated, of the same. Mindless (she is nothing but sexual unconsciousness), un-moral (being incapable of conceptual thought she is incapable of morality), lacking a continuous memory (which is why she is a liar and has no sense of time), and without a will of her own (she is slavishly imitative and impressionable, which is why she can be so easily hypnotized), she has no permanent identity. "The absolute female has no ego." Like the soulless Undine of Fouqué's fairy tale who Weininger characterizes as the "platonic idea" of the female, women have no individuality or character: "Personality and individuality (intelligible), ego and soul, will and (intelligible) character, all these are different expressions of the same actuality, an actuality the male of mankind attains, the female lacks." "Multiplex," "diffused," "undifferentiated," "changeable," "heartless," "nameless," and endlessly double, woman is the bearer of everything that man excludes: she is the absence of "property," of subjective identity—in short, *mimeticism itself.*[60]

---

his rivalry with Freud by adopting or identifying with the persona of "Fiona Macleod," thereby presenting himself as a double (Hale, *James Jackson Putnam,* 323).

60. Otto Weininger, *Sex and Character* (London, 1906), passim. "If a woman possessed an 'ego' she would have a sense of property both in her own case and that of others,"

Just so, like the "half-person" Undine, to whom Prince's friend the neurologist James Jackson Putnam compares her, Sally on this reading is a representative of the "monstrous" danger of mimesis.[61] By the same token, it is mimesis that *produces* the powerful subjectivity effects that Prince and his contemporaries associate with the most fully developed examples of independent personhood. Sally's combative will is a subjectivity effect of exactly this kind; it is both a product of her mimetic identification with Prince, hence gendered male, and it is what is crucial to Prince's (and our) sense of her as the most dynamic, appealing, and autonomous of all the female personalities in the case. Much the same might be said of her internal heterogeneity, which gives to her actions and statements, for all their jejune character, an interest and complexity missing in BI and BIV. Sally is the "self" who is most internally split because mimetic "self-division" is what gives rise to the most forceful and authoritative instatement of the subject. At once lacking a personality (she is merely a morbid, suggestible group of dissociated states) and a fascinating personality in her own right (she has a formidable will, a "sane" (*DP*, 396) intelligence, and a remarkably comprehensive memory), Sally is the site where the mimetic and the antimimetic irreducibly meet.

That is why, like Weininger's woman, she must be made to disappear.[62] A decision must be made against mimesis and in favor of difference and autonomy, although, as William McDougall almost alone among Prince's commentators appears to recognize, Sally's disappearance cannot be motivated within the theoretical terms of Prince's analysis, and although the attempt to master mimesis will be of dubious success.[63] A ritual "exorcism" (*DP*, 137) is required, one that appears to

Weininger writes (205), and: "It is in striking harmony with the ascription to men alone of an ineffable, inexplicable personality, that in all the authenticated cases of double or multiple personality the subjects have been women. The absolute female is capable of subdivision; the male ... is always an indivisible unit" (210–11).

61. James Jackson Putnam, review of Prince's *Dissociation of a Personality* in the *Journal of Abnormal Psychology* 1 (1906): 236–39, where he compares Sally to Fouqué's soulless Undine and describes the case itself as "caricature, a monstrosity."

62. "Woman, as woman, must disappear, and until that has come to pass there is no possibility of establishing a kingdom of God on earth" (Weininger, *Sex and Character*, 343). Weininger's vision of a male homosocial world that would be purified not only of women but of the demands of genital sexuality altogether is consistent with the structure of twentieth-century homosexual panic as described by Eve Kosofsky Sedgwick in her pathbreaking *Between Men: English Literature and Male Homosocial Desire* (New York, 1985).

63. William McDougall, "The Case of Sally Beauchamp," *Proceedings of the English Society for Psychical Research* 19 (1907): 410–31.

depend on the violent imposition of Prince's hypnotic will and that can, therefore, only redouble the mimesis. Worse, isn't Sally a version of Prince, and doesn't killing Sally therefore amount to killing himself?

The Subject of Trauma

One of the ways in which Sally remains occult to Prince—in which she exceeds his interpretive frame—is that she cannot be accommodated within Prince's theory of trauma. In agreement with Janet and other theorists, Prince holds that multiple personality is caused by an external trauma or shock that splits the personality into its component elements. Therapy consists in using hypnosis as an objective instrument to help the patient remember and integrate the dissociated traumatic memory. In the case of Miss Beauchamp, Prince concludes that neither BI, BIV, nor Sally is the "real" or "normal" self; rather, the original self is the one that fell apart under the impact of a dramatic encounter that occurred several years earlier, in 1893, between his patient and the mysterious "William Jones." Jones is an older man who served as spiritual guide to the adolescent Miss Beauchamp after her mother's death and whom the patient revered and adored almost as much as she had previously revered and adored her mother (he is, in those respects, a mother-substitute). It turns out, though, that Jones "unintentionally, and perhaps all unconsciously" (*DP*, 89) is responsible for the "psychical catastrophe" (*DP*, 109) that caused Miss Beauchamp to disintegrate.

Here is Prince's description of Miss Beauchamp's catastrophic encounter with Jones. The scene is set in southern Massachusetts, in a town he calls Providence (actually Fall River, a small industrial town on the Rhode Island border), where Miss Beauchamp has been pursuing a career as a medical nurse:

> One night, while in the nurses' sitting-room conversing with a friend, Miss K., she was startled, upon looking up, to see a face at the window. It was the face of her old friend, William Jones, a man whom with the idealism of girlhood she worshipped as a being of a superior order. He was much older than she, cultivated, and the embodiment of the spiritual and the ideal.

Miss Beauchamp goes downstairs to meet Jones:

> It transpired that he had stopped over in Providence, en route to New York, and had wandered up to the hospital. Seeing a ladder (which had been left by workmen) leaning against the side of the building, he had, in a spirit of fun, climbed

75

up and looked into the window. At the hospital door an exciting scene occurred. It was to Miss Beauchamp of an intensely disturbing nature, and gave her a tremendous shock.. . . . The surroundings, too, were dramatic. It was night, and pitch dark. A storm was passing over, and great peals of thunder and flashes of lightning heightened the emotional effect. It was only by those flashes that she saw her companion. (*DP,* 214–15)

The scene as a whole shares with the two others discussed earlier—the one in which Sally opens her eyes and the mirror scene with BIV—an unmistakable emphasis on seeing and specularity. Like the earlier scenes, this one is virtually cinematic in its intensity, despite the fact that it remains unclear exactly what took place between the patient and Jones at the hospital door. (And despite its inherent implausibility: at one or two o'clock in the morning [*DP,* 219], Jones happens to find a ladder that allows him, in the middle of a violent thunderstorm, to climb to a second floor window in order to play a joke on his young friend.)

Prince repeatedly emphasizes the importance of this scene for his interpretation of the case; in particular the image of Jones's face framed in the window and illuminated by flashes of lightning is the last thing BIV remembers of the episode when she first recalls it in a trancelike state (*DP,* 171–77, 220–22). But there are insuperable difficulties here, starting with the fact that according to Sally's account (to which Prince gives considerable credence) she has existed from earliest childhood, i.e., the patient's dissociation occurred well before the catastrophic encounter with Jones.[64] Furthermore, although in certain passages Prince seems to imply that the restoration of Miss Beauchamp's "real" or "normal" personality involves the reabsorption of Sally into BI and BIV, no proof is given that this is in fact what has taken place.[65]

64. Prince, "Miss Beauchamp: The Theory of the Psychogenesis of Multiple Personality," 150. In a recent reconstruction of the case, Rosenzweig has located the origin of Miss Beauchamp's dissociation in the trauma of the sudden death from sudden infant death syndrome of her newborn brother when the patient was seven years old, a death to which, he suggests, the patient secretly believed she had contributed (hence her sense of guilt). He also posits an additional trauma of paternal physical and sexual abuse (Rosenzweig, "Sally Beauchamp's Career," 5–60). I find Rosenzweig's speculative reconstruction of the case unconvincing, but it has the merit of drawing attention to the inadequacy of Prince's emphasis on the trauma of the scene with Jones.

65. On this basis, McDougall concludes that Sally must be a spirit ("The Case of Sally Beauchamp," 430), a popular turn-of-the-century interpretation of multiple personality that is alluded to but rejected by Prince. Prince used a Dr. Richard Hodgson, a leading Boston spiritualist and James's collaborator in the investigation of the medium Mrs. Leonore Piper, to serve as Miss Beauchamp's physician when he himself was unavailable,

Another, graver source of difficulty concerns the evidence Prince adduces in favor of the reality of the scene with Jones. It is worth stressing in this regard that in his earliest account of the scene at the hospital, Prince gives a different explanation for Miss Beauchamp's fright—her encounter with a mad patient who had tried to seize her; the trauma of seeing Jones's face at the window is added later.[66] Prince never exactly explains why the scene with Jones was not mentioned in the early phases of treatment, merely implying that it subsequently emerged as the case developed (*DP,* 66, 214–15). The problem is that the patient's recollection of the Jones scene is obtained by methods, such as staring into a crystal ball, that are designed to help her visualize the past in a state of abstraction or absorption akin to a light trance but that are known for producing hypnotic confabulation or fictional scenes. The problem of confabulation worries Prince;[67] but it is a problem he either attempts to overcome or simply ignores, for he is committed to the idea that traumas leave "facsimiles"[68] of the original experience in the form of indelible imprints in the brain that can be accurately uncovered in treatment. As he writes: "She saw it all as a *vision,* just as it occurred. She was startled by what she saw and experienced over again all the emotion of the original scene."[69]

---

and there are signs of Hodgson's influence on the patient (*DP,* 356, 377). But, apparently in reaction against his own earlier spiritualist predilections, Prince seems to have limited Hodgson's role in the case, and nowhere in his text does he discuss the spiritualist hypothesis.

66. Prince, "The Pathology, Genesis and Development of Some of the More Important Symptoms in Traumatic Hysteria and Neurasthenia," *Boston Medical and Surgical Journal* 138 (1898): 561; idem, "The Development and Genealogy of the Misses Beauchamp," 146–47.

67. Thus in an 1898 discussion of "crystal ball" gazing and other techniques for obtaining visions of the past in the case of Miss Beauchamp and others, Prince is concerned to refute the charge that the memories solicited in this way might be suggested fabrications (Prince, "An Experimental Study of Visions," *Brain* 21 (1898): 528–46). As far back as 1890, William James had expressed concern that multiple personality might be the product of unconscious suggestion or training, pointing out that the practice of naming the secondary personage tended to encourage the patient's belief in the experience of being double (Prince, "Some of the Revelations of Hypnotism. Post-Hypnotic Suggestion, Automatic Writing, and Double Personality," 55).

68. Prince, "The Unconscious," *Journal of Abnormal Psychology* 3 (1908–9): 342.

69. Prince, "The Development and Genealogy of the Misses Beauchamp," 151. In *The Unconscious* (1914), Prince uses evidence obtained with automatic writing, crystal gazing,

The question of confabulation is especially acute in the Beauchamp case in that the patient's reconstruction of the allegedly traumatic encounter with Jones involves dramatically enacted "reproductions" of the decisive event, in which at a crucial juncture Prince stands in for—plays the role of—the elusive Jones (*DP,* 173). In other words, the patient's reconstruction takes place in the mode of a quasi-hypnotic, emotionally charged acting-in-the-present. (Prince obtains the first, comprehensive account of the scene from Sally and then turns to his patient's other selves for corroboration that the reconstruction is accurate, which could count as persuasive evidence only if the independence as witnesses of all the selves were beyond question.)[70]

Yet despite these problems and considerations Prince never thereafter questions the reality of the scene at the hospital. And I suggest that this is because the function of that scene, like that of the one in which Sally opens her eyes, is to guarantee the status of the subject as a subject. In particular, the specular staging of the scene, Jones's disembodied face framed and frozen in the window, the mutual facing off between the patient and Jones—all serve to validate Miss Beauchamp's separateness as a personality even as she is shattered into multiplicity. Similarly, I suggest that Sally cannot be comprehended within Prince's theory of the trauma because she personifies the production of the subject in and by a mimetic dynamic that the theory of the trauma as a historical, external event coming to the already constituted subject functions to evade. No wonder the scene at the window never ceases to resonate in our minds throughout our reading of Prince's long, convoluted narrative; it is the key, not as Prince believes to the "secret" (*DP,* 214) of Miss Beauchamp's condition, but rather

and dreams in the Beauchamp and other cases of multiplicity to prove that traumatic memories are exact representations of the past.

70. "The veridicality of the facts revealed in this vision was confirmed by the memories of the other personalities BI and BII (the normal reintegrated self) as well as by Sally" (Prince, "Miss Beauchamp: The Theory of the Psychogenesis of Multiple Personality," 184). Yet Prince also admits that one of his patient's visions must have been a distorted reconstruction, since the patient in the visualized scene had her eyes *closed,* making it impossible for the picture to be an exact repetition of the original impression; in other words, at the time of the actual scene she was caught up in it and was therefore unable to be an observer from outside (Prince, "An Experimental Study of Visions," 541–42). In 1899 Freud made our tendency to objectify scenes of childhood by viewing our past selves as if from the outside the basis for commenting on our general disposition to falsify the past by making us spectators of ourselves in accordance with certain dynamic trends (Freud, "Screen Memories" [1899], *Standard Edition,* 3: 321–22).

to his attempt to stabilize that condition even as he struggles to do justice to its protean manifestations.[71]

Postscript: Recent Developments

The recent revival of the multiple personality diagnosis is a striking phenomenon. In *Rewriting the Soul* and other texts Ian Hacking has brilliantly discussed the "memoro-politics" of that revival, noting the ever-growing numbers of cases and alters, the gender imbalance (nearly 90% of reported multiples are female), and above all the battle that has ensued over the reality of the trauma of childhood sexual abuse. The proliferation of cases has been met with considerable skepticism, largely because of the role of hypnosis in diagnosis and treatment. So the same issues that I have raised in relation to Prince's foundational text may also be raised in relation to current developments:

1. Like Prince, present-day theorists of multiple personality tend to assume the existence of an already-constituted female subject comprised of a functional plurality or hierarchy of component parts to which violence comes entirely from the outside to shatter it into dysfunctional multiplicity. As in Prince's text, such a stark opposition between trauma and victim serves to elide the mimetic dimension and to do so on the basis of a normalization of gender roles that represents the female subject as a completely passive victim. This familiar sexual coding finds reverse expression in the often repeated theory that if there are far fewer reported cases of male multiples this is because, being naturally more active or violent than women, males are more likely to end up in jail than in the therapist's waiting room.[72]

---

71. In fact, Prince never really solves the case. In the 1920s, more than twenty years after he first encountered Miss Beauchamp, Prince identifies (mimetically) with Freud by shifting the emphasis of the case towards the etiological role of the *sexual conflicts* aroused in his patient by the scene with Jones. But he also (antimimetically) attempts to reassert his difference from Freud by simultaneously appealing to the patient's *auto-suggestion* in the determination of her psychic splitting. According to Prince, Sally—whose origin as a co-consciousness thus remains unexplained—"erupted" by opening her eyes because in an act of self-assertion she could "will" or hypnotize Miss Beauchamp into letting her emerge. On this basis, Sally goes "back to where she came from" by being repressed by another subject, this time by the "stronger autocratic systems of the real self." But this leaves a remainder in the form of a footnote deferring once again a "fuller explanation" of the case, signalling that the interpretation hitherto given is, if not inadequate, incomplete (Prince, "Miss Beauchamp: The Theory of the Psychogenesis of Multiple Personality," 172).

72. Hacking, *Rewriting the Soul*, 28.

2. The same normalization of gender roles occurs with respect to the question of gender transgression in cases of multiple personality. I have argued that Prince sought to undo the lability of sexual identity implied by his mimetic theory, by positing an antimimetic theory of the self envisaged as a composite of fixed elements that were themselves defined in terms that reinforced heterosexual norms. If according to Prince's own mimetic hypothesis, human sexuality is permeated by mimesis and thus is inherently perverse, the antimimetic turn within Prince's mimetic paradigm explains the perversion of homosexuality as the effect of an abnormal *bisexual or homosexual subject* made up of conventionally, not to say misogynistically, defined female and male selves.

Similarly, today's theorists of multiplicity circumvent the ambiguity and plasticity of sexual identity inherent in a mimetic account of subjectivity by imagining the self as a compound of often fractious and rivalrous male and female "alters" who are defined in conventional ways. Hacking has remarked on the stereotypical nature of the various alters common today, among whom is an aggressive "protector" male personality to balance the lively Sally-like juvenile alter and the compliant Miss Beauchamp-like self.[73] Thus if the modern multiple personality is often considered "bisexual," this is only by imagining bisexuality as made up of a combination of preexisting, self-identical male and female personalities. In other words, the widely held notion that multiple personality is a way for certain women to avoid the gender roles assigned to them by acting out transgressive male roles is undercut by an unexamined tendency to describe those roles in traditional, heterosexist terms. Certain feminists in the field of dissociation portray critics of the multiple personality movement as participants in a general backlash against the liberation of women; nevertheless, their own explanations function to reinforce a conventional view of sexual difference.

3. My account of the Beauchamp case throws light on yet another aspect of the modern multiple personality movement. As we have seen, modern interpretations of dissociation or multiplicity tend to theorize trauma as an external event that shatters the female subject into constituent parts. As in the Beauchamp case, the aim of hypnotic treatment is to help the patient to recover the memory of the trauma and thereby to restore her to a condition of adaptive wholeness. But the use of hypnosis to elicit recollections of past traumatic events and to cure patients is problematic, as the Beauchamp case demonstrates, and as even some of today's most ardent supporters of the multiple personality diagnosis rec-

73. Hacking, *Rewriting the Soul*, 76–78.

ognize. Patients relapse and "true" recollections of the traumatic event are hard to come by. (I say more about the latter point further on.) As one influential architect of the multiple personality diagnosis recently warned: "The therapist must learn to interpret and restructure dissociation rather than try to suppress, ignore, or medicate it. He must remain aware as well that material influenced by intrusive inquiry or iatrogenic dissociation may be subject to distortion. In a given patient, one may find episodes of photographic recall, confabulation, screen phenomena, confusion between dreams or fantasies and reality, irregular recollection, and willful misrepresentation. One awaits the goodness of fit among several forms of data, and often must be satisfied to remain uncertain."[74]

More than any other factor, it is the issue of confabulation that has given rise to skepticism about the attribution of sexual abuse, especially in view of the widespread use of hypnotic techniques to enhance recollection.[75] Hacking has proposed that the basic confrontations between defenders and critics over multiple personality mainly have to do with competing ideologies of the family, ideologies that find expression in the new twentieth-century language of the soul, the science of memory.[76] But he is also aware that sensitivity to the larger cultural meanings of today's multiple personality movement does not obviate the need to ask difficult questions about the role of hypnosis itself.

One response to the charge of hypnotic confabulation is to dismiss the problem of hypnotic suggestion outright, on the grounds that charges of

74. Richard P. Kluft, "Treatment of Multiple Personality Disorder: A Study of 33 Cases," *Psychiatric Clinics of North America* 7 (1984): 13–14.

75. For the claim that the dangers of hypnosis in the treatment of multiple personality have been exaggerated see Richard P. Kluft, "Varieties of Hypnotic Interventions in the Treatment of Multiple Personality," *American Journal of Clinical Hypnosis* 24 (1982): 230–40; idem, "The Treatment of Multiple Personality Disorder (MPD): Current Concepts," *Directions in Psychiatry* 24 (1985): 1–9. For the claim that in the famous case of Sybil the patient's memories of sexual and sadistic abuse were fabricated, see Mikkel Borch-Jacobsen, "An Interview with Dr. Herbert Spiegel," *The New York Review of Books* 54, no. 7 (24 April 1997): 60–63; cf. Hacking, *Rewriting the Soul*, 124. For a useful collection of essays on the relations between hypnosis, memory, and confabulation, including the problems posed by the use of hypnotic visualization and age regression methods to induce hypermnesia, see *Hypnosis and Memory*, ed. Helen P. Pettinati (New York, 1988).

76. Hacking, *Rewriting the Soul*, 126. See also Jean Comaroff, "Aristotle Remembered," in *Questions of Evidence: Proof, Practice, and Persuasion across the Disciplines*, ed. James Chandler, Arnold I. Davidson, and Harry Harootunian (Chicago, 1994), 463–69; and Lawrence Wright, *Remembering Satan: A Case of Recovered Memory and the Shattering of an American Family* (New York, 1994).

the fabrication of "false memories" are simply part of a politically motivated backlash against the women's movement. From this point of view,
there is no problem of suggestion—or if there is one, the problem can be
dealt with by techniques that will supposedly purify the therapeutic relationship of its suggestive elements (see chapter 7 for a discussion of the
views of Judith Herman and Bessel van der Kolk on this point). An opposite response takes the skeptical route of interpreting the entire
phenomenon of multiplicity as an artifact of the contagious effects of
suggestion.[77] A version of the latter is the argument recently offered by
Borch-Jacobsen to the effect that multiple personality is the outcome of a
banal suggestive game between patient and physician, a view that treats
dissociation as an artifact of the hypnotic relationship (see chapter 5).

My own view of the matter, based on a familiarity with the literature on
hypnosis over the past century or more, is that the historian would do well
to stop short of embracing either of the two extreme positions. Hacking
has rightly observed of the often simplistic discussion of hypnotic suggestion by both advocates and critics of the diagnosis of multiple personality, "We have very little grasp of the workings of 'suggestion,' a word
that, as Nietzsche said of 'psychological pain,' so often stands for a question mark rather than for a clear idea. . . . The models of suggestion and
iatrogenesis are cavalierly invoked by skeptics about multiplicity. They
are confidently rejected by advocates. I am urging that the models are
skimpy and superficial."[78] I think this is correct, and I want further to suggest that from the moment when multiple personality was first invented at
the end of the nineteenth century, theories about hypnosis have been
structured in such a way as inevitably to give rise to just such a polarization
between advocates and opponents of the diagnosis of multiple personality disorder. This is a major theme in the chapters that follow.

77. This is the position of Dr. Paul R. McHugh, Director of Psychiatry at the School of
Medicine at Johns Hopkins University, in "Witches, Multiple Personalities, and Other
Psychiatric Artifacts," *Nature Medicine* 1 (1995): 110–14; idem, "Psychiatric Misadven-
tures," *American Scholar* 61 (1992): 497–510. See also McHugh's exchanges with Frank W.
Putnam and David Spiegel, both prominent figures in the multiple personality movement
in Paul R. McHugh and Frank W. Putnam, "Resolved: Multiple Personality Disorder Is an
Individually and Socially Created Artifact," *Journal of Academy of Child and Adolescent Psy-
chiatry* 34 (1995): 957–63; and David Spiegel and Paul McHugh, "The Pros and Cons of
Dissociative Identity (Multiple Personality Disorder)," *Journal of Practical Psychiatry and
Behavioral Health* (September 1995): 158–66.

78. Hacking, *Rewriting the Soul*, 255.

III

Traumatic Cures: Shell Shock, Janet, and the Question of Memory

When soldiers began to break down on a large scale during the First World War and when it became evident to some physicians that, in the absence of physical lesions, their wounds were psychological rather than organic in nature, hypnotic suggestion proved to be a remarkably effective treatment. The use of hypnosis to deal with the war neuroses marked a return to a therapy that, since the time of its flourishing under Charcot's leadership more than twenty years earlier, had been largely abandoned by the medical profession. More precisely, practitioners reinstated Breuer and Freud's early method of treatment of hysteria by hypnotic catharsis or "abreaction," a method whose abandonment by Freud around 1896 had been the decisive gesture by which he had sought to differentiate the "discipline" of psychoanalysis from the "enigma" of suggestion. The revival of hypnosis to cure what was understood as a virtual epidemic of male hysteria during the war was attended by a revival of the many doubts and objections that have repeatedly accompanied the use of hypnosis as a technology of the subject in the West. The nature of those doubts and objections is complex, but I believe they can all be seen to revolve around a single question: How does hypnosis cure?[1]

---

1. For the history of the concept and treatment of shell shock see especially Paul Fussell, *The Great War and Modern Memory* (New York, 1975); Esther Fischer-Homberger, *Die traumatische Neurose: Vom Somatischen zur Sozialen Leiden* (Bern, 1975); Eric J. Leed, *No Man's Land: Combat and Identity in World War I* (Cambridge, 1979); P. Lefebvre and S. Barbes, "L'Hystérie de guerre: étude comparative de ses manifestations au cours des deux derniers conflits mondiaux," *Annales médico-psychologiques* 142 (1984): 262–66; Martin Stone, "Shellshock and the Psychologists," in *The Anatomy of Madness: Es-*

In London in the spring of 1920 that question was the topic of a brief but, I argue, highly significant debate between three well-known doctors who had played significant roles as psychotherapists during the war. The discussion was led by William Brown, who had seen nearly 3,000 cases of war neuroses in France and Britain.[2] Following Breuer and Freud, Brown argued that the characteristic signs of shell shock—mutism, loss of sight or hearing, spasmodic convulsions or trembling of the limbs, anesthesia, exhaustion, sleeplessness, depression, and terrifying, repetitive nightmares, symptoms hitherto associated chiefly (although not exclusively) with female hysteria—were all bodily expressions of obstructed or "repressed" emotions. Brown reasoned that when a soldier was confronted with the need to maintain self-control and army discipline in front-line conditions of unremitting physical and psychological stress, he was likely to respond to any significant trauma by breaking down. Unable to discharge his powerful emotions directly, through action or speech, he unconsciously "materialized" them by converting them into physical or bodily symptoms. Most striking of all, the patient could not remember

---

says in the *History of Psychiatry*, ed. W. F. Bynum, Roy Porter, and Michael Shepherd (London, 1985), 2: 242–71; Edward M. Brown, "Between Cowardice and Insanity: Shell Shock and the Legitimation of the Neuroses in Great Britain," in *Science, Technology and the Military*, ed. Everett Mendelsohn, Merritt Roe Smith, and Peter Weingart, Sociology of the Sciences (Dordrecht, Germany, 1988), 12: 323–45; Ted Bogacz, "War Neurosis and Cultural Change in England, 1914–22: The Work of the War Office Committee of Enquiry into 'Shell-Shock,'" *Journal of Contemporary History* 24 (1989): 227–56; Sue Thomas, "Virginia Woolf's Septimus Smith and Contemporary Perceptions of Shellshock," *English Language Notes* 25 (1987): 49–57; Harold Mersky, "Shell-Shock," in *150 Years of British Psychiatry, 1841–1991*, ed. German E. Berrios and Hugh Freeman (London, 1991), 245–67; Ben Shephard, "'The early treatment of mental disorders': R. G. Rows and Magull, 1914–1918," in *150 Years of British Psychiatry*, vol. 2, *The Aftermath*, ed. Hugh Freeman and German E. Berrios, 434–64; and Chris Fendtner, "'Minds the Dead Have Ravished': Shell Shock, History, and the Ecology of Disease-Systems," *History of Science* 31 (1993): 377–420. As has been noted by several authors, the diagnosis of shell shock provided a useful middle term between the prevailing dichotomy of outright malingering versus insanity or psychosis, allowing a relatively humane approach to the large number of men who broke down in the war.

2. The three doctors were William Brown, Charles S. Myers, and William McDougall. Brown delivered a paper—"The Revival of Emotional Memories and Its Therapeutic Value"—at a meeting of the medical section of the British Psychological Society on 18 February 1920, Myers and McDougall responded, and Brown's rejoinder closed the meeting. The proceedings were published under the title of Brown's paper in the *British Journal of Medical Psychology* 1 (1920): 16–33; hereafter abbreviated "R."

anything about the horrifying events that lay at the origin of his pitiable state. Dissociation or amnesia was therefore the hallmark of the war neuroses.[3] *"Hysterics suffer mainly from reminiscences."*[4] The famous Breuer-Freud formula, according to which hysterics suffered from repressed traumatic memories, served Brown as the basis for a hypnotic therapy designed to restore the victim's memory through the trancelike repetition and abreaction of the shattering event ("R," 16–17).

For Brown the efficacy of hypnosis depended crucially—though as we shall see, not exclusively—on the emotional catharsis involved. What appeared to him to be fundamental was that in the hypnotic or trance state the traumatic event was "reproduced" or "relived" with all the affective intensity of the original experience. Only in this way, he thought, could the pent-up emotion be successfully abreacted: "The essential thing seems to be the revival of the emotion accompanying the memory" ("R," 16). Breuer and Freud had also emphasized the importance of emotional discharge in the cathartic treatment. "Recollection without affect almost invariably produces no result," they had observed. "The psychical process which originally took place must be repeated as vividly as possible; it must be brought back to its *status nascendi* and then given verbal utterance" (*SH*, 6). At military centers just behind the French front line, Brown had obtained such emotional relivings without difficulty. But in cases of longer standing back home in Britain, where the symptoms had had a chance to become more "fixed," he had found it much less easy to obtain the same results. Brown stated that one of his patients, who had suffered from hysterical deafness and loss of speech, had recovered his memories under hypnosis on several occasions but had not regained his voice and hearing until, one night, he had experienced an extremely intense dream and had suddenly tumbled out of bed with his speech and hearing restored. "In the case of deaf-mutes treated in the field such failure never occurred," he observed. "The explanation seems to be that, in this case, I did not produce the emotional revival with sufficient vividness" ("R," 16).

But C. S. Myers and William McDougall, the other participants in the debate, rejected Brown's emphasis on the emotions in hypnotic abreaction. They maintained that what produced the relief of symptoms was

3. William Brown, "The Treatment of Cases of Shell Shock in an Advanced Neurological Center," *Lancet* 2 (August 17, 1918): 197; hereafter abbreviated "TC."

4. Josef Breuer and Sigmund Freud, *Studies on Hysteria* (1895), *The Standard Edition of the Complete Psychological Works of Sigmund Freud*, trans. and ed. James Strachey (London, 1955–74), 2: 7; hereafter abbreviated *SH*.

not the affective catharsis but the cognitive dimensions of the cure. Implicitly embracing the traditional distinction between the lower emotional appetites and the higher functions of rational control, they emphasized not the affective reliving but the conscious reintegration of the dissociated or "repressed" memory into the patient's history. "It is the recall of the repressed scene, not the 'working out' of the 'bottled up emotional energy' . . . which is responsible for the cure," argued Myers ("R," 21). "The essential therapeutic step is the relief of the dissociation," McDougall agreed. "The emotional discharge is not necessary to this, though it may play some part" ("R," 25). McDougall pointed out that in an earlier discussion of his procedure, Brown had insisted to the patient, while the latter was still under hypnosis, that on waking he would remember the scenes that he had just relived. Without such a precaution, the patient on being roused from the trance state characteristically forgot again everything that had just occurred. "In this procedure [Brown] seems to have recognised practically that the emotional excitement was not in itself the curative process," McDougall noted, "but that at the most it was contributory only to the essential step in the process of cure, namely the relief of amnesia or dissociation" ("R," 25). (As will become clear, this did not wholly misrepresent Brown's ideas.) McDougall conceded that the revival of emotion was important as an aid to securing the complete relief of the traumatic experience, both directly, by giving force and vivacity to the recollection, and indirectly, by overcoming the continued tendency to repress or forget the unpleasant memories. But the essential thing in treatment was the reappearance of the traumatic memory in the clear light of consciousness. Indeed, McDougall claimed that it was possible to obtain the recovery of the repressed traumatic event without emotional excitement of any appreciable kind ("R," 25–26).

What is the significance of the debate between Brown and his colleagues? I submit that theirs was not simply a disagreement about a minor point of therapeutic technique. Nor, in spite of McDougall's interest in the topic, was theirs essentially a dispute about the cerebral mechanisms that might underlie the symptoms of the war neuroses. Far more basic issues were at stake. For the force of Myers's and McDougall's denial of the importance of emotional abreaction was to insist that what mattered in the hypnotic cure was to enable the traumatized soldier to win a certain knowledge of, or relation to, himself by recovering the memory of the traumatic experience. The idea was to help the subject achieve an intellectual reintegration or resynthesis of the forgotten memory so that he could overcome his dissociated, fractured state and accede to a coherent narrative of his past life. For this a certain degree of the patient's participation was required. Put more generally, it is as if two competing accounts of the role or position of the patient in medicine opposed one another in the debate. The first account, which I call the *participatory* account, imagined that the collaboration of the patient was an inseparable part of the cure, while the second account, which I call the *surgical* account, imagined that, as in the case of drug therapy or surgery—dominant modes of medical therapy in the West—the collaboration of the patient was irrelevant to treatment. For the psychotherapists of the war neuroses the key question appeared to be this: Did hypnosis heal the patient by soliciting the subject's participation? Or did a suggestive therapeutics achieve its effects by encouraging the patient's "subjection" to the authoritative command of the hypnotist that by-passed the consent and as it were the collaboration of the self? If we rephrase those positions in the light of Foucault's work on discipline and knowledge, we might say that the first account emphasized the active role of a subject understood as constituted through categories of consent and refusal, while the second imagined a subject—but does the term make sense in this context?—who somehow escapes both alternatives.[5]

Now a revealing feature of the 1920 debate in this regard was the fear that, in the absence of cognitive insight, the hypnotic reliving of the trauma might be positively harmful to the patient by reinforcing an emotional dependence on the physician that was held to be incompatible with psychical autonomy and self-control ("R," 24–25). This was also the message of psychotherapist Paul-Charles Dubois, whose influential attacks on hypnosis, starting in 1905, had helped precipitate the rapid decline of hypnotic therapy in the prewar years. Eerily anticipating the equation between the therapeutic value of self-control and the requirement of military discipline that was characteristic of medical discourse in World War I, Dubois had compared neurotics to "army stragglers" whom we are tempted to regard as lazy or simulators: " We do not know whether to believe in their hurts and put them in the infirmary, or to handle them roughly and send them back to the ranks. We are already involved in a problem of liberty and of responsibility, and it is the absence

5. If for Foucault power, discipline, and knowledge line up together, they do so by presupposing the production of a subject capable of consent and resistance. "Power is exercised only over free subjects, and only insofar as they are free," he writes (Michel Foucault, "Afterword: The Subject and Power," in Herbert L. Dreyfus and Paul Rabinow, *Michel Foucault: Beyond Structuralism and Hermeneutics* [Chicago, 1982], 220–21). For Foucault, power and freedom are thus mutually constitutive, as has been emphasized by Mark Maslan, "Foucault and Pragmatism," *Raritan* 7 (Winter 1988): 94–114.

of a clear solution which makes us hesitate which course to follow."[6] Dubois's widely heeded response to that uncertainty had been to demand the abandonment of hypnotic "manipulation" in favor of a moral reha- bilitation of the patient based on "rational persuasion."]Rejecting what he defined as the hypnotist's exploitation of the patient's childish and "ef- feminate" passivity and automatic obedience, he had urged physicians in- stead to increase the soldier's virile self-discipline and autonomy by strengthening his rational and critical powers. "It is our moral stamina which gives us strength to resist these debilitating influences [or sugges- tions]," he had maintained.[7] Even Pierre Janet's scathing condemnation

6. Paul-Charles Dubois, *The Psychic Treatment of Nervous Disorders*, trans. Smith Ely Jelliffe and William A. White (New York, 1908), 35, 45–46, 116. The extent to which those in charge of shell-shock patients in Austrian military hospitals were guilty of abusive or "disciplinary" treatment was the subject of hearings in 1920 at which Freud gave testi- mony. See K. R. Eissler, *Freud as an Expert Witness: The Discussion of War Neuroses between Freud and Wagner-Jauregg* (New York, 1986). For recent discussions of the conflation be- tween the requirements of military discipline and medical therapy in the treatment of the traumatic neuroses in World War I see Stone, "Shellshock and the Psychologists"; Pat Barker, *Regeneration* (New York, 1992), for a discussion of the treatment of the poet Siegfried Sassoon; and Elaine Showalter, "Male Hysteria: W. H. Rivers and the Lessons of Shellshock," in *The Female Malady:* Women, Madness, and English Culture, 1830–1985 (New York, 1985), 167–94.

7. Dubois, *Psychic Treatment*, 116. For objections to hypnosis similar to those expressed by Dubois, see Alfred Binet, *Les idées modernes sur les enfants* (Paris, 1910), 193; and J. Dejerine and E. Gauckler, *The Psychoneuroses and Their Treatment by Psychotherapy* (Philadelphia, 1913). Suggestion was characteristically defined as "the process by which ideas are introduced into the mind of a subject without being subjected to his critical judge- ment. The effect of any suggestion depends on its evading the critical judgement of rea- son" (J. A. Hadfield, "Treatment by Suggestion and Persuasion," in *Functional Nerve Disease*, ed. H. Crichton Miller [London, 1920], 63). By contrast, persuasion was defined as "the form of treatment which appeals to the conscious reason and enforces its claims on logical grounds." Ominously, however, for those who—like Dubois—wanted to maintain an absolute distinction between these two processes, Hadfield went on to remark that "in actual practice the success of persuasion depends on suggestion, especially that derived from the authority of the physician and the expectancy of the patient" (ibid., 82). In effect, Hadfield—like Brown and so many others—attempted to distinguish between "good" suggestion, which helped strengthen the patient's will and freedom, and "bad" suggestion, which weakened psychical autonomy. Suggestion thus conformed to the structure of the *pharmakon* (or *supplement*) in Derrida's sense of those terms, as that which is simultaneously remedy and poison. It is worth noting in this regard that the term "shell shock," introduced by Myers and others early in the war, was officially banned in 1917 on the grounds that it helped spread, by contagion or "suggestion," the very symptoms whose cure *by* suggestion

of Dubois's position—his proposal that hypnosis should be considered no different from drug therapy or surgery, the efficacy of which did not depend on the patient's insight or awareness—could not prevent the reorientation of psychotherapy towards moralization and "rational" analysis that occurred at this time.[8]

All this suggests that for hypnosis to be installed successfully at the core of a medical therapeutics during the First World War it had to be retheorized as exemplifying not the hierarchical, "coercive" model but the consensual or participatory model of treatment. And in fact Brown himself interpreted hypnotic suggestion along these lines. If, for Brown, emotional catharsis had a legitimate place in the treatment of the war neuroses, this was precisely because it avoided the abjection and mechanical automaticity of "direct suggestion." In hypnotic catharsis, Brown had earlier explained, "the patient goes through his original terrifying experiences again, his memories recurring with hallucinatory vividness. It is this which brings about the return of his powers of speech, and not direct suggestion, as in the ordinary method of hypnosis." Catharsis was thus imagined as "free from the defects" attaching to the "ordinary" use of hypnotic suggestion ("TC," 198).[9] As a means toward helping the patient achieve self-mastery and self-knowledge, the emphasis in Brown's treatment fell squarely on the recovery and resynthesis of the forgotten memory ("TC," 198–99). Hypnotic catharsis was theorized not as an apparatus of behavioral manipulation but as a "supplementary

---

was the goal of psychotherapeutic treatment (see C. S. Myers, *Shellshock in France 1914 1918, Based on a War Diary Kept by Charles S. Myers* [Cambridge, 1940], 12–13, 92–97).

8. For Janet's criticisms of Dubois and the turn to rational persuasion see Pierre Janet, *Psychological Healing. A Historical and Clinical Study* (1919; reprint, New York, 1976), 1: 129ff; hereafter abbreviated *PH*. "Influenced by the prevailing fashion," Janet sarcastically remarks of a certain Dr. Levy who had converted to Dubois's rational therapeutics, "he now tells us that hypnotism has fallen into disfavour 'because it is regarded as a special nervous condition.' He, too, wants the patient to participate in the work of the cure, which is, of course, to be 'rational.' The patient must learn to discipline himself morally and physically. In a word, the whole of Levy's therapeutic system depends upon 'rational education and re-education'" (*PH*, 1: 113). In response to this Janet quotes from an article by Max Eastman: "'It is difficult to see why it is any more a suspension of judgment to let a physician you have decided to trust lodge a helpful idea in your mind, than to let him lodge an ominous-looking capsule in your body'" (*PH*, 1: 337).

9. By the "ordinary" method of hypnosis Brown meant Hyppolite Bernheim's method of direction suggestion in which the physician verbally suggested to the hypnotized patient that the symptoms (or their cause) would vanish. See William Brown, "Hypnosis, Suggestion, and Dissociation," *British Medical Journal* (June 14, 1919): 735.

aid" to a medical treatment designed to "discipline" the subject by getting him to accept a certain version of his history and identity.] Brown conceded:

> Psychologically we are forced to recognise the great therapeutic effect produced by the intellect in the analytic review of past memories, especially in the analytic treatment of what have been called "anxiety states," where the patient is helped and encouraged to look at past events from a more impersonal point of view, and so to obtain a deeper insight into their mutual relations and intrinsic values. The method, which might be called the method of *autognosis*, does produce a readjustment of emotional values among the patient's past memories. These memories are all scrutinized from the point of view of the patient's developed personality—or rather of his ideal of personality so far as it becomes revealed in the course of the analysis—and the relative autonomy that some of them had previously enjoyed by virtue of their over-emphasis is withdrawn from them. The progress is from a state of relative dissociation to a state of mental harmony and unity. The "abreaction" of excessive emotion here is no merely mechanical process, but is controlled at every step by the principle of relativity and intellectual adjustment. ("R," 19)[10]

In other words, the disagreement between Brown and the others, basic as it seemed at the time, emerges in retrospect as a matter of emphasis, not of fundamentally opposed viewpoints.

It is significant in this regard that, in order to avoid the perceived dangers of hypnosis, Brown advocated limiting its use to a "very small minority of cases," namely the major hysterias, and preferably to only one treatment ("TC," 199).[11] According to Brown and other commentators, it was owing to the brevity of hypnotic treatment that there were no diagnoses of multiple personality in the war—a fact of considerable interest given the previous success of the multiple personality diagnosis in Europe and the United States. In 1926, Bernard Hart commented on the

---

10. For similar descriptions of the task of hypnotic therapy see Brown, *Talks on Psychotherapy* (London, 1923), 10; Myers, *Shell Shock in France*, 68; W. H. Rivers, "Freud's Psychology of the Unconscious," *Lancet* 1 (16 June 1917): 914; and Simmel's discussion of catharsis in Sándor Ferenczi, Karl Abraham, Ernst Simmel, and Ernest Jones, "Symposium Held at the Fifth International Psycho-Analytical Congress in Budapest, September 1918," in *Psycho-Analysis and the War Neuroses* (London, 1921), 30, 33.

11. See also William Brown, *Talks on Psychotherapy*, 34–35; and idem, *Suggestion and Mental Analysis: An Outline of the Theory and Practice of Mind Cure* (London, 1923), 170.

"remarkable absence" of cases of double personality in the literature of psychoanalysis.[12] In the discussion that followed, Freud's disciple, Ernest Jones, attributed such a lack to the Freudian rejection of hypnotic methods that he, Jones, regarded as especially likely to produce the weakening and dissociation of the ego characteristic of multiple personality. Brown was inclined to agree, pointing out that although functional nervous diseases with amnesias, fugues, and other symptoms, had been produced in the thousands during the war, "no well-marked cases of multiple personality were reported or observed." He ascribed this to the absence of prolonged hypnotic treatment which would otherwise have "trained" the patient to display the disorder.[13] Brown's observations can help us understand multiple personality as a historical-social construct. During World War I, in a context that saw a modest revival of Freud's cathartic methods, the fear of suggestibility and automaticity in the male—the demand for the revirilization of the demoralized soldier—limited the deployment of hypnotic suggestion in such a way as to contain the emergence of the more florid symptoms hitherto associated with the diagnosis of multiple personality.[14] That factor, in combination with Joseph Babinski's assault on the entire hysteria diagnosis and the general neurological-organicist orientation of the psychiatric profession, ensured that the shell-shocked soldier might be regarded as a malingerer or treated as a case of male hysteria, but he would never be seen as an example of multiple personality.[15]

Nevertheless, the war neuroses brought into prominence once again the very phenomenon of dissociation or splitting that had been considered the defining characteristic of Prince's Beauchamp case and of female hysteria generally. "The war neurosis, like the peace neurosis, is the expression of a splitting of the personality," Ernst Simmel wrote.[16] The rediscovery of splitting as the essential feature of shell shock reopened

12. Bernard Hart, "The Conception of Dissociation" (with discussion), *British Journal of Medical Psychology*, 6 (1926): 241–56.

13. Discussion of Hart's paper, *British Journal of Medical Psychology* 6 (1926): 260.

14. Similarly, the diagnosis of multiple personality seems to have been very rare in combat neurosis patients in World War II. I have come across only one reference: Harry C. Leavitt, "A Case of Hypnotically Produced Secondary and Tertiary Personalities," *Psychoanalytic Review* 34 (1947): 271–95.

15. For a contemporary critique of Babinski's views on hysteria by one of the chief architects of the multiple personality diagnosis see Morton Prince, "Babinski's Theory of Hysteria," *Journal of Abnormal Psychology* 14 (1920): 312–24.

16. Simmel, "Symposium Held at the Fifth International Psycho-Analytical Congress at Budapest, September 1918," 33.

the debate, inaugurated by Freud, over the role of sexuality in the production of hysteria. Since for Brown, Myers, McDougall, and many other physicians the notion of sexual conflict seemed inapplicable to the traumas of the war, the threat of annihilation—the feeling of utter helplessness when confronted with almost certain death—rather than sexual repression came to be regarded as the cause of hysteria. Indeed, repression itself was called into question as the mechanism of hysteria, with the result that psychotherapists returned to Breuer's concept that in the traumatic neuroses a "hypnoid" or psychical splitting of the ego occurred prior to, or independently of, any mechanism of repression.[17] Furthermore, with the new emphasis on the trauma of death came a return to a thematics of maternal identification and maternal trauma that we also found in Prince's Beauchamp case. (We shall see that Janet's case of the dissociated and traumatized Irène, discussed in this chapter, fits this model.) The mother, conceived as the mesmerizing "object" of the suggestible child's first passionate identificatory tie, was scapegoated as the source of her son's "feminine" hysteria and lack of virile courage in actual battle.[18] In addition, the war neuroses came to be thematized—notably by psychoanalyst Sándor Ferenczi and by Freud—as a repetition of the child's earliest reaction to the threatened loss or disappearance of the maternal figure.[19] In short, the hysterical dissociation of the traumatic neuroses of both war and peace emerged as the sign of a prior, impossible mourning for the lost mother.

17. For British postwar discussions of the distinction between splitting and repression see W. H. R. Rivers, *Instinct and the Unconscious: A Contribution to a Biological Theory of the Psycho-Neuroses* (Cambridge, 1922); and William McDougall, *Outline of Abnormal Psychology* (New York, 1926), esp. chap. 12.

18. For mother-fixation as a cause of the war neuroses see for example Maurice Nicoll, "Regression," and H. Crichton Miller, "The Mother Complex," in *Functional Nerve Disease: An Epitome of War Experience for the Practitioner,* ed. H. Crichton Miller (London, 1920), 101, 115–28; and Simmel, "Symposium held at the Fifth International Psycho-Analytical Congress at Budapest, September 1918," 31. For general discussions of the gender aspects of the war experience see Margaret Randolph Higgonnet, ed., *Behind the Lines: Gender and the Two World Wars* (New Haven, Connecticut, 1987).

19. In an astonishing note added to his discussion of the war neuroses in 1918, Ferenczi cited the work of Moro on the reflexes of the newborn in order to compare traumatic hysteria to the reflex reactions of the very young infant traumatized by a sudden shock, which Ferenczi interpreted as an atavistic reversion to the young monkey's reflex clasping of the mother (Ferenczi, "Symposium Held at the Fifth International Psycho-Analytical Congress at Budapest, September 1918," 21). Ferenczi's linking of trauma to maternal separation anticipated Imre Herman's study of maternal trauma in *L'Instinct filial* (1943; Paris, 1972) as well as Nicolas Abraham's and Maria Torok's analysis of the traumatic "dual union" between the mother and child in *L'ecorce et le noyau* (Paris, 1978).

Affect, Memory, and Representation

Could hysteria so defined be cured? Here I want to emphasize that Brown's advocacy of hypnotic abreaction as a technique for recovering traumatic memories involved the claim that emotion always "involves a reference, vague or explicit, towards some object" ("R," 17), which is to say that the emotions belonged to a system of representations. That is what made it possible for emotions to persist in the mind with the same continuity and verisimilitude as the images on a movie reel to which Brown compared them, just as the experiences or objects to which the emotions were attached were completely preserved in the memory. And that is why, when emotions were repressed or dissociated, they did not disappear but were lodged in the unconscious in the form of forgotten recollections. For Brown, it was because the affects participated in the same representational system as other experiences that they could be recalled or "reproduced" under hypnosis with all the intensity of the original experience ("R," 17).

Brown's idea that the subject was incapable of forgetting anything— that even if conscious access to such memories was blocked we unconsciously retained a complete record of every single event or experience that has ever happened to us, however insignificant—testifies to the extraordinary importance traditionally attached to memory as—along with volition—*the* defining mark of personal identity.[20] But what if emotional memories were not what they were assumed to be? What if the (often temporary yet) "undeniable successes"[21] of hypnosis in the treatment of the war neuroses depended not on the revival of emotions that had been previously experienced and were now re-presented to the subject as

20. "As one has said for a long time, no fact disappears from memory," wrote the physiologist Charles Richet in 1894: "everything is fixed in it. Everything acts on us and leaves an indelible trace" (cited by Michael Roth, "Hysterical Remembering," *Modernism/Modernity* 3 [May 1996]: 13).

21. Sándor Ferenczi and Otto Rank, *The Development of Psychoanalysis* (1924; reprint New York, 1956), 61. For positive evaluations of hypnosis and suggestion in the treatment of shell shock in World War I, see M. D. Eder, *War-Shock: The Psycho-Neuroses in War Psychology and Treatment* (London, 1917), 128–43; and Frederick Dillon, "Treatment of Neuroses in the Field: The Advanced Psychiatric Centre," and J. A. Hadfield, "Treatment by Suggestion and Hypno-Analysis," in *The Neuroses of War*, ed. Emmanuel Miller (New York, 1940), 119–27, 128–49. For more pessimistic assessments of the value of these methods, especially in chronic cases, see Norman Fenton, *Shell Shock and Its Aftermath* (St. Louis, 1926), and Abram Kardiner, *The Traumatic Neuroses of War* (Menasha, Wisconsin, 1941).

past, but on the repetition of the emotional experience *in the present*, with all the energy of the initial "event"? What if, accordingly, the passionate "relivings" or "reproductions" characteristic of the cathartic cure could not be used to retrieve emotional memories, for the simple reason that the memories in question did not exist? More broadly, what if the emotions defied a certain kind of representational economy?

It is greatly to Brown's credit that he realized that the question of emotional memory, far from being "entirely unreal" as McDougall believed ("R," 24, n. 1), was central to the issue of the nature of the hypnotic-cathartic cure. Moreover, he was also aware that it was a question that in the prewar years had excited the curiosity of many of the best psychologists of the day, with results that did not always support his own position. Brown drew attention to two contributions of particular interest, those of the Swiss psychologist Edouard Claparède and Sigmund Freud.

In a paper of 1911 Claparède had rejected the theory of emotional memory that subtended Brown's analysis. It will help us get our bearings here if we recall that the controversy over emotional memory was part of a wider turn-of-the-century debate over the epistemological foundations of psychology. One consequence of that debate was a general shift away from an atomistic, sensationalist psychology to a more intentionalist or functional-pragmatic approach that called into question the general role of sensation and the image, or representation, in psychical life. Hovering about these prewar developments, and influencing them in ways that have yet to be fully analyzed, was the talismanic figure of Henri Bergson. We will not let Sartre's brilliantly articulated phenomenological critique of Bergson, Claparède, and others that, despite their new orientation and new terminology, they retained the concept of the image in its classical empiricist-materialist guise, prevent us from acknowledging the historical interest of their work in revising the interpretation of the place of the image and representation in mental life.[22]

Although what has chiefly attracted the attention of historians is the debate over "imageless" thought,[23] the role of the image in emotion was also a topic of discussion. A key figure here was William James, discussed by both Brown and Claparède, who had denied the existence of emo-

22. Jean-Paul Sartre, *Imagination: A Psychological Critique* (Ann Arbor, Michigan, 1972).

23. See George Humphrey, *Thinking: An Introduction to Its Experimental Psychology* (New York, 1963); *Thinking: From Association to Gestalt*, ed. Jean Matter Mandler and George Mandler (New York, 1964); and David F. Lindenfeld, *The Transformation of Positivism: Alexius Meinong and European Thought, 1880–1920* (Berkeley, California, 1980), 220–64.

tional memory. When we think of a past feeling, James had argued, what surges up in our consciousness is not the memory of that feeling, but a new feeling experienced in the present. "The revivability in memory of the emotions, like that of all the feelings of the lower senses, is very small," James wrote in a passage cited by Claparède: "We can remember that we underwent grief or rapture, but not just how the grief or rapture felt. This difficult *ideal* revivability is, however, more than compensated in the case of the emotions by a very easy *actual* revivability. That is, we can produce, not remembrances of the old grief or rapture, but new griefs and raptures, by summoning up a lively thought of their exciting cause. The cause is now only an idea, but this idea produces the same organic irradiations, or almost the same, which were produced by its original, so that the emotion is again a reality. We have 'recaptured' it."[24] With respect to the hypnotic treatment of the shell-shock victim, we might put it that, according to James's theory of emotion, it was because the organic conditions of the original experience had been brought back so vividly to the traumatized soldier that they again produced the emotion of fear—but the emotion was an actual, present feeling caused by the visceral sensations aroused during the hypnosis.

Claparède extended James's argument. Specifically, he set out to discover what he himself experienced when he tried to remember a past emotion. Claparède reported that when he attempted to project an emotion into the past—the sadness he experienced at the thought of his dead parents (significantly, an example of mourning)—either he continued to feel the emotion actually, in the present, and hence not as past, or he ceased to experience the emotion altogether and instead merely represented himself to himself as a kind of depersonalized or dead "mannequin-self" whom he saw objectively, at a distance, without any emotion, as if he were a spectator of himself. "For me," he wrote, "it is impossible to *feel* an emotion as *past*." He observed: "Thus I know that I was sad, but I have no consciousness of any state of sadness. In order for these nonaffective images of sadness to renew their original meaning and their life, I am obliged to retranslate them into affective terms; but then I relapse into emotional states in the present, which is to say that it is my present self that is sad, and no longer only my past self." An emotional state and projection into the past are "incompatible facts," Claparède stated. He observed:

> As soon as I project the past moment far from the present moment which fills myself, then it is as a simple spectator, so to speak, that I consider these past

24. William James, *The Principles of Psychology* (1890; reprint Cambridge, 1983), 1087–88.

memories—which is to say that if I represent myself there to myself, I see myself *from outside*, in the same way that I represent other individuals to myself. My past self is thus, psychologically, distinct from my present self, but it is . . . an emptied and objectivized self, which I continue to feel at a distance from my true self which lives in the present. And if, from being a simple spectator, I try to become an actor, if I try to identify myself with this second self [*image-sosie*], then I draw it back to the present in order to reincarnate it; but it attracts with it the ambient images, and then I have the impression of again enjoying in the present the scene that has passed.

He added:

This tendency to experience in the present a previously experienced scene is especially likely to occur when I seek to represent to myself a past emotion: the emotion can only be experienced as a state of myself. It can only be known from within, and not from outside. If I attribute it to my phantom-self [or double] (which is only seen from outside) then in that very moment I see it vanish from my present consciousness. One cannot be a spectator of one's feelings; one feels them, or one does not feel them; one cannot imagine them [image them, represent them] without stripping them of their affective essence.[25]

This is a fascinating description of what might be called the "phenomenology of affect" in that, in his modest yet elegant way, Claparède appears to break with an entire theory of representation according to which—in a genealogy that goes back to the dominant interpretation of Descartes—the certitude of the *Cogito* is conceived as the spectatorial or specular certitude of the self-observing subject or onlooker who sees or represents himself to himself, as if in a mirror or on a stage. On the contrary, in his critique of the concept of emotional memory or affective representation Claparède abandons all such metaphorics of specularity and

25. Edouard Claparède, "La question de la 'mémoire' affective," *Archives de psychologie* 10 (1911): 367–69. Among those Claparède mentioned as supporting the idea of emotional memory were Ribot, Pillon, Pieron, Dugas, Paulhan, Dauriac, Baldwin, Bain, Fouillée, and Patini; among those he cited as opposing the idea of emotional memory were James, Titchener, Höffding, and Mauxion. At the end of his essay Claparède recognized the role of the psychologist Alfred Binet in launching a general critique of the mental image in psychology. In this connection see especially Alfred Binet, "Qu'est-ce qu'une émotion? Qu'est-ce qu'un acte intellectuel?" *L'Année psychologique* 17 (1911): 1–47, cited with approval by Claparède in "Feelings and Emotions," in *Feelings and Emotions: The Wittenberg Symposium*, ed. Martin L. Reymert (Worcester, Massachusetts, 1928), 136. In that essay, Claparède was still asking: "Does a true affective memory exist? (Or do affective memories constitute an *actual* revival of feelings and emotions?)" (125).

spectatorship. This does not mean that he rejected the image or representation as such, but a particular interpretation of the image or representation as always involving a "representative theatricality," implying a specular distance between the subject and object, ego and alter ego.[26] Indeed there is a sense in which he breaks even more decisively with that metaphorics than Freud himself. This becomes clear when we consider the other text on emotional memory, besides Claparède's, to which Brown in his paper on the traumatic neuroses of the war also drew attention—Freud's metapsychological essay on the unconscious of 1915. It is one of Brown's achievements that he pointed to the precise moment in Freud's difficult and disconcerting text where he seemed to posit the absolute irreducibility between affect and representation on which Claparède also insisted. This is the moment when Freud appeared to acknowledge that if there is such a thing as an unconscious idea or representation—since, for Freud, even in the unconscious the drive (or instinct) is known only through its representations—*affect itself* manifests the drive directly, without any intermediary. In the passage cited by Brown, Freud wrote:

> An instinct can never become an object of consciousness—only the idea [*Vorstellung*] that represents the instinct can. Even in the unconscious, moreover, an instinct cannot be represented otherwise than by an idea. If the instinct did not attach itself to an idea or manifest itself as an affective state, we could know nothing about it. . . . We should expect the answer to the question about unconscious feelings, emotions and affects to be just as easily given. It is surely of the essence of an emotion that we should be aware of it, i.e., that it should become known to consciousness. Thus the possibility of the attribute of unconsciousness would be completely excluded as far as emotions, feelings and affects are concerned. But in psycho-analytic practice we are accustomed to speak of unconscious love, hate, anger, etc., and find it impossible to avoid even the strange conjunction, "unconscious consciousness of guilt," or a paradoxical "unconscious anxiety." Is there more meaning in the use of these terms than there is in speaking of "unconscious instincts"?
>
> The two cases are in fact not on all fours. In the first place, it may happen that an affective or emotional impulse is perceived but misconstrued. Owing to the repression of its proper representative it has been forced to become connected with another idea, and is now regarded by consciousness as a manifesta-

26. On this topic see Michel Henry, *The Genealogy of Psychoanalysis* (Stanford, California, 1993); Mikkel Borch-Jacobsen, *The Freudian Subject* (Stanford, California, 1988); idem, *Lacan: The Absolute Master*, trans. Douglas Brick (Stanford, California, 1991), esp. 43–71.

tion of that idea. . . . Yet its affect was never unconscious; all that had happened was that its *idea* had undergone repression . . . . *Strictly speaking . . . there are no unconscious affects.* . . . The whole difference arises from the fact that ideas are cathexes—basically of memory-traces—whilst affects and emotions correspond to processes of discharge, the final manifestations of which are perceived as feelings. In the present state of our knowledge of affects and emotions we cannot express this difference more clearly.[27]

Brown remarked of this passage that "Freud finds great difficulty in coming to a conclusion on the nature of 'unconscious affects' as contrasted with 'unconscious ideas,' and recognises that the problem of the former is different from that of the latter" ("R," 33).[28] Of the same problem of affect in Freud's 1915 text, Mikkel Borch-Jacobsen has commented:

It is no accident that Freud writes . . . "Even in the unconscious, moreover, an instinct cannot be represented (*repräsentiert sein*) otherwise than by a *Vorstellung* [an idea]," despite immediately adding, as though with remorse, that the drive would remain unknowable if it "did not attach itself to an idea or *manifest itself as an affective state.*" In reality, it is only the *Vorstellung* that *repräsentiert* the drive, for the good reason that the affect, for its part, *presents* it immediately, without the slightest mediation. This is attested to by the fact . . . that affect, by Freud's own admission, cannot possibly be unconscious, as if it would short-circuit every distance and every exteriority between the drive and the psyche (between "body" and "soul"). Affect either is or is not. . . . Contrary to the *Vorstellung*, which can be and yet not appear, the affect *is* only in appearing, exists only as manifest. . . . That is why, according to Freud, there cannot be, in all rigor, any "unconscious affects." And so, in speaking of "unconscious anxiety" or, still more paradoxically, of an "unconscious consciousness of guilt" (*unbewusstes Schuldbewusstein*), the psychoanalyst would only mean that the representation to

27. Sigmund Freud, "The Unconscious" (1915), *Standard Edition*, 14: 177–78; emphasis added. Brown's source for Freud's text was Freud's "Das Unbewusste," *Sammlung kleiner Schriften zur Neurosenlehre* (Vienna, 1906–22), 4: 309.

28. In support of his belief in the existence of emotional memory Brown mentioned the case of one of his patients who on two separate occasions experienced exactly the same emotion when recalling under hypnosis the events of his sixth birthday ("R," 18). In 1921 Brown referred to the same experiments on hypnotic age regression as supporting Bergson's theory of memory (Brown, *Psychology and Psychotherapy* [London, 1921], 179–90). It is a sign of the complexity of Bergson's role in these developments that, directly or indirectly, through his influence Brown and Claparède adopted opposed positions on the existence of emotional memory.

which the affect was initially attached has succumbed to repression. But the affect itself would never cease to impose itself on consciousness. In other words, the affect may well be "suppressed" ("inhibited," "blocked," reduced to the state of a "rudiment"), but it can by no means be *repressed*.[29]

Indeed it would be possible to show that, in his writings of the 1920s on the transference and the second topography, Freud simultaneously conceived affect as that which is always and only experienced *in* consciousness and as that which absolutely *resists* coming into consciousness: paradoxically, Freud appeared to undo the very distinction between consciousness and unconsciousness that he elsewhere seemed to enforce. "[T]he affect," Borch-Jacobsen has observed in this connection, "far from being a second psychic *Repräsentanz* of the drive . . . is, rather, its very manifestation. That affect always be 'conscious' means, in effect, that the psyche can never 'distance' it, never flee it (repress it) like an exterior reality, never ob-pose itself to it in the light of the *Vor-stellung*, and thus neither can it ever dissimulate it from itself. In short, this signifies that the opposition of consciousness and the unconscious is not applicable to affect" ("UN," 139).[30] So that—always according to the same logic—the transference, or emotional tie to the analyst, far from dissimulating a prior, repressed Oedipal or pre-Oedipal memory or representation, as Freud continued to argue, rests on an affect that, as Freud also stated, can only be experienced in the immediacy of an acting in the present that is unrepresentable to the subject ("UN," 186–87). Most paradoxically of all, it was hypnosis that, again according to Freud, best exemplified the peculiar workings of the unconscious defined in those terms. Strangely, Freud treated hypnosis as the paradigm of the emo-

---

29. Borch-Jacobsen, "The Unconscious, Nonetheless," in *The Emotional Tie: Psychoanalysis, Mimesis, and Affect* (Stanford, California, 1993), 123–54, 138–39; hereafter abbreviated "UN." Borch-Jacobsen points out in this regard that Freud never uses the expression "affective representation" ("UN," 197, n. 26).

30. Borch-Jacobsen goes on to observe that this does not mean that the unconscious is thereby reabsorbed into consciousness as pure manifestation, presence, or auto-affection, as the phenomenologist Henry maintains, but that—following Freud's own arguments of *The Ego and the Id* and other writings—"the unconscious invades consciousness itself; indeed, here everything depends on that infinitesimal yet decisive difference of accent between a *conscious* unconscious and an *unconsciousness* of consciousness" ("UN," 142). For related comments on the "formidable difficulties" with which Freud surrounded the concept of affect see Philippe Lacoue-Labarthe and Jean-Luc Nancy, "The Unconscious is Destructured Like an Affect (Part I of 'The Jewish People Do Not Dream')," trans. Brian Holmes, *Stanford Literature Review* 6 (Fall 1989): 197–99.

tional transference to or identification with the other at the very moment he sought to exclude hypnosis from the psychoanalytic project.

All this suggests that according to the mimetic theory of affect adopted by Claparède, Freud, and others what was problematic in the use of hypnosis to cure the war neuroses was precisely the attempt to recover past traumatic experiences in the form of emotional representations that could be brought back into the subject's consciousness; for on the theory of affect examined here the passionate relivings or "reproductions" characteristic of hypnotic abreaction preceded the distinction between "self" and "other" on which the possibility of self-representation and hence recollection depended. The same was true of psychoanalysis, defined as the reconstitution of the subject's history through the recovery and analysis of the patient's repressed memories or fantasies, because the existence of such affective memories or affective representations is what Freud in his discussion of affect appeared to call into question. In sum, there was no "subject" of suggestion in the sense of a subject who could see or distance himself from his emotional experience by re-presenting that experience to himself as other to himself: that seemed to be the lesson of Claparède's and Freud's dissection of the emotions.

But that is a conclusion that Freud also resisted, as did Brown and his colleagues. They remained committed to the view that what "disciplined" or cured patients was that they could be made to distance themselves from their traumatic emotional experiences by representing (re-presenting) them to themselves as other to themselves in the form of recollected, "repressed" or "dissociated" experiences. Accordingly, they demanded that the emotional acting out of the hypnotic catharsis be converted into self-representation and narration—that the patient's speech and behavior under hypnosis be interpreted not as a "reproduction" of the traumatic scene in the mode of a "blind" emotional acting in the *present* but as a narrative in full consciousness of that lived experience as *past.*

Yet a scrutiny of the case histories of the traumatic neuroses suggests that this was a demand that could not always or readily be met. The subject in deep hypnosis did not appear to be a spectator of the (real or fantasized) emotional scene but was completely caught up *in* it, as Claparède claimed. If speech or verbalization often accompanied those scenes, it did not do so in the form of a discourse in which the patient narrated the truth of his past to himself or another (the physician or hypnotist), but in the mode of an intensely animated, present-tense miming or emotional reliving of the alleged traumatic scene that occurred in the absence of self-observation and self-representation. As Brown himself reported of the shell-shocked soldier:

[He] immediately begins to twist and turn on the couch and shouts in a terror-stricken voice. He talks as he talked at the time when the shock occurred to him. He really does live again through the experiences of that awful time. Sometimes he speaks as if in dialogue, punctuated by intervals of silence corresponding to remarks of his interlocutor, like a person speaking at the telephone. At other times he indulges in imprecations and soliloquy. . . . *In every case he speaks and acts as if he were again under the influence of the terrifying emotion.* ("TC," 198)

Significantly, it proved difficult for physicians like Brown to obtain certain kinds of information while patients were reenacting the traumatic scene; indeed the victims often became confused to the point of swooning when they were asked to narrate their experiences in the past tense. "In some cases he is able to reply to my questions and give an account of his experiences," Brown related. "In others he cannot do so, but continues to writhe and talk as if he were still in the throes of the actual experience" ("TC," 198). Sometimes, patients responded to the demand for self-narration by alternating between the past and present tense. "One subject . . . whispered to me, 'Did you see that one? . . . It went up on top,'" Myers reported. "'What now?' I asked, 'What did you say?' 'I was talking to my mate,' was the reply. To my question 'What were you saying?' he answered 'get rifles.' He could be made to realise he was in hospital, but explained his inconsistent behavior by the remark: 'Can't help it. I see 'em and hear 'em (the shells).'" "[His thoughts] repeatedly fly to the trenches," Myers noted of another patient's tendency to repeat the traumatic past in the present tense, as if the events were happening all over again. "For a few minutes his attention could be gained, then his answers became absurd; the question 'How old are you?' for example receiving the reply, 'It passed my right ear.' He would often ask me to speak louder when on the point of lapsing into thoughts of trench life. In another case the alternation of states was so marked that on being unduly pressed for his thoughts when in a stuporous condition he assumed an attitude of hostility, rushing about the room with an imaginary rifle in his hands."[31] (Indeed, for Millais Culpin present-tense abreaction became an explicit goal of the cathartic treatment, anticipating in this regard the same emphasis on present-tense abreactions by William Sargant and others in World War II.)[32]

31. Myers, "Contributions to the Study of Shell Shock," *Lancet* 1 (8 January 1916): 67–68.

32. Thus Culpin spoke of the need for the soldier to use the "correct style of description" in reporting his experiences during the hypnotic catharsis, by which he meant that

Breuer and Freud had earlier made similar observations. "In the afternoons she would fall into a somnolent state which lasted till about an hour after sunset," Breuer had stated of his patient, Anna O. "She would then wake up and complain that something was tormenting her—or rather, she would keep repeating in the impersonal form 'tormenting, tormenting.' For alongside of the development of the contractures there appeared a deep-going functional disorganization of speech. . . . It was also noticed how, during her *absences* in day-time she was obviously creating some situation or episode to which she gave a clue with a few muttered words. . . . When she was like this it was not always easy to get her to talk, even in her hypnosis" (*SH*, 2: 24–30). "The words in which she described the terrifying subject-matter of her experience were pronounced with difficulty and between gasps," Freud reported of Emmy von N., adding of the case of Elizabeth von R. that the details of a certain episode "only emerged with hesitation and left several riddles unsolved" (*SH*, 2: 53, 151).

Moreover, as Freud was the first to observe, <u>patients lacked conviction as to the reality of the reconstructed traumatic scenes.</u> "Sometimes, finally, as the climax of its achievement in the way of reproductive thinking," he observed of his "pressure" technique in *Studies on Hysteria*, "it causes thoughts to emerge which the patient will never recognize as his own, which he never *remembers*, although he admits that the context calls for them inexorably, and while he becomes convinced that it is precisely these ideas that are leading to the conclusion of the analysis and the removal of his symptoms" (*SH*, 2: 272). To which he added in a stunning admission of the irretrievability of the traumatic origin:

> The ideas which are derived from the greatest depth and which form the nucleus of the pathogenic organization are also those which are acknowledged as memories by the patient with greatest difficulty. Even when everything is finished and the patients have been overborne by the force of logic and have been convinced by the therapeutic effect accompanying the emergence of precisely these ideas—when, I say, the patients themselves accept the fact that they thought this or that, they often add: "But I can't *remember* having thought it." It is easy to come to terms with them by telling them that the thoughts were *unconscious*. But how is this state of affairs to be fitted into our own psychological views? Are we to disregard this withholding of recognition on the part of patients, when, now that the work is finished, there is no longer any motive for

the patient ought to use the present tense (see Ben Shephard, "'The early treatment of mental disorders': R. G. Rows and Magull, 1914–1918," 449).

their doing so? Or are we to suppose that we are really dealing with thoughts which never came about, which merely had a *possibility* of existing, so that the treatment would lie in the accomplishment of a psychical act which did not take place at the time? It is clearly impossible to say anything about this—that is, about the state which the pathogenic material was in before analysis—until we have arrived at a thorough clarification of our basic psychological views, especially on the nature of consciousness. (*SH*, 2: 300)

The problem of the patient's lack of confidence in the reality of the memory of the trauma—the victim's inability to remember, and hence testify with conviction to, the facticity of the reconstructed event—will haunt not only psychoanalysis but the entire modern discourse of trauma. "There is one feature of the modern that is dazzling in its implausibility: that the forgotten is what forms our character, our personality, our soul," Ian Hacking has recently observed.[33] But one implication of the material we have just explored is that according to the theory of affect endorsed by Clarapède and Freud it was precisely what could not be remembered that was decisive for the subject. To put this slightly differently, the implication was that the emotional trauma could not be lifted from the unconscious because that trauma was never "in" the unconscious in the form of repressed or dissociated representations.

Now it might appear that the "hidden observer" phenomenon of age regression and other hypnotic experiences contradicts the idea that the hypnotic experience involves an immersive or "mimetic" absorption in the traumatic scene to the point of self-forgetfulness. Thus Breuer reported of Anna O. that "even when she was in a very bad condition—a clear-sighted and calm observer sat, as she put it, in a corner of her brain and looked on at all the mad business" (*SH*, 2: 46), and in another passage stated that "many intelligent patients admit that their conscious ego was quite lucid during the attack and looked on with curiosity and surprise at all the mad things they did and said" (*SH*, 2: 228). Similarly, Freud observed of his patient Emmy von N. that she "kept a critical eye upon my work in her hypnotic consciousness" (*SH*, 2: 62, n. 1). According to Breuer, Anna O. even went so far as to subsequently reproach herself with shamming or simulating her entire illness—as if she had never lost awareness of what had been going on. Breuer qualified Anna O.'s claim to simulation by attributing it in part to her retrospective sense of guilt for all the trouble she had caused and to her feeling that, from the perspective

33. Ian Hacking, "Memory Sciences, Memory Politics," in *Tense Past: Cultural Essays in Trauma and Memory*, ed. Paul Antze and Michael Lambek (New York, 1996), 70.

of her reunited personality, she could have prevented it, noting of her that "this normal [or specular] thinking which persisted during the secondary state must have fluctuated enormously in amount and must very often have been completely absent" (*SH*, 2: 46).

Borch-Jacobsen, however, has recently dismissed Breuer's statement as an argument designed to meet the charge of simulation by appealing to the notion of the unconscious—a notion he now rejects. His position depends on an interpretation of hypnosis as always a lucid simulation in which the hypnotized subject remains an observer of the spectacle of her trance. It is an interpretation that allows him to treat the problem of traumatic or hypnotic amnesia as an artifact of false ideas about the "immersive" nature of the trance experience and on that basis to skeptically dismiss the entire theory of traumatic dissociation. In chapter 5 I discuss Borch-Jacobsen's opinions at some length. Here I only want to note that if, on the basis of their own conflicting ideas about the nature of affect and the cathartic cure, Brown, Myers, and McDougall glossed over the problem of the failure of memory in the cathartic cure—that is, the failure of memory defined as self-narration and self-representation—Freud, on abandoning hypnosis altogether, interpreted that failure as an expression of the patient's *resistance* to recollection and narration. Such a strikingly original solution opened up an entire dynamics of unconscious desire and repressed representations and dramatically shifted attention away from the affective reliving of the cathartic cure to the question of corporeal signification and linguistic meaning; but it was a solution that would eventually unravel at the level of practice in the problem of traumatic repetition and at the level of theory in the aporias of Freud's second topography. In short, as Freud himself became increasingly aware, nothing was less certain than whether the cathartic "reproduction" or "repetition compulsion" could be converted into conscious recollection, nothing more ambiguous than the nature and mechanism of what he called "working through."[34]

34. Sigmund Freud, "Remembering, Repeating and Working-Through" (1914), *Standard Edition*, 12: 147–56. In this remarkably complex paper, written at the same time as the Wolfman case and apparently in reference to its theoretical and therapeutic difficulties, Freud, oddly, identified hypnotic catharsis with the "simple" or "ideal" form of remembering, that is with self-representation and self-narration, and psychoanalysis with the compulsion to repeat or the tendency to act out (*agieren*) in the absence of any awareness of the repetition. On this basis Freud compared the process of "working through" the resistances, by which the repetition of repressed affects and representations was to be converted into recollection, with the abreaction of the quota of affect "strangulated by repression—an abreaction without which hypnotic treatment remained ineffective" (155–

The Persistence of Janet

If I mention Pierre Janet at this juncture it is partly because, recognizing in one of Janet's early cures a method analogous to theirs, Breuer and Freud placed Janet at the origin of the cathartic cure (*SH*, 2: 7, n. 1)—an ambiguous gesture, as we shall see. Moreover, Janet's views on dissociation were accepted by Brown, Myers, and other physicians as a useful alternative to Freud's theory of repression in the interpretation of certain cases of shell shock.[35] Janet is also pertinent because Judith Herman, Bessel van der Kolk, Onno van der Hart, and other modern theorists have recently hailed him as a pioneer in developing a fully formulated mnemotechnology for the treatment of the trauma victim, including the victim of combat or PTSD. In particular, returning to Janet's long-neglected meditations on the nature of trauma, memory, and narration, Herman and others have praised Janet for distinguishing between two kinds of memory—"traumatic memory," which merely and unconsciously *repeats* the past, and "narrative memory," which *narrates the past as past*—and for validating the idea that the goal of therapy is to convert "traumatic memory" into "narrative memory" by getting the patient to recount his or her history. "In the second stage of recovery, the survivor tells the story of the trauma," Herman writes. "She tells it completely, in depth and in detail. This work of reconstruction actually transforms the traumatic memory, so that it can be integrated into the survivor's life story. Janet described normal memory as 'the action of telling a story.' Traumatic memory, by contrast, is wordless and static. . . . The ultimate goal . . . is to put the story . . . into words."[36]

---

56). Yet Freud's own discussion of affect in "The Unconscious" and other texts raised the question: Did the affects belong to the scheme of repressed representations apparently posited here? If not, what was the nature and mechanism of "working through"?

35. Shephard, "'The early treatment of mental disorders': R. G. Rows and Maghull, 1914–1918," 447, 449.

36. Judith Lewis Herman, *Trauma and Recovery* (New York, 1992), 175–77; hereafter abbreviated *TR*. For similar appeals to Janet's work, see Bessel A. van der Kolk and Onno van der Hart, "The Intrusive Past: The Flexibility of Memory and the Engraving of Trauma," in *Trauma: Explorations in Memory*, ed. Cathy Caruth (Baltimore, 1995), 158–82, and "Pierre Janet and the Breakdown of Adaptation in Psychological Trauma," *American Journal of Psychiatry* 146 (1989): 1330–42; Onno van der Hart and Rutger Horst, "The Dissociation Theory of Pierre Janet," *Journal of Traumatic Stress* 2 (4) (1989): 397–412; Onno van der Hart, Paul Brown, and Bessel A. van der Kolk, "Pierre Janet's Treatment of Post-Traumatic Stress," *Journal of Traumatic Stress* 2 (4) (1989): 379–95; and Frank W. Putnam, "Pierre Janet and Modern Views of Dissociation," *Journal of Traumatic Stress* 2 (4) (1989): 413–29.

But such an appropriation of Janet on the part of Herman and others involves repudiating that aspect of his psychotherapy that seeks to make the patient *forget*. Take for example Janet's famous cure of Marie (the case cited by Breuer and Freud). Marie was a nineteen-year-old girl whom Janet saw at Le Havre early on in his career when she was hospitalized for hysterical convulsive crises and a delirium that, Janet soon established, always coincided with the arrival of her menstrual periods, periods that, after about twenty hours, would then abruptly cease. During her delirium, Marie sometimes "uttered cries of terror, speaking incessantly of blood and fire and fleeing in order to escape the flames; sometimes she played like a child, spoke to her mother, climbed on the stove or the furniture, and disturbed everything in the room," he wrote in his first description of the case in 1889. The end of each hysterical crisis was accompanied by the vomiting of blood. Marie was completely amnesiac for what had transpired. In between her attacks she suffered from small contractions of the muscles of the arms and chest, various anesthesias, and a hysterical blindness of her left eye.

Positing a connection between the origin of Marie's hysterical symptoms and the onset of her menstrual periods, Janet hypnotized her in order to "bring back" the apparently forgotten memories. Based on Marie's dramatic reenactments in the trance state, Janet was "able to recover the exact memory of a scene which had never been known except very incompletely." Owing to the shame she had felt when, aged thirteen, she had experienced her first menstrual period, Marie had succeeded in interrupting the flow of blood by plunging into a large tub of cold water. The shock had produced shivering, a delirium for several days, and a complete cessation of her periods; when five years later these had recommenced, they had produced the symptoms that had led to her hospitalization.[37]

Janet's "supposition" concerning Marie's originary trauma, "true or false" as he expresses it (*AP*, 412), served as the basis for her cure. "I could only succeed in effacing this [fixed] idea by a unique method. It was necessary to take her back by suggestion to the age of thirteen, to put her back again into the initial circumstances of the delirium, and thus to convince her that her period had lasted for three days and had not been interrupted by any unfortunate incident. Now, once this was done, the next menstrual period arrived on time and lasted for three days, without lead-

37. Pierre Janet, *L'Automatisme psychologique: Essai de psychologie expérimentale sur les formes inférieures de l'activité humaine* (1889; reprint Paris, 1989), 410–12; hereafter abbreviated *AP*.

ing to any pain, convulsion or delirium" (*AP*, 412–13). He treated Marie's remaining symptoms, including her hysterical blindness, as well as other cases of dissociation by the same method.[38]

In other words, according to Janet's first account of the case, and contrary to the ingrained beliefs of many his commentators, Marie was cured not by the recovery of memory but by *the excision of her imputed or reconstructed trauma* (see *AP*, 7). In 1880 the novelist Edward Bellamy imagined an invention for the extirpation of thought processes. "I deem it only a question of time," Dr. Gustav Heidenhoff says to Henry, who loves a woman driven almost to suicide by a guilty sexual past that she cannot forget, when "science shall have so accurately located the various departments of thought and mastered the laws of their processes, that, whether by galvanism or some better process, the mental physician will be able to extract a specific recollection from the memory as readily as a dentist pulls a tooth, and as finally, so far as the prevention of any future twinges in that quarter are concerned."[39] In 1894 Janet himself remarked that one of the most valuable discoveries of pathological psychology would be a sure means of helping us to forget (*NIF*, 404). The same year he criticized Breuer and Freud's account of the cathartic cure on the grounds that what mattered in the treatment of the neuroses was not the "confession" of the traumatic memory but its elimination (*NIF*, 163).[40] Nor did

38. See also Janet's "L'Amnésie continue" (1893), "Histoire d'une idée fixe" (1894), and "Un cas de possession et l'exorcisme moderne" (1895) in *Études expérimentales sur les troubles de la volonté, de l'attention, de la mémoire, sur les émotions, les idées obsédantes et leur traitement* (1898), vol. 1 of *Névroses et idées fixes* (1898; reprint Paris, 1990), 156–212, 375–406; hereafter abbreviated *NIF*.

39. Edward Bellamy, *Doctor Heidenhoff's Process* (1880; reprint New York, 1969), 101. In the novel, Dr. Heidenhoff, who describes his process of memory extirpation as "merely a nice problem in surgery" (104), criticizes traditional notions of moral responsibility, based on our capacity to remember past acts, as grounded in ideas concerning the permanence of identity—ideas he rejects (120–24). But Bellamy does not endorse the solution to the problem of morality adopted by his fictional character, Dr. Heidenhoff. The haunted young woman cannot be absolved of her guilt by being helped to forget her shameful past, and Dr. Heidenhoff's method of erasing memories turns out to be only a dream of Henry brought on by a sleeping powder. Morality, Bellamy seems to be saying, depends on our having permanent remembered identities; the novel ends with the heroine refusing Henry's offer of marriage and choosing instead to kill herself. Only death can bring about the absolute forgetting that she craves.

40. Here as elsewhere Janet added that more than simple suggestion was necessary to cure hysteria and went on to describe the various methods he used to remove or "rub out" (*enlever*) or efface (*effacer*) or otherwise transform the patient's traumatic "memories." These methods included the method of "decomposition" and "substitution" by which

the ethical implications of such "modern exorcism" or "psychological surgery" trouble him (*PH*, 1: 678). As he observed of a cure strikingly similar to that of Bellamy's fictional scenario, that of a hysterical husband whose guilt over his infidelity had driven him into hysteria: "The memory of his fault was transformed in all sorts of ways thanks to hallucinated suggestions. Finally even the wife of Achille, evoked by hallucination at an appropriate moment, came to give a complete pardon to this husband who was more unfortunate than guilty" (*NIF*, 404).[41]

But it is precisely that aspect of Janet's legacy that Herman disowns. "Janet sometimes attempted in his work with hysterical patients to erase traumatic memories or even to alter their content with the aid of hypnosis," she observes. "Similarly, the early 'abreactive' treatment of combat veterans attempted essentially to get rid of traumatic memories. This image of catharsis, or exorcism, is also an implicit fantasy in many traumatized people who seek treatment. It is understandable for both patient and therapist to wish for a magic transformation, a purging of the evil of the trauma. Psychotherapy, however, does not get rid of the trauma. The goal of recounting the trauma story is integration, not exorcism. In the process of reconstruction, the trauma does undergo a transformation, but only in the sense of becoming more present and more real. The fun-

---

traumatic memories were broken down into their component parts—into specific images, words, or even parts of words—and hypnotic suggestion was then deployed, in a lengthy treatment process, to substitute neutral or positive experiences for each of the traumatic component elements.

41. Janet does not seem to have been aware of Bellamy's novel. But, thanks to Peter Swales, we now know that Freud was familiar with the book and that when, in July 1889, during his visit to Bernheim, he wanted to give his sister-in-law, Minna Bernays, an idea of how he was treating his patient "Frau Cäcilie M." (Anna von Lieben), he advised her to consult *Doctor Heidenhoff's Process* (Peter Swales, "Freud, His Teacher, and the Birth of Psychoanalysis," in *Freud: Appraisals and Reappraisals: Contributions to Freud Studies*, ed. Paul E. Stepansky [Hillsdale, New Jersey, 1986], 1: 35–36). Swales's description of Dr. Heidenhoff's procedure as a brilliant anticipation of Freud's cathartic method encapsulates the ambiguities inherent in that method as I have been attempting to describe them. Understanding Dr. Heidenhoff's procedure as involving a "kind of hybrid version of catharsis and ECT [Electroconvulsive Therapy]," Swales reports that Dr. Heidenhoff's patient is liberated from her traumatic memories by being induced to recall and narrate them during the operation (36). But according to the novel, the patient has no difficulty in remembering what troubles her; on the contrary, her problem is that she can't forget. Nor is she asked to tell or narrate the events of her past during treatment; rather, she is asked to concentrate her attention on the memories she finds impossible to forget so that Dr. Heidenhoff can remove them with his electrical machine.

damental premise of the psychotherapeutic work is a belief in the restorative power of truth-telling" (*TR*, 181). What appears to motivate Herman's attitude here is a powerfully entrenched commitment to the redemptive authority of history—even if that commitment is tempered by an awareness of the difficulty of historical reconstruction. For Herman and for the modern recovery movement generally, even if the victim of trauma *could* be cured without obtaining historical insight into the origins of her distress, such a cure would not be morally acceptable. Rather, the victim must be helped to speak the horrifying truth of her past—to "speak of the unspeakable" (*TR*, 175)—because telling that truth has not merely a personal therapeutic but a public or collective value as well. It is because personal testimony concerning the past is inherently political and collective that the narration of the remembered trauma is so important (*TR*, 181). As Herman states: "Remembering and telling the truth about terrible events are prerequisites both for the restoration of the social order and for the healing of individual victims" (*TR*, 1).[42]

42. Even as they emphasize the therapeutic importance of transforming traumatic memories into narrated memories, van der Kolk and his colleagues recognize the difficulty of achieving such a transformation in severe or chronic cases of trauma and indeed acknowledge that the restoration of memories alone doesn't necessarily cure (van der Hart, Brown, and van der Kolk, "Pierre Janet's Treatment of Post-Traumatic Stress," 380). Moreover, they grant that Janet treated some patients not by converting traumatic memories into narration but by hypnotically exorcising the past—for example, by using the trance state to substitute pleasant memories for painful ones (van der Kolk and van der Hart, "The Intrusive Past," 450). These authors even observe that Janet's famous patient, Marie, was cured in this way (van der Hart, Brown, and van der Kolk, "Pierre Janet's Treatment of Post-Traumatic Stress," 388). But, apparently uneasy with the idea of altering or playing with history, they appear to equate Janet's hypnotic manipulation of memory with the patient's voluntary control of the past (van der Kolk and van der Hart, "The Intrusive Past," 450).

It must be acknowledged that the relative merits of forgetting and remembering in any given case are a matter of therapeutic possibilities and tact, as well as of political circumstances. Van der Hart has coauthored an article on the hypnotic treatment of traumatic grief in which he recognizes the value of Janet's "substitution technique." That method, as he shows, has its counterpart in some recent hypnotic procedures designed to help the patient neutralize, revise, or remedy the past (van der Hart, Brown, and Ronald N. Turco, "Hypnotherapy for Traumatic Grief: Janetian and Modern Approaches Integrated," *American Journal of Clinical Hypnosis* 32 [April 1990]: 263–71).

Van der Hart, however, has also stated his awareness of the need in current practice for a more "egalitarian" therapeutic relationship in which, when indicated and possible, the patient rather than the therapist should take the lead in "rewriting" the script. Moreover, he also believes that, because of the secrecy and taboo that have surrounded the topic of

Recently, a few critics have begun to question the assumption that the determination and recuperation of the historical past has an inherent ethico-political value.[43] But what I want to focus on here is the influence exerted by an apparently similar commitment to the importance of historical reconstruction on Janet's representation of his own contribution to psychotherapeutics. As a consequence of a growing emphasis on recollection and narration in psychotherapy, mediated in part by his famous rivalry with (but also implicit dependence on) Freud's model of the "talking cure," Janet came to distort his own record.[44] He did not want to forget that he was the first to discover a technique for the cure of patients by getting them to remember their traumas. But precisely because he did not want to forget his priority, he forgot what his discovery was.

In fact from the start Janet's attitude toward memory appeared am-

<hr />

sexual abuse, the therapeutic task in such cases should be one of bearing witness to the past, an emphasis that valorizes narrative memory (see in this regard van der Hart and Brown, "Abreaction Re-Evaluated," *Dissociation* 5 [1992]: 127–40). In the context of the current recovered/false memory debate, he states that he has become even more wary of therapist-induced changes in memories of sexual abuse. The experience of the Holocaust has also contributed to the reaction against concepts of altering the past. Van der Hart reports that an important author on Holocaust testimonies was shocked by the statement made by van der Kolk and himself to the effect that the patient might be encouraged to imagine alternatives to the traumatic past (personal communication). As I show in chapter 7, the need to acknowledge the reality of trauma among physicians working with concentration camp victims whose claims for compensation were being denied by the German courts has contributed to the recent emphasis on the reality and accuracy—indeed literality—of traumatic memory.

43. For vigorous arguments against the widespread belief that there is an intrinsic relationship between history and ethics, arguments bearing for the most part on the question of collective rather than individual memory, see Steven Knapp, "Collective Memory and the Actual Past," *Representations* 26 (Spring 1989): 123–48. See also Walter Benn Michaels, "Race into Culture: A Critical Genealogy of Cultural Identity," *Critical Inquiry* 18 (Summer 1992): 679–80, "The Victims of New Historicism," *Modern Language Quarterly* 54 (March 1993): 111–20, and "The No-Drop Rule," *Critical Inquiry* 20 (Summer 1994): 758–69. For reflections on the moral dilemmas associated with the often unrealizable demand that veterans of the Vietnam War and other wars be made to remember as a requirement of treatment strategies for post-traumatic stress disorder see Allan Young, *The Harmony of Illusions: Inventing Post-Traumatic Stress Disorder* (Princeton, New Jersey, 1995).

44. For Janet's well-known rivalry with Freud, see especially Campbell Perry and Jean-Roch Laurence, "Mental Processes Outside of Awareness: The Contributions of Freud and Janet," in *The Unconscious Reconsidered*, ed. K. S. Bowers and Donald Meichenbaum (New York, 1984), 9–48.

bivalent. On the one hand, he believed that memory was overvalued. "'One must know how to forget,'" he was fond of quoting Taine as saying, and of remarking that: "One must not be surprised at this forgetfulness, it is necessary that it should be so. How could it be that our own minds, our poor attention, could fix itself constantly on the innumerable perceptions which register in us? We must, as has often been said, forget in order to learn. Forgetting is very often a virtue for individuals and for a people" (*NIF*, 421).[45] On the other hand, memory—continuous, narratable memory—increasingly came to have a privileged status in his texts as that which makes us distinctly human. Thus at almost the same moment Janet discovered the therapeutic value of erasing memory, he began to suggest that, in order to be cured, patients must be helped to dissolve their amnesia by telling the story of the traumatic event. As it did for Brown and his colleagues, the task of psychotherapeutics became one of getting the patient to "*say* 'I remember'" (*NIF*, 137; emphasis added). For Janet in this mode, memory proper was more than dramatic repetition or miming; it involved the capacity to distance oneself from oneself by representing one's experiences to oneself and others in the form of a narrated history. In a statement of 1919 that has recently been cited by van der Kolk and others, Janet observed: "*Memory*, like belief, like all psychological phenomena, is an action; essentially, it *is the action of telling a story* . . . . The teller must not only know how to [narrate the event], but must also know how to associate the happening with the other events of his life. . . . A situation has not been satisfactorily liquidated, has not been fully assimilated, until we have achieved, not merely an outward reaction through our movements, but also an inward reaction through the words we address to ourselves, through the organisation of the recital of the event to others and to ourselves, and through the putting of this recital in its place as one of the chapters in our personal history" (*PH*, 1: 661–62). "Strictly speaking," Janet added, "one who retains a fixed idea of a happening cannot be said to have a 'memory' of the happening. It is only for convenience that we speak of it as a 'traumatic memory.' The subject is often incapable of making with regard to the event the recital which we speak of as a memory; and yet he remains confronted by a difficult situation in which he has not been able to play a satisfactory part, one to which his adaptation had been imperfect" (*PH*, 1: 663).

45. Cf. Michael Roth, "Remembering Forgetting: *Maladies de la mémoire* in Nineteenth-Century France," *Representations* 26 (Spring 1989): 54. We are reminded here of William James's remark that "If we remembered everything, we should on most occasions be as ill off as if we remembered nothing" (James, *Principles of Psychology*, 640).

Janet called the act of narration "presentification," an operation of self-observation and self-representation that he imagined as a feat of internal policing or self-surveillance by which at any moment we are compelled to attend to and communicate our present experiences to ourselves and above all to others—for memory is preeminently a social phenomenon—and to situate and organize those experiences in their proper time and place. "Presentification" thus depends on our ability to constitute the present *as present* and to connect the stories we tell about ourselves with present reality and our actual experiences. Janet conceived "narrative memory" in economic terms as an act of abbreviation that—unlike traumatic memory, which is rigidly tied to the specific traumatic situation, takes place without regard for an audience, and by virtue of its inflexible acting out takes a considerable length of time—can be performed in only a few minutes and, depending on the social context, in a variety ways. For Janet, it is precisely because language is conceived as intrinsically portable—as representing an absent present—that narrative memory can be detached from the occasioning event in this manner. (This suggests that Janet did not attach as much significance to the truthfulness of narrative memory as to its plasticity and flexibility.) The act of presentification is one that animals, primitive people, young children, and hysterics are characteristically unable to perform—animals, because they are incapable of self-knowledge and self-representation; and primitive people, young children, and hysterics because, owing to their undeveloped or degenerate or weakened mental condition, they lack the mental synthesis necessary for paying attention to present reality and hence for locating their narratives in an appropriate temporal order. From this perspective, as Janet made clear, the animated acting out or reliving characteristic of the trauma patient, for all its inclusion of verbalization (Herman is clearly wrong to imply that traumatic memory is "wordless"), did not constitute such a narration precisely because it occurred in the absence of self-representation.[46]

Janet's favorite example of the *failure* of presentification was the case of Irène, a young woman who was traumatized by the death of her mother (a case of maternal mourning, as I have already noted). Unable to

---

46. See especially Janet, *L'Évolution de la mémoire et de la notion du temps* (Paris, 1928). Recently, Monique David-Ménard has used the term *presentification* (*Darstellung*) in a sense opposite to that of Janet—not for the process of self-representation to which Janet attaches the term but for the hysterical acting in the present that occurs precisely in the absence of self-representation and symbolization (Monique David-Ménard, *Hysteria from Freud to Lacan: Body and Language in Psychoanalysis* [Ithaca, 1989], 110).

realize the fact of her loss, Irène instead reenacted the scene of death in a somnambulistic repetition that was completely unavailable to subsequent recall. Irène "has not built up a recital concerning the event, a story capable of being reproduced independently of the event in response to a question," Janet observed: "She is still incapable of associating the account of her mother's death with her own history. Her amnesia is but one aspect of her defective powers of adaptation, of her failure to assimilate the event. . . . In her crises she readopts the precise attitude which she had when caring for her mother in the death agony. This attitude is not that of a memory which enables a recital to be made independently of the event; it is that of hallucination, a reproduction of the action, directly linked to the event" (*PH*, 1: 662–63).

Janet seemed to imply that Irène was cured when she became capable of transforming the traumatic memory of her mother's death into narrative representation. But a careful reading of the text from which the above quotation is taken reveals that something far more interesting and complex is going on. For in this text Janet singled out not the case of Irène but that of *Marie* as exemplifying the cure of hysteria by the recollection and narration of the forgotten trauma. Emphasizing his priority over Breuer and Freud, Janet revised his earlier account of the case of Marie by suggesting that she was cured not by the excision but by the recollection and narration of the traumatic memory. Without referring to his attempts to hypnotically *eliminate* memories in that case, Janet stated:

> In my early studies concerning traumatic memories (1889–1892), I drew attention to a remarkable fact, namely that in many cases the searching out of past happenings, the giving an account [*l'expression*] by the subject of the difficulties he had met with and the sufferings he had endured in connexion with these happenings, would bring about a signal and speedy transformation in the morbid condition, and would cause a very surprising cure. *Marie's case was typical.* . . . In the somnambulist state, this young woman told me what she had never dared to confess [*dire*] to anyone. At puberty she had been disgusted by menstruation, and had dreaded its onset. When the flow began, wishing to check it, she got into a cold bath. . . . After she had made this disclosure, her fits of hysterics ceased, and normal menstruation was restored. . . .
>
> In these earlier writings, I drew the inference . . . that the memory was morbific because it was dissociated. . . . The morbid symptoms disappeared when the memory again became part of the synthesis that makes up individuality. I was glad to find, some years later, that Breuer and Freud had repeated these experiments, and that they accepted my conclusions without modification. (*PH*, 1: 672–74)

In other words, Janet appeared to transform a cure based on the excision of memory into a therapy based on the patient's conscious insight and recollection. And yet in another section of the same book, to which in the passage just quoted the translators (but not Janet) refer the reader, Janet wrote of Marie: "Finally, it was found possible, by modifying the memory in various ways [*en modifiant le souvenir par divers procédés*], to bring about the disappearance or the modification of the corresponding symptom" (*PH*, 1: 591)—a formulation that, in the light of my earlier analysis of his original description of the case, strongly suggests that Marie's traumatic memory was altered or replaced by others, and in that sense eliminated.

But what, then, of the melancholic Irène—the focus of the recovery movement's interest in Janet—who failed to mourn her mother by failing to remember that her mother was dead? How was *she* cured? In this text as in later writings, Janet's attempts to describe and explain the therapeutic process are extraordinarily convoluted, as if the task of characterizing the nature of the cure—of defining what Freud called working through— defied systematic articulation. More precisely, his texts are marked by displacements and slippages such that every attempt he makes to stabilize his account of his various psychotherapeutic methods (for he recognizes the need for many different approaches) necessitates repeated gestures of supplementation. So that far from belonging unproblematically to the category of cure by narration—indeed chiasmatically crossing the case of Marie—Irène's case turns out to depend not entirely or exactly on re-memorization but on an additional procedure or set of procedures that Janet called both "assimilation" and "liquidation" and that appears to have much in common with—no surprise here—hypnotic suggestion. "Irène's case is of special interest because her absurd behaviour was so out of place in the circumstances, and because of the lacunae in her interior assimilation which found expression in her amnesia," Janet wrote in the same text of 1919. "After much labour I was able to make her reconstruct the verbal memory of her mother's death. From the moment I succeeded in doing this, she could talk about the mother's death without succumb- ing to crises or being afflicted with hallucinations; the assimilated hap- pening had ceased to be traumatic" (*PH*, 1: 680–81). But in citing this same passage, van der Kolk, van der Hart, and other modern trauma theorists fail to acknowledge that Janet's claim left a remainder.[47] For in a passage omitted by them he immediately added:

---

47. Van der Hart, Brown, and van der Kolk, "Pierre Janet's Treatment of Post- Traumatic Stress," 388.

Doubtless so complex a phenomenon cannot be wholly explained by such an interpretation. Assimilation constitutes no more than one element in a whole series of modified varieties of behaviour which I shall deal with in the sequel under the name of "excitation." Irène, under the influence of the work which I made her do, threw off her depression, "stimulated" herself, and became capable of bringing about the necessary liquidation. . . . Irène was cured because she succeeded in performing a number of actions of acceptation, of resignation, of rememorisation, of setting her memories in order, and so on; in a word, she was able to complete the assimilation of the event. (*PH*, 1: 681)[48]

Under Janet's authoritative "influence" Irène was "excited" to give up her melancholic attachment to her dead mother and adapt to the needs of the present. And in general for Janet the process of cure did not necessarily depend on the recovery and narration of memory. As he wrote in 1923:

The well-known expressions one repeats without cease, "to act, forget, pardon, renounce, resign oneself to the inevitable, to submit," seem always to designate simple acts of consciousness. . . . In reality these expressions designate a complicated ensemble of real actions . . . which liquidate the situation and make one resigned to it. A woman is very gravely ill since the rupture with her lover. You will say this is because she cannot resign herself; no doubt, but this absence of resignation consists of a series of actions which she continues to make and which it is necessary for her to cease making. The physician must help this woman stop carrying out these absurd actions, teach her to make others, give her another attitude. *To forget the past is in reality to change behavior in the present. When she achieves this new behavior, it matters little whether she still retains the verbal memory of her adventure, she is cured of her neuropathological disorders.*[49]

The process of cure thus required *both* assimilation and liquidation. It demanded a discharge or "demobilization" of psychical energies that Janet linked to Freud's method of cathartic abreaction and that in relation to the cure of Irène he described in the following terms:

[I] have already often remarked that it is necessary to . . . use all the resources of rhetoric in order to make a patient change a shirt or drink a glass of water. This is what I especially emphasized in my earliest researches. "The treatment

48. If in this text Janet claims that Irène recovered, her cure cannot have been a simple matter for in 1927, apparently referring to the same case, he observed that "patients act out *indefinitely* the scene of rape or the scene of the death of their mother *for years after the event*" (Janet, *De l'angoisse à l'extase* [1927; reprint Paris, 1975], 2: 334, emphasis added; see also 2: 322).

49. Janet, *La médecine psychologique* (1923; reprint Paris, 1980), 126; emphasis added.

which I imposed on the patient is not only a suggestion, but moreover an excitation. In psychological treatments, one has not always distinguished between
the role of suggestion and the role of excitation which tries to increase the
mental level. I demand attention and effort on the part of Irène, I demand
clearer and clearer consciousness of her feelings, everything that helps to augment the nervous and mental tension, to obtain, if you will, the functioning of
the higher centers. Very often I have observed with her as with so many other
patients that the truly useful séances were those where I was able to make her
emotional. It is often necessary to reproach her, to discover where she has remained suggestible, to support her morally in all sort of ways to raise her up and
to make her recover memories and actions." . . . When the higher functioning
is obtained, the subject feels a modification of his consciousness that translates
into an increase in perception and activity.[50]

If Janet's notion of "assimilation" appears analogous to the recovery
movement's notion of "integration" based on the recovery and narration
of memory, it is nevertheless the case that for Janet narrated recollection
was insufficient for the cure. A supplementary action was required, one
that involved a process of "liquidation" that, terminologically, sounded
suspiciously like "exorcism" or forgetting. Moreover "liquidation" didn't
just supplement "assimilation"; the mutual entanglement of the two operations was so intense that the entire chapter in which Janet in 1919
expounded his understanding of the therapeutic process is called "Treatment by Liquidation"—not "Assimilation." Perhaps most important,
the supplementary procedures necessary for Irène's cure manifestly involved the physician's deliberate manipulation of the patient by processes that Janet himself understood as suggestive in nature (see *PH*, 1:
145). In Janet's mnemo-technology, hypnotic suggestion was discovered
to be not external to the process of cure but internal to its effectiveness. In
sum, Janet's extensive writings bear witness to the impossibility of sus-
taining theoretically or practically the opposition between forgetting and
remembering upon which so much of the edifice of modern psychotherapeutic thought has been made to depend.

*

In 1920, the "daemonic" compulsion to repeat painful experiences—a
phenomenon long familiar to physicians as the "fixation" to the trauma in
the case of female hysterics but appearing during World War I as the rev-

50. Ibid., 129–30.

elation of something new and remarkable now that it was seen to apply to a large number of males—led Freud to posit the existence of death instincts that lay "beyond" pleasure and that seemed to pose a virtually insuperable obstacle to remembering. If Freud never completely abandoned his belief in the curative power of recollection, this is not the case for one major school of his successors—the linguistic-rhetorical school of Lacan and his followers—for whom the failure of memory in traumatic experience exemplifies the need for a structural or formal version of psychoanalysis, conceived (or reconceived) as a discipline that on the one hand invests patient narratives with decisive significance but on the other hand maintains that those narratives are characteristically, perhaps inherently, discrepant with the (themselves often unknowable) "facts" of the case. As Lacan emphasized, in the Wolf Man case of 1918 Freud himself attempted to resolve the tension between forgetting and remembering by proposing just such a structural treatment of the problem of psychoanalytic narrative. Put more strongly, psychoanalysis as Lacan and the Lacanians define it is committed to the project of formalizing memory by eliciting and analyzing narratives whose fidelity to individual experience is no longer of central importance.[51]

More provocatively, it might be argued that Janet's psychotherapeutic work also may be understood as committed to such a project. As we have seen, what mattered according to Janet in the treatment of hysteria was that, through the use of techniques of liquidation and assimilation, patients acquired the ability to produce an account of themselves that conformed to certain requirements of temporal ordering but that did not necessarily entail a process of self-recognition. The distinction between forgetting and remembering thus virtually collapsed in the demand that, whether or not they remembered the traumatic "event," patients became capable of developing a coherent narrative of their lives the importance of which lay not so much in its adequation to personal experience as in its bearing on present and future actions. Viewed in this perspective, Janet's well-known disagreement with Freud over the sexual content of psychoanalysis seems less significant than their agreement that, if narration cures, it does so not because it infallibly gives the patient access to a primordially personal truth but because it makes possible a form of self-understanding even in the absence of empirical verification. In short, what I earlier described as Janet's ambivalence with respect to the problem of memory emerges as more apparent than real. For his seemingly

51. Jacques Lacan, *Freud's Papers on Technique, 1953–1954*, vol. 1 of *The Seminar of Jacques Lacan*, ed. Jacques-Alain Miller (New York, 1988), 12–14.

conflicting claims that memory is overvalued and that memory is fundamental turn out to resolve themselves into the noncontradictory propositions that <u>memory conceived as truth-telling is overestimated but that</u>
<u>memory conceived as narration is crucial.</u>

But this account of the convergence between the views of Janet,
Freud, and Lacan overlooks certain significant differences. In the first
place, <u>notions of speech and narration</u> by no means play an indispensable
role in Janet's assessment of the totality of the methods of psychotherapy.
In Janet's writings the opposition between remembering and forgetting
dissolves in the requirement that the patient learn to make an appropriate "adaptation" to the past, present, and future, but narrative self-
understanding is not always essential for such adaptation and other forms
of adjustment may serve the purposes of cure. For Janet, the physician's
rhetorical and suggestive skills are directed at improving the traumatized
subject's mental synthesis by producing modifications in conduct and behavior, modifications that do not necessarily depend on acts of conscious
self-representation and self-enunciation. (For Lacan, of course, psychoanalysis has nothing to do with what is ordinarily meant—or what Janet
presumably meant—by adaptation.)

Moreover, my discussion of the role of rhetoric and suggestion in
Janet's work implies a second difference between his view of psychical
treatment and the views of Freud and Lacan. Throughout his career
Janet defended the use of hypnosis in psychotherapy and regarded the
emotional rapport between physician and patient as fundamentally suggestive. Such an interpretation of the relationship between patient and
physician was alien to Freud and especially to Lacan who, developing various themes in Freud's thought, configured psychoanalysis as a rhetorical
enterprise but one from which the persuasive arts of hypnotic suggestion
were strictly excluded. But recent discussions of the Freudian corpus
have shown that the problem of suggestion in psychoanalysis cannot be
disposed of so easily. On the contrary, that problem resurfaced in Freud's
texts of the 1920s at precisely those junctures where it appeared that the
best way to understand the nature of the bond between the ego and the
other was by comparing that bond to the unconscious hypnotic rapport
between subject and physician. That comparison threatened to unsettle
the dynamics of repressed emotional and desiring representations on
which the very identity of psychoanalysis—like that of Lacan's subsequent "return to Freud"—ultimately depended. As we see in the next two
chapters, during the same years Freud's disciple, Sándor Ferenczi,
through his revival of the Breuer-Freud theory of trauma and the cathar-

tic method, made those issues a matter of urgent debate. But the history of trauma is a history of forgetting and it is not at all obvious even today, when the traumas of war and sexual abuse, the diagnosis of dissociation, and the deployment of hypnosis for the recovery of traumatic memories are commonplaces of psychiatric and psychotherapeutic practice, that we have grasped the scandalous nature of the traumatic cure.

IV

Imitation Magic: Sándor Ferenczi
and Abram Kardiner on Psychic Trauma

In December 1931 Freud learned that Sándor Ferenczi, arguably his most gifted and attractive disciple, had allowed a patient to kiss him during her analytic sessions. The news reached Freud through the indiscretion of the patient herself, a young American psychiatrist named Clara Thompson, future pioneer of "interpersonal psychiatry" and contributor to psychoanalytic feminism, who was then in a training analysis with Ferenczi. Ferenczi had indulged Thompson's desire to kiss him because he had come to believe that the authoritarian approach characteristic of classical analytic technique was rightly interpreted by the patient as a sign of the analyst's lack of genuine interest; the result was that, since the patient was now as dependent on the analyst as she had been dependent on her parents during childhood, she preferred to doubt her own judgment rather than challenge the analyst's authority. Consequently, she doubted the reality of those sexual traumas that the analytic reconstruction indicated lay at the origin of her hysterical symptoms, just as she had been made to doubt the reality of those same sexual traumas by her parents' hypocritical denials of wrong-doing in the past. Rejecting the standard psychoanalytic approach, Ferenczi had introduced a method of "maternal indulgence" or "relaxation" in which the patient in the analytic situation was encouraged to reexperience or relive the traumatic experience with almost hallucinatory intensity, even to the point of falling into a trance state. As Ferenczi recognized, the procedure was strikingly similar to the earlier hypnotic-cathartic technique of Breuer and Freud, which he thus reinstalled at the center of psychoanalytic practice along with Freud's largely abandoned emphasis on the traumatic origins of hysteria.[1]

1. See especially Sándor Ferenczi, "The Principle of Relaxation and Neocatharsis"

But the kissing episode with Thompson taught Ferenczi something more: so long as the analyst was dishonest with the patient, mere "relaxation" was insufficient. For if the patient sensed that the analyst, in his very passivity, was for whatever reason dissimulating his true feelings, she would stop cooperating with the analysis. That is precisely what happened in Thompson's case when, hypersensitive to Ferenczi's unacknowledged anger over her gossiping about his indulgence towards her, she began to resist her analysis by "acting out" sexually and in other ways. Only when Ferenczi was able genuinely to confess to her his annoyance at the trouble she had caused was trust restored. The analyst's natural and sincere behavior was thus conceived by Ferenczi as the antidote necessary to give the patient the protection and healing she had lacked at the time of her original wound.

In the immediate wake of the kissing episode, Ferenczi began keeping an extraordinary *Clinical Diary* in which he set down a detailed record of the development of his therapeutic experiments and theoretical ideas. In the first entry he jointly credited Thompson and his friend the psychoanalyst George Groddeck with contributing to his discovery of the importance of analytic honesty. Recognizing, though, that his efforts to counteract the rigidity of psychoanalysis by techniques of sincerity and permissiveness could also be counterproductive, he observed:

> See the case of Dm. [Clara Thompson], a lady who, "complying" with my passivity, had allowed herself to take more and more liberties, and occasionally even kissed me. Since this behavior met with no resistance, since it was treated as something permissible in analysis and at most commented on theoretically, she remarked quite casually in the company of other patients, who were undergoing analysis elsewhere: "I am allowed to kiss Papa Ferenczi, as often as I like." I first reacted to the unpleasantness that ensued with the complete impassivity with which I was conducting this analysis. But then the patient began to make herself ridiculous, ostentatiously as it were, in her sexual conduct . . . . It was only through the insight and admission that my passivity had been unnatural that she was brought back to real life, so to speak.

Citing the sexual trauma at work in Thompson's case, he added:

(1929), *Final Contributions to the Problems and Methods of Psychoanalysis*, ed. Michael Balint (New York, 1955), 119–20. For the therapeutic failures within psychoanalysis that led to Ferenczi's technical innovations see especially Leon Chertok et al., *Hypnose et psychanalyse* (Paris, 1987); Leon Chertok and Isabelle Stengers, *La coeur et la raison: l'hypnose en question, de Lavoisier à Lacan* (Paris, 1989); and André Haynal, *The Technique at Issue* (London, 1988).

Simultaneously it became evident that here again was a case of repetition of the father-child situation. As a child, Dm. [Thompson] had been grossly abused sexually by her father, who was out of control; later, obviously because of the father's bad conscience and social anxiety, he reviled her, so to speak. The daughter had to take revenge on her father indirectly, by failing in her own life.[2]

Ferenczi remarked that a logical extension of his principle of analytic sincerity was a more democratically conceived *mutual* analysis—the idea that not only ought the patient to be given the right to analyze the analyst himself but that both patient and analyst should sink into a kind of mutual, hypnotic exchange. Only in that way, he thought, could the tendency of the neurotic to repeat rather than remember the traumatic past be cured (*CD*, 3).[3]

Ferenczi died in 1933 at the age of 59, and for more than half a century his *Clinical Diary* was held by many of his psychoanalytic colleagues to be too revisionist, indeed too *reactionary*, to be published.[4] It is only in the wake of the present-day revival of interest in psychic trauma that many have come to regard Ferenczi's work as timely and significant (his *Clinical Diary* was first published in French in 1985, in German and English in 1988).[5] Yet throughout the recent discussion of Ferenczi's late work one

2. *The Clinical Diary of Sándor Ferenczi*, ed. Judith Dupont (Cambridge, Massachusetts, 1988), 1–2; hereafter abbreviated *CD*. The identification of "Dm." as Clara Thompson is discussed on page 3. I have also consulted the German edition of Ferenczi's diary, *Ohne Sympathie keine Heilung: Das klinische Tagebuch von 1932*, Herausgegeben von Judith Dupont (Frankfurt am Main, 1988).

3. For Ferenczi's reservations about the use of mutuality see Haynal, *The Technique at Issue;* for a recent discussion of mutual analysis see François Roustang, *Influence* (Paris, 1990) and idem, *Qu'est ce que l'hypnose?* (Paris, 1994).

4. Freud strongly objected to Ferenczi's late work and the disagreement between the two men acted, as Balint put it, like a "'trauma on the psychoanalytic movement'" (cited by Haynal, *The Technique at Issue*, 33). The debate between Freud and Ferenczi can be followed in their correspondence during these years.

5. Jeffrey Moussaieff Masson set the stage for Ferenczi's rehabilitation by praising him for challenging psychoanalytic orthodoxy on the role of fantasy and emphasizing instead the reality of the external sexual trauma; others have made similar efforts of reclamation. See Jeffrey Moussaieff Masson, *The Assault on Truth: Freud's Suppression of the Seduction Theory* (New York, 1984); Bessel A. van der Kolk, Alexander C. McFarlane, and Lars Weisath, *Traumatic Stress: The Effects of Overwhelming Experience on Mind, Body, and Society* (New York, 1996), 56; Lewis Aron and Adrienne Harris, eds., *The Legacy of Sándor Ferenczi* (Hillsdale, New Jersey, 1993); Alex Hoffer, "The Freud-Ferenczi Controversy—a Living Legacy," *International Review of Psychoanalysis* 18 (1991): 465–72; Benjamin Wolstein, "The Therapeutic Experience of Psychoanalytic Inquiry," *Psychoanalytic Psychology*, 7 (4)

crucial fact has been largely glossed over or ignored: as Ferenczi privately admitted, his therapeutic innovations *did not work*.

For Ferenczi, the goal of treatment was to abreact the trauma under conditions of analytic honesty and sympathy such that the patient's neurotic tendency to repetition—a tendency Freud attributed to the radical unleashing of the death drive—could be converted into a conscious recollection of the repressed or dissociated event. Like Breuer, Freud, Prince, Brown, and others before him, Ferenczi thus envisaged catharsis as a mnemotechnology designed to convert traumatic forgetting into self-representation and recollection.[6] In his last published paper, his famous (or infamous) "Confusion of Tongues between Adults and the Child," Ferenczi claimed success in this regard. *"It is this confidence that es-*

(1990): 565–80; Arnold Wm. Rachman, "Confusion of Tongues: The Ferenczian Metaphor for Childhood Seduction and Emotional Trauma," *Journal of the American Academy of Psychoanalysis* 17 (2) (1989): 181–205; Judith Dupont, introduction to *The Clinical Diary of Sándor Ferenczi*, xi–xxvii.

In a large general literature on Ferenczi I have also consulted the following: *Sándor Ferenczi*, ed. T. Bokanowski, K. Kelley-Laine, and G. Pragier (Paris, 1995); Michèle Bertrand, Thierry Bokanowski, Monique Déchaud-Ferbus, Anouk Driant, Madelaine Ferminne, and Nagib Khouri, *Ferenczi: patient et psychanalyste* (Paris, 1994); Eva Brabant-Gero, *Ferenczi et l'école hongroise de psychanalyse* (Paris, 1993); *The Correspondence of Sigmund Freud and Sándor Ferenczi*, vol. 1, 1908–1914, vol. 2, 1914–1919, ed. Eva Brabant, Ernst Falzeder, and Patrizia Giampieri-Deutsch (Cambridge, Massachusetts, 1993); Martin Stanton, *Sándor Ferenczi: Reconsidering Active Intervention* (Northvale, New Jersey, 1991); Ilse Grubrich-Simitis, "Six Letters of Sigmund Freud and Sándor Ferenczi on the Interrelationship of Psychoanalytic Theory and Technique," *International Review of Psychoanalysis* 13 (1986): 259–77; Marie Torok, "La correspondance Ferenczi-Freud: La vie de la lettre dans l'histoire de la psychanalyse," *Cahiers confrontation* 12 (1984): 79–99; Barbro Sylwan, "An Untoward Event, Ou la guerre du trauma de Breuer à Freud, de Jones à Ferenczi," *Cahiers confrontation* 12 (1984): 101–15; Wladimir Granoff, "Ferenczi: faux problème ou vrai malentendu," *Le block-notes de la psychanalyse* 4 (1984): 35–62; Pierre Sabourin, *Ferenczi: paladin et grand vizir secret* (Paris, 1985); Claude Lorin, *Le jeune ferenczi: premiers écrits 1899–1906* (Paris, 1983); Bela Grunberger, "From the 'Active Technique' to the 'Confusion of Tongues': On Ferenczi's Deviation," in *Psychoanalysis in France*, ed. Serge Lebovici and Daniel Widlocher (New York, 1980), 127–52; Ilse Barande, *Sándor Ferenczi* (Paris, 1972); Izette de Forest, *The Leaven of Love: A Development of the Psychoanalytic Theory and Technique of Sándor Ferenczi* (London, 1954).

6. The classical formulation of this mnemotechnology, which served as the starting-point for Ferenczi's subsequent therapeutic innovations, was Freud's "Remembering, Repeating and Working-Through (Further Recommendations on the Technique of Psycho-Analysis)" (1914), in *The Standard Edition of the Complete Psychological Works of Sigmund Freud*, trans. and ed. James Strachey (London, 1953–1974), 12: 145–56.

tablishes the contrast between the present and the unbearable traumatogenic past," he observed, "the contrast which is absolutely necessary for the patient in order to enable him to re-experience the past no longer as hallucinatory reproduction but as an objective memory."[7] But Ferenczi's dazzlingly speculative and often moving attempt to clarify the theoretical and therapeutic meaning of his findings tells a different story: his patients failed to remember the traumatic events in question. "By no means," he admitted, "can I claim to have ever succeeded, even in a single case, in making it possible for the patient to *remember* the traumatic processes themselves, with the help of symptom-fantasy submergence into dreams, and catharsis" (*CD*, 67).

Why did Ferenczi's project reach such an impasse? A close reading of his *Clinical Diary* demands that we confront at least two possibilities: (1) The nature of trauma is such that it can never be consciously experienced and hence remembered; (2) Hysterics tend to lie, which is why the truth about their pasts can never be known with confidence. (If this sounds suspiciously like the traditional claim that hysterics are *simulators*, that is not wrong. But wait.) Is there a relationship between these two possibilities, implicit in Ferenczi's speculations, which would explain his theoretico-therapeutic setback? Does his predicament throw light on current stalemates in the controversy over the nature and treatment of trauma?

Imitation Magic

Let me start by emphasizing Ferenczi's idea that in the moment of trauma the victim's psyche is split apart in such a way as to lose all psychic coherence. Splitting was imagined by him as the disintegration or fragmentation of the psychical apparatus: the effect of shock was to destroy or dissociate all mental associations and synthesizing functions. Ferenczi placed the problem of *imitation* or *mimicry* at the center of his conceptualization of trauma when he suggested that in the traumatic

7. Sándor Ferenczi, "Confusion of Tongues Between Adults and the Child" (1933), in *Final Contributions to Psychoanalysis*, 160; hereafter abbreviated "CT." Freud and his followers made every effort to prevent Ferenczi from presenting this paper to the International Congress of Psycho-Analysis held at Wiesbaden in September, 1932. The details can be followed in *The Complete Correspondence of Sigmund Freud and Ernest Jones, 1908–1939*, ed. R. Andrew Paskauskas (Cambridge, Massachusetts, 1993). Although Ferenczi's paper was published in 1933 in the *International Zeitschrift für Psychanalyse*, it was not translated into English until sixteen years later.

moment the victim's best solution to the crisis was not to resist but to give in to the threatening person *by imitating or identifying with him (or her)*. For Ferenczi, the process of imitation or identification with the aggressor—the mimetic acceptance of unpleasure, including the incorporation or introjection of the aggressor's guilt—held the key to the victim's split or fragmented mental state. According to him, hypnosis provided the paradigm for this unconscious process of mimetic yielding or identification.

What enormously complicated Ferenczi's work, however, is that he posited *two different models* of traumatic splitting and of the imitative mechanisms connected with them. On the first model of trauma, which I call the *originary* model, splitting was imagined as producing the separation of the ego from the object; it was the very process which constituted the subject *as* a subject by the splitting apart of the ego from the objective world. Trauma and splitting on this model were absolutely originary for the subject-object opposition and hence were normal and inescapable. But on the second model of trauma, which I call the *postoriginary* model, splitting was imagined by Ferenczi as taking place on the basis of an already existing ego. Trauma and splitting on the postoriginary model were pathological and exceptional processes that happened to the already constituted subject—the human being in the postoriginary condition. Throughout his late writings on trauma Ferenczi did not keep the two models distinct but continually moved between them or superimposed the one on the other. Several recent commentators, tempted to read Ferenczi in the light of their own, often reductive accounts of trauma, fail to recognize the difficulty of associating Ferenczi definitively with one or other contemporary position, with the result that they fail to acknowledge the complexities and corrosive implications of his work.

A series of brief entries in Ferenczi's *Clinical Diary* will help launch my analysis. In those entries he addressed a fundamental problem in psychoanalysis, one that Freud had already discussed in several important texts and that Ferenczi himself had also grappled with in the past, namely, the origin of intellectual judgment. The question posed was: How does the nascent ego come to form judgments about the world? More specifically: How does the infant come to have a knowledge of the reality of objects? Ferenczi began with the following assertion: "Repression of the self, annihilation of the self, is the precondition for objective perception" (*CD*, 111). His formulation appears to be a restatement of the German philosopher Arthur Schopenhauer's famous evocation of esthetic perception or contemplation, which is described as involving the disappearance of the ego and its desires and its absorption into the object of its apprehension.[8] Ferenczi first gave a psychoanalytic gloss on Schopenhauer's thesis by bringing out the egoistic motives involved in the selflessness associated with objective perception, namely, the *selfish* wish that there be no other consciousness or will to impede the harmonious relations between oneself and the world. Thus Ferenczi wondered, "What is the motivation for such selflessness?" and replied, "really only the experience that, through it, the self will be helped in another, better way. I disappear for a moment, I do not exist, instead things outside of me exist" (*CD*, 111). As for the child imagined as on the verge of knowledge of the world, "The original wish is: *nothing* should exist that disturbs me, nothing should stand in my way" (*CD*, 111). To which he added:

> But certain wicked things will not obey me and force themselves into my consciousness. So: there are *other* wills besides my own. But why does a sort of *photograph* of this external body appear in me as soon as, aware of my weakness, I vanish by withdrawing? (Why does the horror-struck person in his anxiety imitate the features of the horrifying thing?) The *memory mask* develops, perhaps always at the cost of the temporary or permanent dying away of a part of the ego. Originally an effect of the shock. *Imitation magic?*
>
> Memory is thus a collection of *scars of shocks* in the ego. Fear dissolves the rigidity of the ego (resistance) so completely that the material of the ego becomes as though capable of being molded *photochemically*—is in fact always molded—by external stimuli. Instead of my asserting *myself*, the external world (an alien will) asserts itself at my expense; it forces itself upon me and *represses the ego*. (Is this the primal form of "repression"?) (*CD*, 111)

8. The following passage from Arthur Schopenhauer resonates with Ferenczi's idea that the ego withdraws or disappears during objective perception. In discussing the process by which we move from a knowledge of particular things to objective knowledge, Schopenhauer writes: "Raised up by the power of the mind, we relinquish the ordinary way of considering things. . . . [W]e do not let abstract thought, the concepts of reason, take possession of our consciousness, but, instead of all this, devote the whole power of our mind to perception, sink ourselves completely therein, and let our whole consciousness be filled by the calm contemplation of the natural object actually present, whether it be a landscape, a tree, a rock, a crag, a building, or anything else. We *lose* ourselves entirely in this object, to use a pregnant expression; in other words, we forget our individuality, our will, and continue to exist only as a pure subject, as clear mirror of the object, so that it is as though the object alone existed without anyone to perceive it, and thus we are no longer able to separate the perceiver from the perception, but the two have become one, since the entire consciousness is filled and occupied by a single image of perception" (Arthur Schopenhauer, *The World as Will and Representation* [New York, 1966], 2: 178–79). Although Ferenczi does not cite Schopenhauer at this juncture, many of his writings testify to his interest in Schopenhauer's work.

These passages are amazingly interesting, not least because they are at odds with Freud's account of the psychology of the infant. In several texts well known to Ferenczi, Freud had argued that mental life begins with a state of psychical rest that is disturbed by the pressure of unmet internal needs—needs that, in the absence of the desired object (the mother's breast), are at first satisfied by wishes of a pleasurable, hallucinatory kind. Freud's aporetic attempts to explain how the child, inevitably disappointed when mere wishes fail to bring satisfaction, masters the necessary but difficult task of substituting painful reality for pleasurable hallucinations lie behind Ferenczi's speculations concerning the origin of the sense of reality.[9] Like Freud, Ferenczi in his *Clinical Diary* posited an original state of infantile psychic repose whose disturbance by excitation starts the process of reality judgement. But unlike Freud, Ferenczi tended to focus less on the disturbing force of the infant's unmet internal needs (the role of desire) than on the disrupting power of external objects—especially the will of another human being.[10]

In addition, by the time of his *Clinical Diary* Ferenczi had begun to question Freud's assumption that the reactions of young children are identical to those of adults and that from the start those reactions are regulated by the wish to obtain satisfaction through the hallucination of libidinally desired objects. Instead, he had begun to stress the idea that the infant's primordial mode of response is not a *libidinal wish* for the desired object but a *mimetic-identificatory response* that, in the absence of an ego, occurs at a stage preceding object relations and the advent of desire. Ferenczi wrote: "In one psychic process the importance of which has perhaps been insufficiently appreciated, even by Freud himself, namely that

9. The key texts are Sigmund Freud, *Project for a Scientific Psychology* (1895), *Standard Edition*, 1: 281–387; idem, *The Interpretation of Dreams* (1900), *Standard Edition*, 5: 565–7; idem, "Formulations on the Two Principles of Mental Functioning" (1911), *Standard Edition*, 12: 213–26; idem, "Negation" (1925), *Standard Edition*, 19: 233–39; Sándor Ferenczi, "Introjection and Transference" (1909), in *Contributions to Psychoanalysis* (Boston, 1916), 30–79; idem, "Stages in the Development of the Sense of Reality" (1913), *Contributions to Psychoanalysis*, 181–203; idem, "The Problem of the Acceptance of Unpleasant Ideas: Advances in Knowledge of the Sense of Reality" (1926), *Further Contributions to the Theory and Technique of Psychoanalysis* (London, 1950), 366–79.

10. "We may suppose that to the new-born child everything perceived by the senses appears unitary, so to speak monistic," he had observed. "Only later does he learn to distinguish from his ego the malicious things, forming an outer world, that do not obey his will" (Ferenczi, "Introjection and Transference," 41). Cf. Ferenczi, "The Problem of the Acceptance of Unpleasant Ideas: Advances in Knowledge of the Sense of Reality," 366, 371–72.

of *identification as a stage preceding object relations*, we have until recently not sufficiently appreciated the functioning in it of a mode of reaction already lost to us, but one that nevertheless exists; although perhaps we are faced with the functioning of a quite different kind of reaction principle, to which the designation *r*eaction can no longer be applied; that is, a state in which any act of self-protection or defense is excluded and all external influence remains an impression without any internal anti-cathexis." He continued:

> The most concise summary of this situation was perhaps given by Dr. [Clara] Thompson, when she said that people at the beginning of their lives have as yet no individuality. Here my view on the tendency to fade away (falling ill and dying in very young children) and the predominance in them of the death instinct: their extreme impressionability (mimicry) may be also just a sign of rather weak life and self-assertive instincts: indeed it is perhaps already an incipient, but somehow delayed, death.[11] But if this is true, and this kind of mimicry, this being subject to impressions without any self-protection, is the original form of life, then it was rash, even unjustified, to ascribe to this period, still almost bereft of motility and of course also probably intellectually inactive, the only self-protective and hallucinatory mechanisms we know and are accustomed to (wish impulses). *The hallucinatory period, therefore, is preceded by a purely mimetic period. (CD*, 147–48; emphasis added)[12]

In the series of *Clinical Diary* entries that concerns us here, it is precisely this mimetic, identificatory process of being "subject to impressions without any self-protection" that Ferenczi likened to the imprint-

11. Ferenczi was referring here to his paper "The Unwelcome Child and his Death Instinct" (1929), *Final Contributions to Psychoanalysis*, 102–7, on the tendency of unwanted children to commit suicide or fade away because they are abused or maltreated or unloved. Ferenczi interpreted such a tendency as an example of the unleashing of the death drive, and hence conceptualized trauma as, at the limit, a kind of suicide. Krystal and other recent trauma theorists likewise stress that massive traumatization can cause "psychogenic death" (Henry Krystal, "Trauma and the Stimulus Barrier," *Psychoanalytic Inquiry* 5 [1985]: 135).

12. Ferenczi repeated these ideas in "Confusion of Tongues" ("CT," 163). The "automatization of the ego" or puppet-like obedience and surrender, which occurred in concentration camp victims as one of the phases of response before the final collapse into the so-called Muselmann stage, exemplifies this mimetic-hypnotic process of imitation, as was recognized by Primo Levi (*Survival in Auschwitz* [New York, 1961]). Cf. Bruno Bettelheim, "Individual and Mass Behavior in Extreme Situations" (1940), in *Survival and Other Essays* (New York, 1980), 48–83; William G. Niederland, "Psychiatric Disorders among Persecution Victims," *Journal of Nervous and Mental Disease* 139 (1964): 463–64; and Henry Krystal, "Trauma and Aging," in *Trauma: Explorations in Memory*, ed. Cathy Caruth (Baltimore and London, 1995), 80–81.

ing of a photographic plate or the photochemical substance of the eye. This extraordinary analogy, repeated elsewhere, radicalized his ideas about identification and imitation. Imagining the horror-struck adult confronted with danger as reverting to the condition of the helpless infant having to come to terms with the existence of other objects or persons, Ferenczi suggested that the adult and infant alike experience the moment of trauma mimetically, as though the mimesis in question were comparable to the literal or material alteration of the sensitive cells (photons) of the retina. According to Ferenczi, it is as a result of the marks ("scars") left by the mimetic impression that the ego is magically split off from the rest of the world and becomes conscious of itself and of other objects. In his words:

> The traumatic *mimicry impressions* are utilized as memory-traces, useful to the ego: "dog" = bowwow, bowwow. When I am frightened of a dog, I become a dog. After such an experience, the ego consists of the (undisturbed) subject and the part that has become the object through the influence of the trauma = memory-traces = permanent imitation (speech is telling the story of the trauma). (*CD*, 113)[13]

Here Ferenczi theorized speech itself as the product of an imitative mechanism which, by permitting objects to be represented to the subject, brings about the differentiation of the ego from the external world. *"What does 'becoming conscious' mean?"* Ferenczi asked, and answered: "Becoming aware of being torn apart into ego and *environment* (dog). The part of inner experience that can be represented in gesture and speech is separated from one's own ego, as external world. Simultaneously I become *conscious of myself*: conscious of the existence of an *external world*" (*CD*, 113). For Ferenczi, however, neither speech nor gestures are necessary for the process of becoming conscious of the separation between ego and the rest of the world—only the mimetic operation of imprinting itself:

> Actually it may be that no reproduction is necessary here—the *photochemistry of the retina imitates pictorially* the external world (or the external world *takes possession* of the specifically traumato-philic substance of the retina). This picture

13. As Ferenczi also observed, danger "produces traumatic 'photo-hyperesthesia' (sensitivity to light and sound); chemotropic change in structure, in which self-assertion is relinquished to some extent (perhaps only temporarily) and the external world is able to shape the ego. But a part of the ego remains undestroyed, indeed, it seeks to profit from this demolition (scars)" (*CD*, 112–13). Ferenczi here took as his example the situation of the child missing its mother, made famous by Freud in *Beyond the Pleasure Principle* (1920).

of the external world, which is forced upon the organism . . . is used for orientation in space. In this way *what remains of the ego* acknowledges the rule of the *reality principle.* (*CD*, 113)

For Ferenczi the "primal" form of repression (*CD*, 111)—the same primal repression which interests Cohen and Winston in their recent return to the problem of trauma (see chapter 1)—*is* this mimetic process of subject-object differentiation. From an original state of undisturbed unity or monistic harmony between self and other, the infant becomes aware of difference.

There are numerous problems, tensions and paradoxes in the above passages that cry out for analysis, all of them stemming from Ferenczi's continued vacillation between an originary and a postoriginary model of traumatic splitting. One result of that vacillation is that Ferenczi's account of the ego is radically unstable. In some of the passages just cited Ferenczi speaks of the temporary or permanent "withdrawal" of the ego in the traumatic moment as if he conceives of traumatic splitting as befalling an already constituted ego. But as many other passages suggest, from the perspective of the child coming to terms with reality there is no ego prior to the traumatic mimicry or identification that brings the ego into being by separating it off from the world of objects. Indeed, the more Ferenczi extends his analysis of trauma back in time to the very origins of subjectivity, the more trauma and splitting are held to occur prior to any properly intersubjective encounter. From this point of view, it is not the accidental imposition of an alien *human will* that traumatizes the individual but the mere existence of *objects as such*. We might describe this as a normalization of the idea of trauma, or better, as a *traumatogenesis of normality*, as traumatic splitting emerges on Ferenczi's first or originary model as a normal and necessary stage in psychic development.

There are two immediate consequences of such an originary model of trauma. First, its logic seems to preclude the possibility of remembering the trauma. Before the trauma there is no ego that can represent the traumatic event to the subject; while consequent upon the trauma the ego comes into existence too late to record the traumatic moment that brought it into being. In short, on the originary model the traumatic "event" is defined as that which, precisely because it triggers the "trauma" that is mimetic identification, cannot strictly speaking be described as an event at all since it does not happen to a pregiven subject. Put slightly differently, traumatic imitation precedes the very distinction between "self" and "other" on which the possibility of self-representation and hence recollection depends.

A second consequence of the originary model is that, since traumatic-mimetic splitting is the condition of the ego's coming into consciousness of itself and the world, the traumatic experience is one to which the notion of cure scarcely seems to apply. If Ferenczi thought that such a condition of splitting *ought* to cured, it was only by imagining the possibility of a complete regression to the *pre*traumatic, monistic state of union between the individual and the world, a state he characterized as a condition of total unconsciousness. "Complete restitution would therefore be possible only at the level of complete *unconsciousness*," Ferenczi wrote, "that is to say: with a return to that which is still unconscious (an as yet undisturbed state of the ego)" (*CD*, 111). Or again: "Sleep is regression to a primal unity, as yet unsplit. (Without consciousness and, when completely *without objects*, dreamless.) Regression to the pretraumatic" (*CD*, 113). What kind of cure is it that it can only be conceived as a yielding to a permanent condition of dreamless sleep?

But what I want to emphasize is that the *same* impossibility of recollection and hence of cure also structures Ferenczi's postoriginary model of trauma. This is because, in disorientingly close proximity to the first model, Ferenczi's second or postoriginary model pictured the individual shattered by trauma as dissociated into two distinct psychical systems, a *subjective* emotional system which feels the emotions of a trauma that it cannot represent, and an *objective* intellectual system which perceives a trauma that it cannot feel (as if the trauma were happening to another person) (*CD*, 6, 103–4). Just as on the first or originary model of trauma, so on the second or postoriginary model of trauma, mimicry or identification was a central dimension of the traumatic experience. However, imitation no longer took place, as it did on the originary model, prior to the ego-object divide but happened to the already constituted human being. On the postoriginary model of splitting, the victim therefore preserved a distance or inner differentiation from the person with whom she identified (*CD*, 32). Thus according to Ferenczi, in the moment of trauma the objective system that splits off from the rest of the psyche assesses the situation of danger and calculates the psychic advantages of imitatively yielding to the unpleasure while remaining a conscious spectator of the scene. "It really seems as though, under the stress of imminent danger, a part of the self splits off in the form of a self-observing psychic instance wanting to give help, and that possibly this happens in early—even the earliest—childhood," he wrote in 1931.[14] In the person-

14. Sándor Ferenczi, "Child Analysis in the Analysis of Adults," *Final Contributions to Psychoanalysis*, 136 (translation modified).

ifying language Ferenczi deployed in this context, the objective fragment achieves the status of a clever philosopher, "guardian angel" (*CD*, 105), "psychiatrist" (*CD*, 165, 172) or "wise baby" (*CD*, 82) who impersonally diagnoses the peril and adapts to the situation by imitating or identifying with the enemy.[15] Ferenczi attributed occult powers to the dissociated or split-off fragment or self which, capable of such "super-performances," functions telepathically or clairvoyantly, like a maternal principle of self-sacrifice, to aid the victim by mimetically yielding to the aggressor in times of danger.[16]

Meanwhile, on the postoriginary model of traumatic splitting the suffering and feeling part of the self that is destroyed by the trauma is cut off or dissociated from the perception or representation of the scene and is driven underground, to be experienced corporeally as hysterical conversion symptoms. Representation of the traumatic scene is thus severed from the emotional response, with the result that the objective system or personality that can represent the event is radically disjoined from the

15. The concept of the "wise baby" first appeared in a brief paper of 1912 in which Ferenczi described a not uncommon dream. In it, a newly born or young child appears who is able to "talk or write fluently, treat one to deep sayings, carry on intelligent conversations, deliver harangues, give learned explanations, and so on" (Ferenczi, "The Dream of the 'Clever Baby,'" *Further Contributions to Psychoanalysis*, 349). Ferenczi suggested that the superficial layer of the dream interpretation pointed to an ironical view of psychoanalysis and its claims for the importance of experiences of childhood. But he also proposed that the irony served as a medium for "deeper and graver memories" of childhood (350).

16. Ferenczi's ideas here were connected not only to the concepts of trauma, dissociation, and multiple personality associated with the work of Janet, Binet, and Morton Prince, with whose work he was undoubtedly familiar, but also with the spiritualist ideas of the period. One of Ferenczi's principal patients, Elizabeth Severn (identified by the initials R.N. in the *Clinical Diary*), an early American practitioner of mental healing, especially hypnotic therapy, was closely identified with the spiritualist movement and, from the evidence of the *Clinical Diary* and other texts, clearly influenced Ferenczi's formulations. She was the author of several books, including *Psychotherapy: Its Doctrine and Practice* (1914), in which she discussed Prince's "famous case of Sally Beauchamp" (63); *The Psychology of Behavior: A Practical Study of Human Personality and Conduct with Special Reference to Methods of Development* (1917), which she gave to Ferenczi for his birthday in 1925; and *The Discovery of the Self* (1933), which reflected Ferenczi's last ideas on trauma. For Severn's biography and influence see Christopher Fortune, "The Case of 'RN': Sándor Ferenczi's Radical Experiment in Psychoanalysis," in Aron and Harris, *The Legacy of Sándor Ferenczi*, 101–20; idem, "Sándor Ferenczi's Analysis of 'R. N.': A Critically Important Case in the History of Psychoanalysis," *British Journal of Psychotherapy* 9 (1993): 436–43.

subjective system or personality that feels the trauma's effects. As Ferenczi observed of one of his patients:

> [W]hile her emotional life vanishes into unconsciousness and regresses to pure body-sensations, her intelligence, detached from all emotions, makes a colossal but . . . completely unemotional progression, in the sense of an adaptation-performance by means of identification with the objects of terror. . . . The trauma made her emotionally embryonic, but at the same time wise in intellectual terms, like a totally objective and unemotionally perceptive philosopher. . . . One could thus take the view that after a shock the emotions become severed from representations and thought processes and hidden away deep in the unconscious, indeed in the corporeal unconscious, while the intelligence goes through the progressive flight [of imitative identification]. (*CD*, 203)

At first sight the posited split between affect and representation consequent on trauma appears to conform to Freud's classical account of repression; Ferenczi at times presented his ideas in these terms (*CD*, 159). It is as if in this mode Ferenczi theorized trauma as a purely external event that, dissociated from the affects involved and repressed into the unconscious in the form of mnemonic traces, ideally could be brought into the victim's consciousness during the neocathartic treatment. From this only a slight step was required, which he took, to attribute neurosis to the reality of a trauma that assaulted the passive and innocent child in the form of a brutal, erotic aggressivity (*CD*, 167). But it is also the case that, just as on Ferenczi's originary model of traumatic splitting, so on his postoriginary model, the traumatic experience did not meet the requirement that it be remembered and cured—at least if by remembering was meant that the patient recollected the traumatic experience in an emotionally integrated and subjectively convincing way. In terms reminiscent of late nineteenth-century descriptions of dissociation and dual personality, Ferenczi viewed the split-off psychical systems as distinct personalities that were mutually amnesiac (*CD*, 105; "CT," 165). He hoped to remove the splits by a therapeutics of analytic sincerity, and throughout the *Clinical Diary* there are moments when he claimed to be on the verge of success (*CD*, 59–60, 65, 103, 204). But despite his celebrated therapeutic optimism, the goal of recollection eluded him (see especially *CD*, 168–69, 179–81). As he recognized, by the very terms of his own analysis, a patient who attempted to heal her split state by verbally narrating or representing the event to herself and others could achieve a conscious, *intellectual* knowledge of the event, but such knowledge necessarily lacked the *affective* experience required to give it validity. By the same token, if she attempted to reexperience the emotions associated with the trauma by sinking into a trance state, she felt the suffering involved but on "waking" had no confidence in the reality of the reexperienced or reconstructed trauma (*CD*, 203).

Representation and affect on this model were mutually incompatible, which meant that the split in the subject could not be healed. As Ferenczi stated: "The task of the analysis is to remove this split, although here a dilemma arises. Reflecting on the event and reconstructing it by one's own reasoning—or even the fact that one perceives the need to reflect [*Denknotwendigkeit*] on it—represents the preservation of the splitting into two parts: one that is destroyed, and one that sees the destruction" (*CD*, 39). His neocathartic methods could not solve that dilemma because, as he noted, the emotional reliving occurred in the mode of a trancelike acting out in the present that appeared to be unavailable for subsequent representation and narration:

> If in catharsis the patient sinks into the experience phase, he feels the suffering in this trance, but still does not know what is going on. Of the series of object and subject sensations, only the subject side is accessible. If he wakes from the trance, the direct evidence disappears immediately; once again the trauma will be grasped only from the outside, by reconstruction, without any feeling of conviction. (*CD*, 39; see also *CD*, 28, 130–31, 180–81)

The result was that affect and representation could not be reconciled and that, according to the logic of Ferenczi's second model of trauma, a certain kind of forgetting of the traumatic experience was ineluctable.

Medusa's Mirror

There is more to be said about the relationship between Ferenczi's two models of traumatic splitting and the ways those models undermined the project to cure the patient by the narrative integration or recollection of the traumatic origin. At first glance, in Ferenczi's postoriginary model the victim's split-off ego appeared to remain intact, thus preserving a difference from the person with whom she identified. "The early-seduced child adapts itself to its difficult task with the aid of complete identification with the aggressor," Ferenczi wrote in one of several similar passages. "[S]uch identificatory love leaves the ego proper unsatisfied" (*CD*, 190). In other words, the traumatic-mimesis leaves a resisting, self-observing, spectatorial fragment in place. "A third personal fragment is a kind of substitute mother, who keeps a permanent watch over the other two fragments," he remarked (*CD*, 64). Or as he also observed of a patient, "[S]he only pretended to accept social conventions, but deep down she remained convinced that modesty is senseless (insane) and a lie" (*CD*, 163), a formulation that linked the traumatic experience to the problem of hypocrisy and lying, an important theme in his thought, as we shall see, and implicitly equated mimicry with deliberate simulation. In still another passage he commented: "Peculiar experiences in one case make it appear not impossible that the 'wise baby,' with his wonderful instinct, accepts the deranged and insane as something that it forcibly imposed, yet keeps his own personality separate from the abnormal right from the beginning." To which he added:

> The personality component expelled from its own framework represents this real, primary person, which protests persistently against every abnormality and suffers terribly under it. This suffering person protects himself, by forming wish-fulfilling hallucinations, against any insight into the sad reality, namely, that the evil, alien will is occupying his entire psychic and physical being (being possessed). (*CD*, 82)

It is not surprising in this context that Ferenczi was drawn to the idea of traumatic imitation as ironic caricature or grimace, as if in identifying with the aggressor the traumatized person, at least at first, preserved intact a defiant, observing ego:

> Now, in situations where protest and negative reaction, that is, all criticism and expression of discontent, are forbidden, criticism can find expression only in an indirect form. For example, the opinion, "You are all liars, idiots, lunatics, who can't be trusted," is illustrated indirectly on oneself through exaggerated, crazy behavior and nonsensical productions, rather like the child who in grimacing distorts himself but only to show the other how he looks. . . . The child recognizes at an early age the absurdities in the behavior of those in authority over him, yet intimidation precludes the exercise of criticism. Ironic exaggeration, the nature of which is not recognized by the environment, remains the only means of expression. The question remains of how and when the irony of the expressions becomes unconscious for the child as well. The insane "superego," being or becoming imposed upon one's own personality, transforms the previous irony into automatism. (*CD*, 50)

Ferenczi here imagined traumatic mimicry as a kind of conscious simulation which only becomes unconscious and involuntary over time. But a model of mimicry as ironic caricature or simulative grimace was clearly at odds with Ferenczi's originary model of imitation defined as an "alteration" of the subject that occurred prior to, or in the absence of, conscious ego functions and ordinary self-observation. For the originary model suggested that if the subject imitated the aggressor, by assuming his contemptuous and aggressive behavior, she was unaware that she was mirroring him in that way. Ferenczi's interpretation of splitting was ambiguous on just this point, hovering between an ideal of self-observation and a notion of blind immersion in the violence and irrationality of the threatening other.

Indeed at more than a few junctures the two models were collapsed into each other with a force that made them mutually inextricable. "The most potent motive of repression, in almost all cases, is an attempt to make a sustained injury not have happened," Ferenczi stated in an astonishing passage in the *Clinical Diary:*

> Another, perhaps even more potent motive is identification out of fear—one must know the dangerous opponent through and through, follow each of his movements, so that one can protect oneself against him. Last not least: an attempt will be made to bring to his senses even a terrifying, raging brute, whose behavior suggests drunkenness or insanity. When the Medusa, threatened with decapitation, makes a horrible angry face, she is actually holding up a mirror to the bestial attacker, as though she were saying: This is how you look. In the face of the aggressor one has no weapons; and no possibility exists of instructing him or bringing him to reason in any other way. Such deterrence by means of identification (holding up a mirror) may still help at the last moment (*ta twam asi:* this art thou). (*CD*, 177)[17]

But of course in the classical myth Medusa is killed not by a terrifying, raging brute but by the hero Perseus, which is to say that Ferenczi's retelling of the myth gives us a Perseus already infected by (already insensibly "mimicking") the monster he is seeking to kill. Moreover, insofar as a mirrored surface figures in the myth—and it does, centrally, in the form of Perseus's bright, reflecting shield—the mirror is held up not by Medusa, as Ferenczi writes, but by Perseus, who deflects the Medusa's monstrous gaze back at her, thereby (presumably) paralyzing her so that

17. The Sanskrit phrase *ta twam asi*, "this are thou," reveals Ferenczi's interest in eastern philosophy. The phrase was something of a cultural commonplace. As Eugene Taylor has shown, it was familiar to William James, whose knowledge of Asian thought may have come from Emerson or the transcendentalist philosophy of his father (Taylor, *William James on Consciousness Beyond the Margin* [Princeton, New Jersey, 1996], 61). The phrase also occurs several times in Schopenhauer's *The World as Will and Representation* as an expression of the experience of pity or sympathy in which the egoistical distinction between self and other breaks down (Rudiger Safranski, *Schopenhauer and the Wild Years of Philosophy* [Cambridge, Massachusetts, 1990], 234–35).

he is able to cut off her head. And this implies in turn that Medusa in Ferenczi's version insensibly assumes Perseus's role (or identity), as if the mutual identification of the two antagonists was already complete before Ferenczi's mini-narrative got under way.

It might seem that Ferenczi's distortion of the Medusa myth is inexplicable, all the more so in the light of what we know to have been Ferenczi's role in instigating Freud's extraordinarily brilliant though mythically "orthodox" short paper on "Medusa's Head" (1922).[18] But in fact Ferenczi's version of the myth amounts to an inspired instance of the very effects of identification that he eccentrically adduces that myth to epitomize. It is as if, in his revision of the Medusa story, Ferenczi so thoroughly identifies with the traditional aggressor (in order to know the dangerous opponent through and through) that he comes to inhabit with considerable pathos the position of the Medusa herself, and hence to recast Perseus not as a hero but as a threatening wild beast. Ferenczi's revision of the myth thus amounts to his enacting the traumatic process of mimetic identification that he is using the figure of the Medusa to metaphorize. In the process he figures the Medusa's originarily Medusizing expression as itself the mirror response to the menace embodied by Perseus and indeed, presumably, by the reflection of her features in Perseus's mirror shield.

Moreover, Ferenczi implicitly de-Oedipalizes the Medusa myth by ignoring the theme of castration that is fundamental to Freud's analysis and by staging the myth outside the triangular erotics of the child-mother-father relationship. In so doing, he not only interprets imitation as involving a dual relationship between subject and other that occurs prior to the subject-object relationship of Freud's libidinal scenario, but also forestalls the narratives of recuperation that Freud's reading of the myth make possible. By this I mean that in Freud's essay the Medusa myth becomes a story in which the little boy, as the spectator who sees the horrifying decapitated head of Medusa, is able to compensate for his fear by becoming erect in terror, thereby consoling himself that he still possesses the penis whose lack the Medusa represents. In Ferenczi's *Clinical Diary* entry, in contrast, no such dialectical "resolution" of the scenario of threat is possible.

---

18. In his short, posthumously published paper, originally dated May 14, 1922, Freud credited Ferenczi with being the first to interpret the Medusa's head as a symbol of castration (Freud, "Medusa's Head" [1922; 1940], *Standard Edition*, 18: 273–74, citing Ferenczi's "On the Symbolism of the Head of Medusa" [1923], *Final Contributions to Psychoanalysis*, 360).

Finally and crucially, Ferenczi's revision of the Medusa myth prevents us from unproblematically assimilating the mimetic scenario to a scene of conscious observation and visualization. That is, Ferenczi's description of the Medusa as holding up a mirror may seem to imply a certain specular distance between the victim and the aggressor. But what I have tried to show is that the primordial (or at least prior) identification of the two antagonists implies an altogether different visual and affective dynamic. (This is especially apparent when we consider Ferenczi's understanding of the hypnotic rapport as the prototype of mimetic identification. I say more about that in the next chapter.)

Ferenczi and Kardiner on Tic and the War Neuroses

Another key text in which the instabilities in Ferenczi's account of traumatic imitation surface is his magnificent, if somewhat reckless, essay "Psychoanalytical Observations on Tic" (1921). The latter, often cited in the *Clinical Diary*, has been almost completely neglected in the secondary literature.[19] Yet it deserves to be placed at the center of any discussion of Ferenczi's late work on trauma, not least because of its importance to the American psychoanalyst Abram Kardiner's contemporaneous attempt to theorize the traumatic neuroses of war.

"Psychoanalytical Observations on Tic" reflected Ferenczi's interest in Breuer and Freud's original theory of trauma. More precisely, it continued the discussion of issues that he had begun to address in two previous essays on the war neuroses, many cases of which he had examined while serving as a medical officer during the Great War. It also announced Ferenczi's interest in the "pathoneuroses," a term coined by him to describe neuroses, such as tic, that he regarded as somatic in origin, the result of disease or injury to those parts of the body that are crucial to life, or highly prized by the ego (e.g., the face or genitals, so often mutilated in the trench combat of the First World War). The effects of both physical and psychical blows to the *ego* were thus the particular focus of Ferenczi's inquiry, which is to say that his paper constituted a contribution to the new psychoanalytic interest in the ego initiated by Freud in the 1920s.

Ferenczi understood victims of tics as people who are abnormally sensitive to any physical stimulus, which they cannot endure without a de-

---

19. I know of only two brief references: Adele Covello, Roger Gentis, Maria Torok, "Entretien autour de Sándor Ferenczi," *Le bloc-notes de la psychanalyse*, no. 2 (1982): 43–58, and Claude Barrois, "S. Ferenczi et les névroses traumatiques," *Psychiatrie française* 17 (1986): 29–39.

fensive reaction because they are pathologically in love with themselves. In other words, tics were narcissistic disorders: they were not transference neuroses, like the conversion hysterias, but were allied instead to the psychoses, such as schizophrenia and paranoia, which Freud had placed among the narcissistic disorders because they were characterized by a profound narcissistic withdrawal from the external world. According to Ferenczi, in such disorders there was a pathological damming up of excitation in an organ or part of the body—either because the injury threatened life, or because the injury affected a highly libidinally cathected part of the body, or because the patient was a constitutional narcissist. He suggested that the various stereotyped motor symptoms, including the grimaces and mannerisms associated with tic and the war neuroses, or the negativism, rigidity, and spasms associated with catatonia, were all expressions of the attempt to rid the hypersensitive, narcissistic individual of the accumulated tension.

But what complicated Ferenczi's discussion of tic was the notion of narcissism itself, especially Freud's notoriously problematic notion of *primary narcissism*. How were we to understand the state of primary narcissism to which the *tiqueur* and catatonic were said by him to be regressed? In a discussion that stresses the incoherences internal to Freud's notion of primary narcissism, Laplanche and Pontalis in a preliminary definition describe primary narcissism as an "early state in which the child [or ego] cathects its own self with the whole of its libido."[20] But as they make clear, precisely the status of the ego is problematic in such a formulation. On the one hand, as a state in which the ego takes *itself* as its love-object, primary narcissism corresponds to the first emergence of a unified subject or ego. On the other hand, Freud also conceptualized primary narcissism as a primitive state of the infant that occurs *prior* to the formation of an ego, a state epitomized by life in the womb. On the latter account primary narcissism is a strictly "objectless" or undifferentiated state, implying no split between the subject and the external world. As Laplanche and Pontalis comment, from a topographical standpoint it is difficult to know just what is supposed to be cathected in primary narcissism thus conceived.

Moreover, to confuse matters further, Ferenczi's deployment of the concept of narcissism was deeply entangled with the same dynamic of imitation that governed his account of trauma. Thus Ferenczi repeatedly emphasized that the *tiqueur* and the psychotic shared a strong tendency to

---

20. J. Laplanche and J.-B. Pontalis, *The Language of Psycho-Analysis* (Baltimore, 1976), s.v. "primary narcissism, secondary narcissism."

mimicry, a tendency that made them prone to imitate the words and actions of others. The "waxen flexibility" of the limbs, characteristic of
catatonics and people with tics, was a familiar example of that imitative
tendency; such persons passively allowed their limbs to be placed in every
sort of position without resistance and were prepared to hold those positions for considerable lengths of time. Like the deeply hypnotized subject to whom Ferenczi also compared them, both the *tiqueur* and the
catatonic were thus highly susceptible to the suggestions of other persons, with whom they identified. But as Ferenczi recognized, mimicry of
that sort contradicted the claim that tic and related conditions involved a
regression to the stage of primary narcissism, *both* versions of which assumed the absence of any relation to external objects: according to the
first version of narcissism the ego loved only itself, while according to the
second version the organism existed in an objectless state prior to the formation of an ego. How was the contradiction to be resolved?

Ferenczi tried to resolve it by treating the imitation in question as a
mode of traumatic defense that protected the victim against danger by allowing him to only outwardly identify with or imitate the aggressor while
preserving his own ego intact. That is, he assumed the existence of an
ego, in accordance with the first version of primary narcissism, and argued that the imitative subordination of the patient to every opposing
will was carried out in the mode of an egoistic "irony" that betrayed the
victim's profound detachment from the disturbances of others. To the
man suffering from catatonia, he wrote:

> [E]verything is of equal value: his interest and libido are concentrated on his
> own ego; he only desires that the outside world shall leave him in peace. In spite
> of complete automatic subordination to every opposing will, inwardly he is
> actually independent of his disturbers. . . . [In] catalepsy . . . the patient ac-
> quires[s] that degree of fakir-like concentration on the inner ego when even his
> body appears as . . . a part of the environment, whose fate leaves its owner ab-
> solutely cold. Catalepsy and mimicry therefore would be regressions to a much
> earlier primitive method of adaptation of the organism, an auto-plastic adap-
> tation (adaptation by means of alteration of the organism itself), while flight
> and defence aim at an alteration in the environment (allo-plastic adaptation).[21]

21. Sándor Ferenczi, "Psychoanalytic Observations on Tic" (1921), *Further Contributions to Psychoanalysis*, 163–64; hereafter abbreviated "POT." According to Ferenczi, an
"autoplastic" response to trauma is a reaction to unbearable stimuli that occurs when the
organism cannot defend itself by flight or defense aimed at altering the environment ("alloplastic" adaptation); instead, through recourse to the more primitive method of defense
associated with identification the organism responds "autoplastically" by making alter-

For Ferenczi, there was no conflict between mimicry thus conceived and primary narcissism:

> Now the fact that the dement and the *tiqueur* both possess such a strong tendency to imitate everyone in word and action, taking them as it were for an object of identification and ideal, seems to be in opposition to the assertion that they have regressed to the stage of primary narcissism or have never advanced beyond it. This opposition is, however, only apparent. Like other blatant symptoms of schizophrenia, these exaggerated expressions of the identification-tendency serve the purpose of concealing the lack of real interests; they act, as Freud would express it, in the struggle for healing, the struggle to regain the lost ego-ideal. But the indifference with which every action, every form of speech, is simply imitated, stamps these identification-displacements as a caricature of the normal search for an ideal; they often operate in an ironical sense. ("POT," 164–65)

The "inner ego" Ferenczi posits in these passages thus stands in a relation of spectatorial opposition to objects to which it is indifferent but that it nevertheless sees and imitatively distantiates in an ironical and caricatural manner. Yet the narcissistic subject ironically mimicking others is also so absorbed in its own ego that it has no relation to objects at all—or rather it relates only to the self or "other" of the ego to which it is autistically cathected. We might put it that *tiqueurs* have broken off all genuine communication with external objects and are madly absorbed in their own subjective or internal processes, which is to say that Ferenczi in his deployment of the notion of narcissism here preserves the ego-object relationship while simultaneously denying it.

Indeed, it was precisely the *absence* of any relation to external objects that allowed Ferenczi to make a contribution to the theory of trauma by proposing instead a new mode of defense in tic and related disorders, including the traumatic neuroses. Thus Ferenczi argued that although the repetitive movements of tic were linked to the psychosexual apparatus, they were not primarily conversion symptoms symbolic of repressed and displaced libidinal desires. Rather, they were essentially actions by which the narcissistic ego attempted, however inadequately, to respond to the local damming up of excitation consequent on a trauma. Like the hysterical fixations on the trauma described years before by Breuer and Freud,

---

ations in itself. Cf. Laplanche and Pontalis, *The Language of Psycho-Analysis*, s.v. "autoplastic, alloplastic"; and Martin Stanton, *Sándor Ferenczi: Reconsidering Active Intervention*, 81–85, 194.

the ceremonial tic was an "incompletely mastered shock affect"—an attempt by the patient to "abreact" the emotion that had been incompletely discharged at the time of the trauma itself. But unlike hysterical fixations, tic involved no libidinal relation to an object, with the result that the withdrawn cathexes could not be displaced elsewhere (in phobias, substitutive formations, regressions, and so on). Put more strongly, the cathexes, having nowhere else to go, remained dammed up in the ego itself, drastically impairing its functions: "Hysteria is a transference neurosis in which the libidinal relation to the object (person) is repressed and appears as a conversion-symptom, as it were an auto-erotic symbolization in the body of the patient himself. In tic, on the contrary, it would seem that no relation to the object is hidden behind the symptom; in this case the memory of the organic trauma itself acts pathogenically" ("POT," 155). As Ferenczi interpreted them, the erotic symptoms frequently accompanying tics and related narcissistic disorders were attempts at libidinal satisfaction superimposed on a more primordial mode of "autoplastic" response unconnected to the repression of a desired object. He therefore distinguished the mode of defense in tic from that of repression itself: "Abreaction is a more archaic method of relieving accrued stimulation: it approximates more closely to the physiological reflex than does the still primitive method of control, e.g., repression. It is characteristic of animals and children" ("POT," 153).

Although in his paper on tics Ferenczi was careful to remain within the orbit of Freud's libido theory by endorsing the concept of narcissism, Freud's disciple Karl Abraham immediately recognized the potentially subversive character of Ferenczi's ideas. "Ferenczi says that a tic does not seem to contain any relation to an object," he protested in a discussion of Ferenczi's paper at the Berlin Psychoanalytic Society. "In my analyses, however, I have found a double relation to the object, namely, a sadistic and an anal one. . . . On the basis of my material . . . the tic seems to me to be a conversion symptom at the anal-sadistic stage."[22] But far from accepting Abraham's criticisms, Ferenczi extended his analysis of tic to a variety of other conditions, including the severe ego disturbances associated with the late stages of syphilis (or general paralysis of the insane, one of the so-called pathoneuroses) and, crucially, the traumatic neuroses of war ("POT," 156).[23] In each instance, he attributed the vari-

22. Karl Abraham, "Contribution to a Discussion on Tic" (1921), *Selected Papers of Karl Abraham* (1927; reprint, London, 1988), 324–25.

23. Ferenczi characterized the symptoms of the war neuroses as a mixture of conversion hysteria and narcissistic phenomena akin to pathoneurotic tics ("POT," 156).

ous symptoms observed to a direct wound to the ego and to the subsequent damming up of the cathexes. In his view, the erotic symptoms frequently accompanying such ego injuries were attempts at libidinal satisfaction superimposed secondarily on a more primordial or archaic mode of defense unconnected to the repression of a desired object.

In 1932 the American psychoanalyst Abram Kardiner, in a neglected but highly intelligent and ambitious monograph-length paper, tried to produce a general theory of the traumatic neuroses by drawing on Ferenczi's ideas. Kardiner had studied chronic cases of war neuroses after World War I while serving as an attending physician at a United States veterans' hospital. Dissatisfied with Freud's libido theory, in the early 1930s he turned to Ferenczi's work for inspiration. What struck Kardiner most about the war neuroses, and what motivated his criticisms of libido theory, was the relative absence of symptom elaboration or displacement: it seemed to him that unlike the classic conversion hysterias or transference neuroses, whose symptoms could be understood as a series of substitutive libidinal gratifications, the traumatic neuroses were marked by a profound mental paralysis or psychic inhibition that prevented the substitution and displacement of symptoms. Adopting as "by far the most valuable idea ever advanced about the nature of the traumatic neurosis" Freud's suggestion in *Beyond the Pleasure Principle* that such neuroses were the result of a breach in the individual's protective shield against stimuli, Kardiner identified the essential deficit in trauma as a sudden paralysis of the functions of the *ego* by which perceptual recognition, orientation, and mastery were achieved.[24] In economic terms, the ego's inhibition could be understood as due to a withdrawal of cathexes, not unlike the withdrawal of cathexes from disagreeable or unacceptable instinctual impulses by which repression had been defined. But the novelty of Kardiner's approach lay in his insistence on the difference between the wounded ego's mode of defense and that of the psychosexual apparatus. In short, Kardiner claimed that the traumatic neuroses could not be understood in terms of libido and repression but involved a different mode of reaction.[25]

24. Abram Kardiner, *The Traumatic Neuroses of War* (Washington, D.C., 1941), 137.

25. It was Kardiner's criticism of libido theory that unnerved his psychoanalytic colleagues. Even so intelligent a critic as Otto Fenichel who, unlike most of Kardiner's American psychoanalytic colleagues, welcomed his contribution as "very thrilling," felt it necessary to warn: "Doesn't Kardiner's essay revive the old reproach made by our opponents that the traumatic neurosis contradicts Freud's theory of the libidinal nature of the neuroses?" (Otto Fenichel, review of Kardiner's paper in *Internationale Zeitschrift für Psychanalyse* 1934: 124). See also Fenichel's respectful citation of Kardiner in "The Con-

Critical to Kardiner's argument, and serving as his point of departure, was Ferenczi's paper of 1921 on tic, whose expansive and potentially disruptive theoretical arguments Kardiner shrewdly commandeered. In his own article, with its unperspicuous yet signature Ferenczian title, "Bio-Analysis of the Epileptic Reaction," Kardiner summarized Ferenczi's controversy with Abraham:

> In his article on tic, Ferenczi revived the concept of trauma. This neurosis was originally regarded as a conversion hysteria, to which Abraham added the qualification "pre-genital." Ferenczi's views brought down considerable adverse criticism, because the principle of reflex defense was inharmonious with libido theory. How could the tic as a defense be reconciled with the finding that it is a masturbatory substitute and that there is unquestionable evidence of pre-genital fixations? The symptom, one felt, must be in the direct line of pursuit of the repressed cravings. Ferenczi later studied a condition with manifest organic basis—general paralysis—and there again discovered the reaction to a traumatic factor; he found a profound disturbance in the discharge of somatic tensions based upon an actual impairment of the executive weapons of the ego; and discovered a systematic effort on the part of the ego to heal the breach by means of secondary processes which involved the libido and other parts of the ego.[26]

cept of Trauma in Contemporary Psychoanalytic Theory," *International Journal of Psychoanalysis* 26 (1945): 33–44. The hostile responses of Kardiner's American colleagues are described by him in his "Reminiscences of Abram Kardiner" (interviews presented by the Psychoanalytic Project, Oral History Research Office, Columbia University, New York, 1965, 232, 235, 238).

26. Abraham Kardiner, "Bio-Analysis of the Epileptic Reaction," *Psychoanalytic Quarterly* 1 (1932): 380; hereafter abbreviated "BER." In his paper Kardiner followed a tradition stemming from the time of Charcot by treating the symptoms of traumatic neuroses and epilepsy as a continuous series of reactions. Kardiner also compared the traumatic neuroses to Freud's "anxiety" or "actual neuroses," a category that according to Freud's earliest formulations also lacked the symbolic transposition, substitutability, and fantasmatic elaboration of symptoms characteristic of the so-called "psychoneuroses," as well as being characterized by irritability and a tendency to syncope ("BER," 377). Freud himself made the same connection in *Inhibitions, Symptoms and Anxiety*, 141, a text Kardiner cites. Kardiner's 1932 paper was destined to miss its mark: not only did he himself immediately become dissatisfied with it, but it vanished from the literature virtually without a trace, with the result that nine years later he was compelled to observe that the conclusions he had reached about the traumatic neuroses of the First World War had simply been forgotten (*The Traumatic Neuroses of War*, v, 240). See chapter 6 for a discussion of Kardiner's later revisions.

Applying Ferenczi's ideas to the traumatic neuroses of war he added:

> There are [sic] a group of hysteroid phenomena, such as blindness, deafness, aphonia, paralyses and epileptoid manifestations of many kinds, which arise under traumatic conditions, and in which the essential metapsychological standards of conversion hysteria or hypochondria fail to apply. That is, they do not stand in relation to love objects, and the gain is not in the interests of genitality or pregenitality by any of the standards we know of. Here we meet the same challenge that Ferenczi encountered in connection with tic. Ferenczi accepted the challenge and said that the tic is primarily a defensive reflex that has outlived its original function; that the eroticization that we invariably find— for the tic is a masturbatory equivalent—is the result of a secondary process; and that as a neurosis it has no primary relation to libido objects. The disturbance is primarily in certain fixed organs or in the psychic functions with which they are inevitably accompanied; and such libido realignments as follow upon it are *secondary* processes. ("BER," 380)

On this Ferenczian basis, Kardiner argued that the sensory, motor, intellectual, and other inhibitions characteristic of the war neuroses now made sense—not as symptoms of repressed libidinal desires, but as the manifestations of a more primordial mode of reaction by which the injured ego tried to cope with the dangerous world. Just as in states of fatigue or sleep—normal states of disorganization, in which mastery was temporarily suspended or destroyed—an effort was made by the psyche to reduce the stimulation of the external world, so in the traumatic neuroses an attempt was made to handle the situation of danger by fleeing from the traumatic excitation. The ego's defense thus approximated more closely to a reflex defense against pain or injury than to a defense based on psychosexual repression. In such a scenario, if the traumatized soldier could be characterized as regressed, his was not a regression to an earlier stage of psychosexual development but to an even more elementary phase in the child's development. As Ferenczi had replied to Abraham: "[T]he regression of the ego is far more extensive in this form of neurosis than in hysteria or obsessional neurosis (obsessional neurosis regresses to the 'omnipotence of thought,' hysteria to 'magic gestures,' tic to the plane of defence reflex)."[27] In Kardiner's words: "It is not enough for us to designate what happens in these cases as 'regression'. . . . The technique of the regression is not partial and gradual, as when regressive erotic cravings are made compatible with an intact sensorium; it consists

---

27. Sándor Ferenczi, "Discussion on Tic," *Final Contributions to Psychoanalysis*, 350.

in a total and actual reproduction of the infantile adaptation to the world *en bloc*" ("BER," 419).

It is not surprising in this context that Kardiner should have defined the reaction characteristic of the war neuroses as a reproduction of the infant's earliest traumatic experiences—specifically, the trauma of birth. In one case he regarded as especially instructive, a soldier reported that, after experiencing a frightful shelling that had knocked him completely unconscious, he had woken up in a hospital six days later completely deprived of all body-consciousness. Although he had retained certain sensations, including some cutaneous sensations, he had been unable to localize his feelings or to move his limbs voluntarily. Incapable of differentiating between sounds, he had been powerless to understand what people were saying to him. Deprived of the ability to swallow, he had had to be tube-fed. Day and night had been the same to him, for he had been unaware of the passage of time. Most remarkable of all had been the patient's profound loss of perceptual knowledge and skills. "He was unaware of any relation to . . . objects," Kardiner observed, "hence they were meaningless. . . . The optical pictures of reality were destroyed" ("BER," 402).[28] Taking his analysis to its logical conclusion, Kardiner argued that in the traumatic moment the subject's mastery of the outer world "ceased momentarily; he was torn forcibly from a hitherto friendly world. . . . This is . . . a reproduction of the birth situation where the primitive ego is abruptly torn from the mother. . . . In the state that follows severe trauma . . . we have a regression to the birth situation" ("BER," 416, 461). In short, Kardiner defined the moment of traumatic suffering as a repetition of the trauma of the child's loss of the protective mother.

It will of course be recognized that Kardiner's interpretation of the traumatic experience as prototypically a birth trauma traced its origins to the investigations launched by Freud and appropriated especially by Otto Rank concerning the nature of the child's earliest emotional ties to the maternal figure—another example of the thematization of the mother that we have already detected at the heart of the discourse of traumatic dissociation. What is striking, though, is that Kardiner understood the

28. In his paper Kardiner mentioned a similar case, one of four such cases described by William McDougall in "Regression," in *Outline of Abnormal Psychology* (New York, 1926), 281–98. The soldier's retrospective reconstruction of his reactions and experiences immediately after the shelling needs to be treated with caution. For a valuable discussion of confabulation in hypnotic age regression experiments, see *Hypnosis and Memory*, ed. Helen M. Pettinati (New York, 1988), 21–63, 128–54.

traumatic experience as involving the same imitative process that Ferenczi placed at the center of the traumatic experience. Specifically, Kardiner interpreted trauma as involving the repetition of an infantile mechanism of affective identification with the mother that entailed such a hypnotic fusion between the self and other as to preclude all possibility of cognition and self-representation. Thus for Kardiner, as for Ferenczi, trauma produced an unconscious imitation of the traumatic scene or person that occurred in a state akin to hypnosis and at the cost of the ego's cognitive integrity and control.[29]

For both authors in this mode, trauma thus involved a fundamental fragmentation, dissociation, or splitting of the ego, a mode of defense more primordial than repression, of which it was the prototype. (In other words, like Ferenczi, Kardiner understood his work on the traumatic neuroses of war as a contribution to the study of *primal repression* ["BER," 446, 464–65].) As an "autoplastic" adaptation, traumatic defense entailed an inhibition, deformation, or fading away of the ego consequent on a hypnotic incorporation of the traumatic situation or person. Kardiner linked the process of fragmentation as a mode of defense to Freud's concept of disavowal. Thus at the conclusion of his 1932 paper on the war neuroses Kardiner compared the traumatized soldier's dissociated condition to the deformation or splitting of the ego by which, according to Freud in "Neurosis and Psychosis" (1924), the psychotic refuses or disavows the traumatic perception, thereby creating a gap or rent in the psyche. Intuitively anticipating the importance Freud himself would increasingly come to attach to the notion of the splitting of the ego, and deliberately extending its application beyond the psychoses and the disavowal of castration to which, not without hesitation and self-contradiction, Freud had tended to restrict it, Kardiner focused attention on the fundamental role played by the cleavage of the ego as a general method of psychical defense ("BER," 464).[30]

---

29. For Kardiner's ideas about hypnotic imitation and his debt in this regard to the work of psychoanalyst Siegfried Bernfeld (1892–1953), who in several publications had addressed the relationship between imitation and hypnosis, see Ruth Leys, "Death Masks: Kardiner and Ferenczi on Psychic Trauma," *Representations* 53 (Winter 1996): 44–73.

30. Freud's paper, "Neurosis and Psychosis" (1924), *Standard Edition*, 19: 149–53, cited by Kardiner, adumbrates Freud's growing concern with the splitting of the ego as a mode of defense, a concern that found expression in "Fetishism" (1927), *Standard Edition*, 21: 152–57 and other texts, and was generalized in his posthumously published paper "Splitting of the Ego in the Process of Defence" (1940), *Standard Edition*, 23: 275–78.

Traumatic Memory

If trauma entails a breakdown of cognitive mastery, how is the traumatic experience registered or fixated? If trauma involves a different kind of defense than repression, is the memory of the trauma recorded in a memory system different from that of ordinary memory? Since the establishment of the diagnosis of PTSD, an answer to those questions has emerged according to which the experience of trauma is inscribed in a special memory system in the brain that is held to be radically different from ordinary memory. Indeed, as I show in chapters 7 and 8, van der Kolk and Caruth have materialized and literalized traumatic memory in ways that make it seem as if trauma stands outside all knowledge and all representation. In retrospect, Kardiner's work can be seen to lend itself to those ideas. Ferenczi may also be associated with the same development because he, too, corporealized traumatic memory. He did not literalize traumatic memory in the ways van der Kolk and Caruth do, nor did he understand the traumatic experience as lying "beyond" all representation—for him the traumatic symptoms with which he dealt always remained modes of "autosymbolic" representation.[31] But he did hypothesize that

31. Ferenczi used the term "autosymbolism" to refer to the various modes of symbolic representation the mind employs to represent its own mental processes and mechanisms, independent of content (Ferenczi, "Development of the Sense of Reality," 221). Autosymbols thus represented inner, unconscious thought mechanisms. "*Autoplastic*" autosymbols used the material of the body for representational purposes of this kind, hence they were modes of material representation. In 1919, in a paper on hysterical materialization, Ferenczi defined autosymbolism as the "realization of a wish, as though by magic, out of the material in the body at its disposal and—even if in primitive fashion—by a plastic representation, just as an artist moulds the material of his conception or as the occultists imagine the 'apport' or the 'materialization' of objects at the mere wish of a medium" (Ferenczi, "Hysterical Materialization" (1919), *Further Contributions to Psychoanalysis*, 96). Hysterical materialization of the kind seen in the war neuroses was thus a kind of corporeal symbolism or figuration analogous to artistic representation. At the time, Ferenczi acknowledged that many investigators regarded as a form of self-deception claims by mediums to have literally materialized thought and professed himself unable to express an opinion (96, n. 1). But in his later *Clinical Diary*, perhaps owing in part to the influence of his patient, Elizabeth Severn, he was willing to give greater credence to the occult idea that at the limit—for example, in extreme states of trance induced by trauma—the mind could really alter matter, such as a photographic plate (see for example the posthumously published fragment entitled "Fakirism," *Final Contributions to Psychoanalysis*, 251; cf. *CD*, 117). Nor was he alone in his belief in spiritual photography; on this point see Michael Roth, "Hysterical Remembering," *Modernism/Modernity* 3 (May 1996): 14–19. The fact, moreover, that Ferenczi likened the traumatic impressioning of the mind to a kind of retinal photography and theorized that the fragmented ego becomes a kind of attenuated matter

the mimetic "defenses" occurring in trauma depended on a special memory system that, in the absence of object-perceptions and repressions, registered the trauma as it were bodily. In short, for Ferenczi traumatic memory was *incarnated memory*.

Thus in his essay on tic Ferenczi appropriated Freud's suggestion that the psychical apparatus consisted of several reflex arcs in the form of unconscious, preconscious, and conscious memory-systems (or M-systems) interpolated between the afferent (sensory) and the efferent (motor) systems. Freud had assumed the existence of a plurality of such memory systems, each subserving different principles of mental association. Pursuing Freud's ideas, Ferenczi suggested the existence of one particular memory system, the "ego-memory system" (or "M" system), whose task was to continuously register the individual's subjective physical and mental processes. According to Ferenczi, as a structure designed to retain traces of experiences that pertained only to the subject's own narcissistic ego, the ego-memory system functioned completely independently of the system subserving the memory of *things*, or external objects. He applied his ideas to the traumatic neuroses of war when he observed: "[A]n unexpectedly powerful trauma can have the result in tic, as in traumatic neurosis, of an over-strong memory fixation on the attitude of the body at the moment of experiencing the trauma, and that to such a degree as to provoke a perpetual or paroxysmatic reproduction of the attitude. . . . The 'ego-memory system,' as well as the system of memory for things, belongs in part to the unconscious and in part extends into the preconscious or into consciousness" ("POT," 156).

In two characteristically brilliant articles on the traumatic neuroses of war to which Ferenczi's essay on tic referred at this juncture, he had described several examples of traumatic fixation of this kind. In one patient, the peculiar contractions of the man's shoulder and elbow had exactly mimicked the position of his arm at the moment of trauma: "The man whose right arm is contracted at an obtuse angle, was concussed by the

---

capable of telepathically and clairvoyantly traversing space and time (*CD*, 33, 81) on the model of "phonography" (*CD*, 70) or "telephony" (*CD*, 159) of the soul—all this inscribed his work in the material technologies that emerged at the turn of the century, or what Kittler has called the "discourse network" of 1900 (Friedrich Kittler, *Discourse Networks 1800/1900* [Stanford, California, 1990]). But Ferenczi did not posit "a level of material deployment that is prior to questions of meaning" or representation in the sense of Kittler's recent discussion (David Wellbery, summarizing Kittler's ideas in the foreword to the latter's book, xii). For Ferenczi, the various figures he deployed to characterize the workings of traumatic memory were just that—figures, or modes of representation.

shell just as he was sliding *his rifle into the 'stand easy' position*. This position corresponds exactly with that imitated by the contracture." Of another patient Ferenczi had stated: "The . . . one who has his shoulder pressed to his side and his elbow fixed at an acute angle perpetuates in the same way the situation in which he was caught by the explosion; he was lying down at the time *with the rifle at his shoulder, and taking aim*—for this he had to press his arm to his side and bend the elbow at an acute angle. . . . [I]n these cases *we are dealing with fixation of the innervation predominating at the moment of the concussion (of the shock)*." And of yet another case he had commented: "This soldier with the permanent contracture of the left calf recounts how he was cautiously descending a steep mountain in Serbia, and, while *stretching his left foot downwards* to find a support, was concussed by an explosion and rolled down. Here . . . therefore, there is a 'petrifaction' due to shock, in the attitude that had just been adopted." For Ferenczi, these cases illustrated the fact that "[t]he sudden affect that could not be psychically controlled (the shock) causes the *trauma;* it is the innervations dominant at the moment of trauma that become permanently retained as morbid symptoms and indicate that undischarged parts of the affective impulses are still active in the unconscious."[32]

In his later work on trauma Ferenczi described hysterical fixation in the same terms and connected it to the hypothetical ego-memory system. "Hysterical symptoms seem to be mere autosymbolisms," he wrote in a posthumously published note, "that is, reproductions of the ego-memory system, but without any connection with the causative moments. One of the main methods of making something unconscious seems to be the accentuation of the purely subjective elements at the cost of knowledge about external causation."[33] The hypothesis of the ego-memory system thus helped explain that dissociation between affect and representation which Ferenczi regarded as characteristic of the traumatic experience by locating the subjective-affective response to trauma in a memory system completely divorced from the memory system concerned with knowledge of the external traumatic origin. As he observed in his *Clinical Diary:*

> My earlier hypothesis of a double memory sequence—subjectively narcissistic and objective, emphasizing alternately one and then the other—offers insight into the formation of hysterical symptoms. If one succeeds in focusing all at-

32. Sándor Ferenczi, "Two Types of War Neuroses" (1916–17), *Further Contributions to Psychoanalysis,* 127–29.

33. *Final Contributions to Psychoanalysis,* 221.

tention on the subjective process while the affects run their course, the object side of the perceptual system is totally empty, uncathected. Great pain, in this sense, has an anesthetic effect: pain without ideational content is not accessible to the consciousness. . . . Such subjective experiences can be attained in a state of trance: a feeling of suffocation, of subjective auditory and visual perceptions without content or form, pain of the most varied types. The feeling of fading away, exploding, etc. (*CD*, 30–31)

In this passage, Ferenczi suggested that at the moment of trauma the victim regresses to the state of an ego so narcissistically absorbed in itself that it has no relation to or knowledge of external objects. But in other passages he implied that the victim regresses to a state prior to the very constitution of the ego and hence prior to the ego-object divide. In one vertiginous statement, in a reference to his theory of the ego-memory system, he combined the two propositions:

[F]rom the moment when bitter experience teaches us to lose faith in the benevolence of the environment, a permanent split in the personality occurs. The split-off part sets itself up as a guard against dangers, mainly on the surface (skin and sense-organs), and the attention of this guard is almost exclusively directed toward the outside. It is concerned only with danger, that is to say, with the objects in the environment, which can all become dangerous. Thus the splitting of the world, which previously gave the impression of homogeneity, into subjective and objective psychic systems; each has its own way of remembering, of which only the objective system is actually completely conscious. (See here the relevant hypothesis in my article on tics.) . . . Actual trauma is experienced by children in situations where no immediate remedy is provided and where adaptation, that is, a change in their own behavior, is forced on them—the first step toward establishing the differentiation between inner and outer world, subject and object. From then on, neither subjective nor objective experience alone will be perceived as an integrated emotional unit (except in sleep or in orgasm). (*CD*, 69)

According to this remarkable, and remarkably slippery, declaration traumatic imitation simultaneously becomes the ground of possibility of knowing objects *and* the ground of impossibility of integrating the memory of those objects with the affects that belong to them. The result is to make the dissociated traumatic experience inherently unavailable for a certain kind of conscious knowledge and recollection.

Towards the end of his life, Ferenczi acknowledged the implications of his own theorizing. "It is unjustifiable to demand in analysis that something should be *recollected consciously* which has never been conscious," he wrote in a very late note in apparent recognition of the impossibility of a certain kind of memory and representation in trauma. "Only *repetition* is possible with subsequent objectivation for the first time *in the analysis*. *Repetition* of the trauma and *interpretation* (understanding)—in contrast to the purely subjective *'repression'*—are therefore the double task of analysis. . . . Now it is the time for 'encouragement' to the 'tasks of life'— future happiness, instead of pondering and digging in the past."[34] This is a formulation that led him to retheorize the role of hypnosis in psycho-analysis as a technique not for recovering or reconstructing the memory of an event that had been repressed, but rather for countermanding or countersuggesting (through "encouragement") the effects of a primarily repressed, or "sequestrated"[35] or "encapsulated" (*CD*, 181) traumatic experience that was incarnated in the body or ego-memory system and that was, as it were, logically unavailable for recollection and self-representation.

Unavoidably, however, as Ferenczi in his *Clinical Diary* and related texts came to realize, such a dependence on hypnotic suggestion raised concerns about the truth-status of his patient's cathartic experiences. For the question arose: Was it possible that patients failed to remember the trauma not because of the inherently inaccessible nature of the traumatic-mimetic experience, as Ferenczi at times seems to have be-lieved, but because the cathartic relivings were simply confabulations performed under the influence of the hypnotic relationship with the physician? Were Ferenczi's trance reenactments, as Freud himself feared, merely suggested performances? Even more drastically, were they deliberate simulations? It is to those questions that I turn in the next chapter.

34. These quotations come from the posthumously published notes entitled "Notes and Fragments," *Final Contributions to Psychoanalysis*, 260–61.

35. Stefan Hollos and Sándor Ferenczi, *Psychoanalysis and the Psychic Disorder of General Paralysis* (New York, 1925), 47; Ferenczi, *Final Contributions to Psychoanalysis*, 270.

V

The Hysterical Lie:
Ferenczi and the Problem of Simulation

The twin problems of suggestion and simulation have haunted the history of trauma from the beginning. Victims of railway accidents were often accused of malingering, especially when they sought compensation in the courts of law. Hysterics were frequently suspected of feigning the symptoms they so dramatically displayed in their spontaneous or hypnotically induced performances. And the shell-shocked soldier of World War I was repeatedly accused of fabricating his symptoms in order to avoid combat. On the face of it, it seems surprising, however, that Ferenczi himself subscribed to a notion of simulation in his own approach to the problem of trauma. Yet a close study of the *Clinical Diary* reveals that both of his models of trauma and splitting were implicated in an account of the individual as destined to deceive, or lie.

Thus in his *Clinical Diary* Ferenczi repeatedly suggested that more fundamental than any other traumatogenic factor, including the actual sexual assaults to which he gave such weight, it was the lies and hypocrisy of adults that, by forcing the child to doubt her own judgment about the reality of her experience, fragmented and hystericized her and made her a liar and hypocrite in turn. Ferenczi actually went so far as to propose that sexual seduction was not inherently traumatizing, for even a quite young child might experience sexual penetration by an adult without undue harm, even with pleasure. He asked, "*What is traumatic: an attack or its consequences?*" and answered, "The adaptive potential 'response' of even very young children to sexual or other passionate attacks is much greater than one would imagine. Traumatic confusion arises mainly because the attack and the response to it are denied by the guilt-ridden adults, indeed, are treated as deserving punishment."[1] The adult distorted the truth to protect his reputation and the child also dissembled the truth by identifying with, or conforming to, the frightening, guilt-ridden adult (*CD*, 155–56). The mother's complicity in the father's hypocritical behavior also played a crucial role (*CD*, 18).[2] Victimized by adult hypocrisy, alone and helpless, the child split off her emotions from her thoughts or representations, dissimulating her own feelings by identifying with her aggressor in the mode of a conscious simulation. The analyst's hypocrisy further exacerbated the problem by encouraging the patient to repeat her dissembling ways. Ferenczi's therapeutics of sincerity was thus conceived by him as an antidote to the problem of simulation, or human mendacity.

But what did it mean for Ferenczi to place the problem of simulation at the center of a psychoanalytic account of trauma? As he knew, psychoanalysis was founded on the assumption that the so-called imaginary illnesses of the hysteric were not fictions in the sense of conscious imitations, and that the traditional opposition between truth and lying must be refused by dismantling the philosophy of consciousness and conscious intention on which the notion of deceit had classically depended. The hysteric was saved from the accusation of conscious simulation in order to found a theory of unconscious fantasy and desire.[3] Yet through-

1. Sándor Ferenczi, *The Clinical Diary of Sándor Ferenczi*, ed. Judith Dupont (Cambridge, Massachusetts, 1988), 178 (cf. 50, 174, 191); hereafter abbreviated *CD*. The effect of Ferenczi's emphasis on the early genitality of the child was to challenge Freud's account of the infant as inherently perverse in favor of a normalized account of desire and sexual difference. Ferenczi in this mode downplayed the importance of psychosexual developmental stages in order to treat homosexuality and the other "perversions" as merely accidental states traumatically induced on an authentic and original heterosexuality (*CD*, 172). Ferenczi's claims, repeated in his paper, "Confusion of Tongues Between Adults and the Child," and amounting, as he recognized, to a rejection of the general significance of the Oedipus complex, were criticized by his psychoanalytic colleagues. The work of tracing the influence of Ferenczi's normalizing views about sexuality and sexual difference on the American school of interpersonal psychiatry, including the feminist theories of Clara Thompson among others, remains to be done. But see Elizabeth Lawrence Capelle, "Analyzing the Modern Woman: Psychoanalytic Debates About Feminism 1920–1950" (Ph.D. diss., Columbia University, 1993).

2. Cf. Ferenczi, "Child-Analysis in the Analysis of Adults" (1931), in *Final Contributions to the Problems and Methods of Psychoanalysis*, ed. Michael Balint (New York, 1955), 138.

3. "All neurotics are malingerers; they simulate without knowing it, and this is their sickness" (Freud's testimony to the "Commission for the Investigation of Derelictions of Military Duty" [1920], cited in K.R. Eissler, *Freud as an Expert Witness: The Discussion of War Neuroses between Freud and Wagner-Jauregg* [Madison, Connecticut, 1986], 62).

out his life Ferenczi was drawn to the problem of simulation or lying as *the* problem of human relations and human society.[4] The result was a tendency to assimilate psychoanalysis to the traditional theory of consciousness from which Freud had attempted to free it. (It is worth stressing that for Ferenczi what was at stake in his reflections on lying was the issue of human speech: what he found problematic about humans is that their speech may dissimulate the truth, which is to say that the distinction that was crucial to his thought was less that between truth and fiction than that between speaking the truth—veracity—and not speaking the truth—lying.)[5]

Mendacity, Trauma, and Splitting

The stakes for Ferenczi of a thematics of lying as the primordial traumatogenic factor can be seen in an especially interesting yet perplexing article, "The Problem of the Termination of Analysis" (1927). In that essay, Ferenczi took as his point of departure his discovery that for more than eight months one of his patients had been deceiving him about an important financial matter. Ferenczi reported that at first he had been embarrassed to find he had been misled in this way but—characteristi-

4. According to Ferenczi, unlike objects, which are unalterable and consistent, human beings are incalculable and deceitful. It is the fact that human beings lie which makes the human infant first realize that there is an external world, separate from him: thus the development of the ego—its splitting off as an independent entity from the objective world—is caused by the mendacity of other human beings (for an early statement of this view see Ferenczi, "Belief, Disbelief, and Conviction" [1913], in *Further Contributions to the Theory and Technique of Psychoanalysis* [London, 1950], 442–43). So that the "wicked" or "malicious" things which come to disturb the infant's initial psychic repose are crucially other human beings whose capacity for lying makes the child susceptible either to a *credulity* that Ferenczi likens to the submissive abdication of critical judgment characteristic of the hypnotic state (for the child depends on the authority of the adults who surround her), or, in a child with a more developed sense of reality, to a *skepticism* which makes her doubt the word of adults. It is of course the child's sexual life about which adults lie most, either by hypocritically denying the importance of the child's instinctual erotic pleasures (Ferenczi's earlier formulation) or by lying about their own sexual actions and assaults (Ferenczi's later formulation).

5. There are obvious connections between Ferenczi's ideas here and Laplanche's concept of sexual trauma as a problem of (mis)communication between parent and child (Jean Laplanche, *New Foundations for Psychoanalysis*, trans. David Macey [London, 1989]). Cf. Jean Laplanche, *Seduction, Translation and the Drives*, ed. John Fletcher and Martin Stanton (London, 1992), 33.

cally wishing not to have to give up in defeat—had decided to try to understand his patient's mendacity. He remembered that before the lie was detected the patient had failed to appear for an analytic session and, when taxed with that absence, had declared that he was completely unaware that he had failed to show up. It then emerged that the patient had forgotten not only his appointment with Ferenczi but the events of the whole day in question. In short, the patient had developed a memory disturbance, and it turned out that he had experienced similar memory disturbances before. On the basis of his patient's amnesia, Ferenczi concluded that he was a split personality. But instead of arguing, as might have been expected, that the patient had lied about his financial affairs because he was a split personality, Ferenczi suggested that the patient was split because he was a conscious liar. "When I obtained incontrovertible evidence of his conscious mendacity I was convinced that the split personality, at any rate in his case, was only the neurotic sign of his mendacity, a kind of indirect confession of his character-defect," Ferenczi wrote, and went on to generalize his finding: "I have no hesitation in . . . interpreting all cases of so-called split personality as symptoms of partially conscious insincerity which force those liable to it to manifest in turn only parts of their personality."[6]

Ferenczi noted that the problem of lying during analysis had come up for discussion before. He observed that in an earlier paper he had suggested that in infancy all hysterical symptoms were originally produced as "conscious fictional structures" ("PTA," 78).[7] Moreover, Ferenczi reported that Freud, too, used occasionally to tell his followers that it was a sign of approaching cure if the patient suddenly expressed the conviction that during the whole of his illness he had really been only shamming. In Ferenczi's words, "it meant that in the light of his newly acquired analytic insight he was no longer able to put himself back into the state of mind in

6. Sándor Ferenczi, "The Problem of the Termination of the Analysis" (1927), *Final Contributions to Psychoanalysis*, 78; hereafter abbreviated "PTA."

7. Although in his 1927 paper Ferenczi did not give a reference, he was probably referring to an earlier article in which he had observed that the unconscious tics and grimaces characteristic of hysterical and other neurotic states had their roots in childish tricks and games which had once been conscious and deliberate but which had later become unconscious and automatic through the influence of habit. "The curious idea which so many neurotics ventilate at the end of treatment," Ferenczi had written, "viz. that the illness which had tormented and incapacitated them so much was after all only 'a pretense' [or simulation] is quite sound in the sense that they have produced in adult life symptoms which during childhood they aimed at and sought after in play" (Ferenczi, "Psychoanalysis of Sexual Habits" [1925], *Further Contributions to Psychoanalysis*, 284).

which he had allowed his symptoms automatically to appear without the slightest intervention of his conscious self" ("PTA," 78). But Ferenczi's use of Freud amounted to a curious misappropriation of the latter's ideas. For it was one thing for Freud to interpret the patient's claim to simulation as a sign of a newly won awareness of the unconscious motives and intentions previously driving him into illness; it was another matter for Ferenczi to interpret the same claim to simulation as proof that hysteria originated in a conscious "fiction" or lie.

The dislocating implications of Ferenczi's argument are evident in what he said about the goals of treatment. In 1897, at a decisive moment, Freud had famously revised his theory of sexual trauma in large part because he had discovered that, as he put it, "there are no indications of reality in the unconscious, so that one cannot distinguish between truth and fiction that has been cathected with affect."[8] "Psychical reality," defined as an order of reality that is the domain of unconscious wishes and their associated fantasies, emerged thenceforth as the specific field of psychoanalytic inquiry. But Ferenczi in 1927 proposed to reverse Freud's position by suggesting that it was not enough for psychoanalysis to uncover the structure of fantasy without regard to its basis in reality; rather, the patient could not be considered fully cured until he had achieved a sense of the absolute difference between reality and fantasy—that is, between truth and lies:

> We had come to the conclusion that the laying bare of the fantasy, which could be said to possess a special kind of reality of its own (Freud called it a psychical reality), was sufficient for a cure, and that it was a secondary matter from the point of view of the success of analysis how much of the fantasy content corresponded to reality, i.e., physical reality or the recollections of such reality. My experience had taught me something else. I had become convinced that no case of hysteria could be regarded as cleared up so long as a reconstruction, in the sense of a rigid separation of reality and fantasy, had not been carried out. . . . No bad way of ferreting out such fantasy nests is detecting the patient in one of those distortions of fact, however insignificant, which so often appear in the course of analysis . . . Associations which proceed from such distortions frequently lead back to similar but far more important infantile events, i.e., to times when the now automatic deception was conscious and deliberate. ("PTA," 78–79)

8. *The Complete Letters of Sigmund Freud to Wilhelm Fliess, 1887–1904*, trans. and ed. Jeffrey Moussaieff Masson (Cambridge, Massachusetts, 1985), 264.

We know, of course, why Ferenczi's patient was a liar: because when he was a child hypocritical adults had lied to him by denigrating his sexual feelings. In short, on this version of trauma everything began in conscious mendacity. The goal of psychoanalysis was to trace the tendency to lie back to its moment of genesis in childhood: the termination of therapy was in sight when patients were made aware of the deliberate duplicity of childhood in which their propensity to deceive originated.[9]

It is striking that in his *Clinical Diary* Ferenczi recounted an incident concerning his analysand Clara Thompson that appears structurally identical to that of his mendacious patient of 1927. Ferenczi reported learning that, on an outing at a swimming pool, two of Thompson's acquaintances had made insulting remarks about her in such a way that she must have overheard them. Yet at the next analytic session Thompson said nothing about the incident. She merely stated that on getting home from the outing she had experienced an "epileptic fit." Because he suspected Thompson of "an intentional failure to hear" (*CD*, 143), Ferenczi confronted her with the truth of the situation. She denied knowing anything about the episode: "She knows *nothing* about it; she was not listening" (*CD*, 144). Ferenczi then proposed the following interpretation. Thompson had lied about the incident. In actual fact, she had "heard everything" (*CD*, 144). But—aided by "her capacity to *swallow* the most unjust accusations" (*CD*, 144)—she had swallowed (or incorporated) the knowledge of what she had heard as a way of coping with the insult. "*She fails to hear* nonsense, lies, and injustices—in order not to explode (kill)" (*CD*, 144). In the process she dissociated her emotions from her conscious knowledge of the event. Thus one part of her psychical system reacted to the trauma by identifying with or imitating the aggressor, while the rage she felt at the injury she had experienced was split off from the "ego consciousness" (*CD*, 144) or the representation and was displaced into the organic in the form of corporeal symptoms (the epileptic fit). The infantile prototype for this deliberate parapraxis, or lie, emerged in one of Thompson's dreams: "C. [Clara Thompson] recounts a dream of the same night: *someone* (mother?) *says the words* 'The man must have been crazy to do a thing like that.' Interpretation: trauma . . . caused by man *is true*: mother's doubts make the child consciously deny her own self. Thus she learns suggestibility, she has no confidence in me, neither in *her own* judgment, nor in her friends. *Postscript to the dream:* persons dead, includ-

9. Similarly, Ferenczi substituted the language of lying for the language of fantasy when he attributed Freud's alleged lack of love for his patients to the latter's discovery, during the seduction theory phase of his career, that hysterics *lie* (*CD*, 93).

ing also (mother) and myself (Dr. F.)" (*CD*, 145). As in the case of the mendacious patient described in 1927, so Thompson's split condition, as manifested by her failure to hear her friends' unjust speech, was a sign or confession of her tendency to lie. And again as in the case of the earlier patient, we know why Thompson tended to lie: it was because her mother had lied to her about the reality of her father's sexual abuse. The mother's skeptical words in the dream ("The man must have been crazy to do a thing like that")[10] explained why the child had learned to consciously deny the truth to herself while helping to confirm—in the mode of a denegation—the truth of the trauma. Thus on Ferenczi's interpretation, if Thompson suppressed her knowledge of the incident at the swimming pool—if she lied about what had happened to her—it was because years ago her mother had lied to her daughter about the latter's traumatic experience. In delayed revenge Thompson now murdered her mother in her dream, along with her analyst—maternal substitute—who repeated the mother's hypocritical ways.

But what are the implications of this line of thought? Ferenczi's interpretation of Thompson's dream confirmed the infantile trauma but it also introduced (or reintroduced) a thematics of doubt that threatened to undermine his entire endeavor. How could it be otherwise? Once Ferenczi had introduced the theme of simulation into his analysis of trauma, it became difficult if not impossible to determine the boundaries between truth and lying. For if it is true that, as he asserted, hysterical fantasies "do not lie" when they tell us that "parents and other adults do indeed go monstrous lengths in the passionate eroticism of their relation with children,"[11] it is also true that, according to the terms of his own theorizing, the hysteric becomes a liar as a consequence of the hypocrisy of the parents on whom she once depended. The testimony of the hysteric is therefore always suspect and—without independent confirmation, always difficult to achieve—it is not clear how the trauma can ever be established with certainty.[12] In this context, of all the paradoxical implications of

---

10. These words can be interpreted differently. They are not obviously skeptical, but Ferenczi took them that way (i.e., as implying that the man *would have had to be crazy* and therefore was unlikely to have done the thing).

11. Ferenczi, "The Principle of Relaxation and Neocatharsis," *Final Contributions to Psychoanalysis*, 121.

12. In *Truth Games: Lies, Money, and Psychoanalysis* (Cambridge, Massachusetts, 1997), John Forrester has remarked in reference to the topic of lying in Ferenczi's 1927 paper that the problem of the reality of infantile scenes is intimately linked to the possibility of the patient's lying to the analyst, and that whenever one topic is raised the other comes with it

Ferenczi's work on trauma, none was more so than his claim that the end of therapy was in sight when the patient admitted that she had been lying all along. And yet one can sense Ferenczi's satisfaction in his account of a critical moment in his handling of Thompson's case. When Ferenczi had acknowledged his anger with Thompson over the kissing episode (see chapter 4), she had made an admission in turn. In Ferenczi's words: "The end of the session in a conciliatory mood; she retained the feeling that she had regained my trust. That I do not treat her as she had been treated in the past, by her father and also by that teacher, who never confessed their offenses toward her. Out of revenge, she then described some of these incidents in much more crude and dreadful terms than was objectively justified. *The hypocrisy of the adults gives the child justification for exaggeration and lying.* If those in authority are more sincere, the child will then come forward on its own with confessions and proposals for good behavior" (*CD*, 57 emphasis added). Are we to imagine that Thompson meant her vengeful misrepresentations to be taken in this way? Or was the psychical legacy of her father's hypocrisy such as to disbar her from admitting the exaggerations that Ferenczi claimed to recognize? The passage leaves us uncertain as to what if anything Thompson "confessed" to Ferenczi. In any case, it scarcely seems surprising that she herself remained doubtful about the role of sexual abuse in her own case, and that the problem of trauma was entirely absent from her own psychoanalytic writings.[13]

Simulation

Now it seems to me significant in this context that in discussing Ferenczi's contributions to psychoanalysis Thompson later singled out his technical procedures for special criticism. In Ferenczi's neocatharsis the reconstruction of the traumatic scene occurred on the basis of a quasi-hypnotic, emotionally charged, childish process in which the patient acted out certain dramatic scenes apparently related to the decisive event

---

(88). But he ignores the question of hypnosis and hypnotic simulation, the focus of this chapter.

13. Susan Shapiro has recently commented on Thompson's failure to address the problem of sexual abuse in her writings on women. Because she is convinced of the reality of abuse in Ferenczi's patients, Shapiro finds Thompson's silence puzzling, even deplorable; she attributes it to a variety of factors, including unresolved transference issues (Susan A. Shapiro, "Clara Thompson: Ferenczi's Messenger with Half a Message," in *The Legacy of Sándor Ferenczi*, ed. Lewis Aron and Adrienne Harris [Hillsdale, New Jersey, 1993], 159–73). Obviously, my discussion goes in a very different direction.

and in which Ferenczi participated by taking on various roles. Ferenczi recognized that the trance relivings included elements of play-acting. Nevertheless, for Ferenczi they were not merely suggested performances: he was committed to the belief that beneath the playful images, hallucinations, and (re)enactments the patient's fundamental traumas and shocks of childhood lay buried and concealed (*CD*, 58).[14]

His procedures were not new. They were modeled on the earlier Breuer-Freud cathartic method, as he himself recognized. His dialogic, play-acting technique also bore a distinct resemblance to the hypnotic-cathartic methods of the kind used by Morton Prince to reconstruct the trauma in his Beauchamp case, or by William Brown and other British therapists to treat victims of shell shock during World War I. Thompson, however, came to reject Ferenczi's procedures precisely on the grounds that the "relivings" seemed to many patients, perhaps including herself, a kind of insincere play-acting. "In order to win the analyst's approval [patients] will go through the reliving activities but they have no feeling about them," she stated in an unpublished assessment of Ferenczi's therapeutic work. "It is acting. . . . If the acting is admitted and the analyst is fully cognizant that it is but an imitation of feeling no harm is done and possibly the demonstration may make something clearer. If however the pt. [patient] fears to admit that he is not feeling what he is doing then of course he is not saying everything that comes in his mind and the analytic situation is blocked by it."[15] Ironically, then, from Thompson's perspective, Ferenczi's techniques exacerbated the problem of insincerity or lying they were designed to cure. They could not be relied upon to enhance the recollection of the trauma because they belonged instead to the domain of acting defined as deliberate feigning, or simulation. The tendency of Thompson's criticism of Ferenczi's techniques was thus to cast doubt on the reality of the scenes "reenacted" in the hypnotic-analytic situation, including presumably her own.[16]

14. Ferenczi, "Child-Analysis in the Analysis of Adults," 129–30.

15. Clara Thompson, "Evaluation of Ferenczi's Relaxation Therapy," unpublished ms, Clara Thompson Papers, William Alanson White Institute, New York; my emphasis. In another paper Thompson characterized Ferenczi's play-acting method as "phantastic," observing that if the patient had a weak sense of reality to begin with, the risk of Ferenczi's technique was that it could drive the patient even further away from contact with reality, i.e., into psychosis. On the other hand, a more integrated patient would simply feel insincere (unpublished lecture on Ferenczi, 55–56, Clara Thompson Papers, William Alanson White Institute).

16. In his *Clinical Diary* Ferenczi acknowledged the problem of insincerity inherent in his neocathartic treatment and expressed his dilemma in the following terms. Either the

Interestingly, Thompson's concerns have something in common with the critique of Borch-Jacobsen, who in several recent publications has argued that the relivings characteristic of hypnotic-cathartic methods are in principle incapable of producing evidence of traumatic origins because they belong entirely to the order of *hypnotic simulation*, that is, to the order of fictive games played out between the patient and hypnotist.[17] According to Borch-Jacobsen's dissident reinterpretation of Freud's original seduction theory, the suggestibility of the patient produces the hallucinatory enactments in compliance with the hypnotist's expectations about sexual etiology. For Borch-Jacobsen, it is not surprising that, as Freud himself reported, patients fail to experience their productions as real memories, for the "memories" in question are nothing but confabulations performed under the influence of the hypnotic relationship. In 1896 Freud treated the failure of recollection in the cathartic treatment—the patient's lack of conviction as to the reality of the traumatic event apparently being reexperienced in the cathartic performance—as a resistance to remembering, a resistance that proved the authenticity of the founding sexual scene or traumatic origin.[18] But according to Borch-Jacobsen no such trauma or scene lies behind the patient's performance, which is simply a fictional or invented scenario acted out by the patient in the hypnotic *rapport* ("N," 33).

His argument can be extended to Ferenczi's approach to trauma. On Borch-Jacobsen's interpretation, the patient's lack of convincing memories of the traumatic scene, so often reported in Ferenczi's *Clinical Diary*,

---

analyst participated in the patient's dramatic performance as if it were a present reality, in which case he was dishonest and insincere, since, insofar as the incident really had occurred, it was not now taking place; or, right from the beginning the analyst presented the events of the past as memories that were not really happening again, in which case the patient remained unconvinced of their truth (*CD*, 24–25). Ferenczi had confronted the same issue of insincerity earlier (see Ferenczi, "On Forced Phantasies" [1924], *Further Contributions to Psychoanalysis*, 68–77).

17. Mikkel Borch-Jacobsen, "Neurotica: Freud and the Seduction Theory," *October* 76 (Spring 1996): 15–43, hereafter abbreviated "N"; idem, *Remembering Anna O: A Century of Mystification* (New York, 1996), hereafter abbreviated *R;* and idem, "L'Effet Bernheim (fragments d'une théorie de l'artefact généralisé)," *Corpus*, No. 32 (1997): 147–73, hereafter abbreviated "L'Effet Bernheim." Thompson has something in common with Borch-Jacobsen, but she renders virtually invisible the hypnotic-suggestive dimension of Ferenczi's play-acting methods that is central to Borch-Jacobsen's critique.

18. Freud, "The Aetiology of Hysteria" (1896), *The Standard Edition of the Complete Psychological Works of Sigmund Freud*, trans. and ed. James Strachey (London, 1953–74), 3:204.

must be attributed not to the inherent tendency of the traumatic experience to fall outside normal recollection, as Ferenczi on both the originary
and the postoriginary models of trauma proposed, but to the simple
absence of any sexual-traumatic event. Even more provocatively, Borch-
Jacobsen believes that Freud realized he had "extorted" his patient's
sexual confessions through suggestion ("N," 16, 41) and thus knew that
his theory of sexual trauma, or seduction theory, was a "suggestive illusion" ("N," 25)—which is why he abandoned it. Borch-Jacobsen is also
convinced that Freud recognized the same problem of suggestive illusion
at work in Ferenczi's use of the hypnotic-cathartic treatment. Citing a
letter from Freud to Max Eitington shortly after Ferenczi's death, Borch-
Jacobsen quotes Freud's condemnatory assessment of Ferenczi's hypnotic techniques: "'His source is what the patients tell him when he man-
ages to put them into what he himself calls *a state similar to hypnosis*. He
then takes what he hears as revelations, but what one really gets are the
phantasies of patients about their childhood, and not the real story. *My
first great etiological error also arose in this very way. The patients suggest something to him, and he then reverses it*'" ("N," 41; Borch-Jacobsen's emphasis).
What this says is that Ferenczi's errors were the result of the patient's
auto-suggestions. But on Borch-Jacobsen's caustic reading, Freud in this
letter also recognized that the scenes of seduction were equally the direct
result of the physician's hypnotic influence—"the veritable *folie-à-deux*"
between the physician and his patient ("N," 41–42). In spite of Freud's
repeated claims that he had not forced or suggested the seduction-fan-
tasies on his patents, Borch-Jacobsen believes that he knew otherwise,
which is why Freud condemned Ferenczi's return to his own earlier theory of sexual trauma as ineluctably tainted by the problem of hypnosis-
suggestion.[19]

Borch-Jacobsen deserves credit for focusing attention on the problems of interpretation inherent in the use of suggestive and hypnotic-
cathartic therapies. In particular, he suggests the naiveté of Ferenczi's
expectation that in every case it ought to be possible to infer the existence
of real memories of sexual traumas from the patient's hypnotic acting
out. He also suggests the naiveté with which hypnosis is currently being
used to produce evidence of sexual abuse in patients diagnosed with multiple personality or other dissociative disorders, without regard to the

19. Borch-Jacobsen thus agrees with those of Freud's contemporaries, such as Robert
Gaupp, Albert Moll, Leopold Löwenfeld, Eugen Bleuler, and others who at the time dismissed the stories of sexual assault told by Freud's patients as merely suggested ("N,"
23–25).

problems of hypnotic confabulation and false memory that are so well documented in the literature. But Borch-Jacobsen has larger goals in mind. Taking his argument in a radically revisionary direction not envisaged by Thompson, who remained loyal to the psychoanalytic project as she understood it, he goes on to condemn the modern obsession with sexual trauma, including Ferenczi's, as a gigantic mistake, a mistake that he traces all the way back to its origins in Freud's suggestive procedures. More contentiously still, he believes that Freud's concepts of the Oedipus complex, infantile sexuality, and wish-fantasy were "arbitrary constructions" cynically designed to account for the patient's stories of incest and perversion while explaining away the problem of simulation posed by the hypnotic acting-out ("N," 43). In other words, for Borch-Jacobsen, the concept of the Oedipal unconscious is fallacious, and the intention of his critique is therefore to challenge the entire validity of psychoanalysis conceived as a theory of unconscious mental states.

It is important to understand that the terms in which Borch-Jacobsen assesses the problem of hypnotic confabulation in psychoanalysis reflect a widespread post–World War II reinterpretation of the nature of hypnotic suggestion itself. That reinterpretation dismisses as naive the traditional, late nineteenth-century view of hypnosis, associated with the work of Charcot and shared by Freud, Ferenczi, and the majority of their contemporaries, that hypnosis is a special or altered state of consciousness. Instead, hypnosis has come to be viewed as a condition that the subject enters into on a conscious and voluntary basis. Far from being a spontaneous and "objective" phenomenon, as Freud and Ferenczi among others always maintained ("N," 27–29), hypnosis on this reinterpretation is the result of a consensual agreement between the hypnotist and subject. "Good hypnotists," Borch-Jacobsen writes, "have always known that initially suggestibility is nothing but the sheer acceptance of the hypnotic contract: the subject must accept the hypnotic game, failing which the game cannot begin. In this sense, suggestibility is always auto-suggestion or, more precisely, a consensual suggestion *negotiated* with the hypnotist. In fact, it starts the very instant the patient decides to visit the 'medicine man,' turning himself over to the man (or his theory), in order to make his ills disappear" ("N," 33). He attributes similar views to Freud's contemporaries, Hippolyte Bernheim and Joseph Delboeuf, whom he thus presents as admirably lucid precursors of the modern movement to demythologize hypnosis ("L'Effet Bernheim," 168). To use Borch-Jacobsen's language, hypnosis is a performative game in which the hypnotized person responds to the hypnotist's suggestions "in every sense of the word" ("N," 29), by which he means that the hypnotist's suggestions are acted out in a conscious, consensual and voluntary fashion (R, 87). The hypnotic acting is not a reproduction of a prior sexual scene or trauma, but an imaginative, invented performance deliberately carried out according to the hypnotist's demands and the subject's expectations.

It would take me too far afield to review the social and cultural factors that have determined such a postwar reorientation in the field of hypnosis. In the United States, that reorientation has been decisively shaped by the effort to repudiate the well-entrenched influence of behaviorism and its mechanistic assumptions in order to reconceptualize the hypnotized person not as a biological machine or puppet-like automaton, but as an active, cognizing subject capable of pursuing goals and intentions within an ongoing social context. The rise of a psychology of personality, especially the psychoanalytic emphasis on the individual's drives and motivation, has played an important role in that development. But there have been other influences, notably the authority of the role-playing theories of George Herbert Mead, Ernest Goffman, and others in sociology, the impact of pragmatist-functionalist-cognitivist conceptions, and the desire to experimentally operationalize the concepts of hypnosis. The result has been a suspicion of all those researchers, clinicians, and others who continue to regard hypnosis as an altered state of consciousness, or who, like today's neodissociationists, do not accept the social-psychological model of hypnosis.[20]

20. In tracing these developments through a large literature I have found the following texts especially useful: Robert W. White, "A Preface to the Theory of Hypnotism," *Journal of Abnormal and Social Psychology* 36 (1941): 477–505; idem, "Hypnosis and the Concept of Dissociation," *Journal of Abnormal and Social Psychology* 37 (1962): 309–28; J. P. Sutcliffe, "'Credulous' and 'Sceptical' Views of Hypnotic Phenomena: A Review of Certain Evidence and Methodology," *International Journal of Clinical and Experimental Hypnosis* 8 (1960): 73–101; Theodore R. Sarbin and William C. Coe, *Hypnosis: A Social Psychological Analysis* (New York, 1972); Theodore R. Sarbin, "Contextualism: A World View for Modern Psychology," *Nebraska Symposium on Motivation* 24 (1977): 1–41; William C. Coe, "The Credibility of Posthypnotic Amnesia: A Contextualist's View," *International Journal of Clinical and Experimental Hypnosis* 26 (1978): 218–45; Nicholas P. Spanos and Theodore X. Barber, "Toward a Convergence in Hypnosis Research," *American Psychologist* 29 (1974): 500–11; Nicholas P. Spanos, "A Social Psychological Approach to Hypnotic Behavior," in *Integrations of Clinical and Social Psychology*, ed. Gifford Weary and Herbert L. Mirels (New York, 1982); idem, "Hypnotic Behavior: A Cognitive, Social Psychological Perspective," *Research Communications in Psychology, Psychiatry and Behavior* 7 (1982): 199–213; Nicholas P. Spanos, John R. Weekes, and Lorne D. Betrand, "Multiple Personality: A Social Psychological Perspective," *Journal of Abnormal Psychology* 94 (1985): 362–76; Nicholas P. Spanos, H. Lorraine Radtke, and Lorne D. Betrand, "Hypnotic Amnesia as a

Basic to the postwar effort to reconceptualize hypnosis has been the attempt to come up with alternative models to that of an altered mental state for the hypnotic experience. What especially interests me here is that the alternative model of choice has been a dramaturgical one: the hypnotized subject is conceived as an actor who performs or mimes her hypnotic "role" in conformity with various contextual cues, social demands, or "plots."[21] In short, what is at stake in the reconceptualization

Strategic Enactment: Breaching Amnesia in Highly Susceptible Subjects," *Journal of Personality and Social Psychology* 47 (1985): 1155–69; Nicholas P. Spanos, "Hypnotic Behavior: A Social-Psychological Interpretation of Amnesia, Analgesia, and 'Trance Logic,'" *Behavioral and Brain Sciences* 9 (1986): 449–502; idem, "Hypnosis, Nonvolitional Responding, and Multiple Personality: A Social Psychological Perspective," *Progress in Experimental Personality Research* 14 (1986): 1–62; Ernest H. Hilgard, "A Critique of Johnson, Maher, and Barber's 'Artifact in the "Essence of Hypnosis": An Evaluation of Trance Logic,'" *Journal of Abnormal Psychology* 79 (1972): 221–33; idem, *Divided Consciousness: Multiple Controls in Human Thought and Action* (New York, 1986); Alan Gauld, *A History of Hypnotism* (Cambridge, 1992).

A key issue throughout the current debate over hypnosis is the dichotomy between voluntary and involuntary behavior. For Nicholas Spanos and other advocates of the nonstate social-psychological approach to hypnosis, hypnotic subjects may subjectively experience their acts as involuntary but they are in fact voluntarily performing their hypnotic role. They are therefore behaving "as if" they have lost control of their behavior, "as if" they are anesthetic to pain, and "as if" they are posthypnotically amnesiac. In other words, the hypnotic self retains control in order to act the role of the helpless, anesthetic, and amnesiac hypnotized subject. From this perspective, all posthypnotic amnesias are reversible, since they are simply the product of contextual cues and hypnotic expectations. The historically older, so-called "special process" or "special state" theory of hypnosis takes the subject's automaton-like appearance and subjective reports of involuntariness and amnesia at face value, so that he or she is held to be unaware or unconscious of the processes required to produce the experience of involuntariness. On this alternative approach, hypnosis is defined as a special mental state, different in kind from ordinary consciousness and volition, one that helps explain the long-term amnesias due to psychical traumas.

One focus of controversy is the existence of a residual group of highly hypnotizable experimental subjects who fail to breach or lift their posthypnotic amnesia even when requested to do so. Spanos's explanation is that the individual's desire to play the role of the good hypnotized subject who expects to forget his hypnotic experiences outweighs the desire to conform to the experimental injunction to remember. But because that explanation appears to cover all contingencies it seems incapable of disproof. Another problem is that since those who adopt an operational approach are leery of subjective self-reports, the question of the specificity of the patient's subjective experiences of the trance state and trance logic becomes a contested issue.

21. See especially Theodore R. Sarbin, "The Concept of Role-Taking," *Sociometry* 7

of hypnosis that largely informs Borch-Jacobsen's reappraisal is precisely the same problem of mimesis or imitation that was central to Ferenczi's late work on trauma. In this context, Borch-Jacobsen makes a deliberate and telling decision when he compares the hypnotized subject to Diderot's image of the great actor in his *The Paradox of Acting* (R, 89). By this he means that the hypnotized or hysterical person is always aware of the imitation she is performing so dramatically, a performance which in effect she sees and represents to herself, as if she were an observer of the enacted scene. Like Diderot's ideal actor, who remains a detached spectator of the emotions and passions she represents on the stage, the hypnotized subject is imagined as miming her role in full awareness that it is merely a game—and a "banal" one at that (R, 85). In short, for Borch-Jacobsen the hypnotic or hysterical enactment is a *simulation*.

Borch-Jacobsen thus renews the debate over the nature of hypnotic and hysterical imitation that has been a factor in the genealogy of trauma from the start. In favor of his views he cites the claim, made by researcher Ernest Hilgard, that there is always a "hidden observer" who remains a witness to the theater of the hypnotic trance. Moreover, as Borch-Jacobsen notes, and as I observe in chapter 3, Breuer's famous patient, "Anna O." (Bertha Pappenheim) made a similar claim when, as Breuer reported, she asserted that even when she was in a very bad state "'a clear-sighted and calm observer (*Beobachter*) sat, as she put it, in a corner of her brain and looked on at all the mad business'" (R, 89). In fact, as Breuer also stated, at the end of her treatment "Anna O." reproached herself with the idea that her entire performance had been simulated. We have seen in chapter 3 that, according to Breuer, "Anna O.'s" claim to simulation was a retroactive illusion. This is the same argument Freud used when, as Ferenczi reported in his paper of 1927, his patients at the end of treatment sometimes expressed the view that they had been shamming all along. It is an argument that treats the patient's claim to simulation as a sign of a newly won reintegration or consciousness. As such it is an argument based on the theory of dissociation or the unconscious—an

---

(1943): 273–85; idem, "Contributions to Role-Taking Theory: I. Hypnotic Behavior," *Psychological Review* 57 (1950): 255–70; William Coe and Theodore R. Sarbin, "An Experimental Demonstration of Hypnosis as Role Enactment," *Journal of Abnormal Psychology* 71 (1966): 400–406; Sarbin and Coe, *Hypnosis: A Social Psychological Analysis*; Spanos, "A Social Psychological Approach to Hypnotic Behavior." Sarbin acknowledges a debt to Mead for his understanding of hypnosis as role-playing but ignores Mead's own struggles with hypnosis. On this point see Ruth Leys, "Mead's Voices: Imitation as Foundation, or, The Struggle Against Mimesis," *Critical Inquiry* 19 (Winter 1993): 277–307.

argument that Borch-Jacobsen treats as erroneous. Instead, taking "Anna O.'s" testimony at face value, he contends against Breuer and Freud: "We can recognize this fallacious argument, also used today by American theorists of Multiple Personality Disorder: it is the argument for 'dissociation of consciousness'—in other words, the argument for 'the unconscious.' The argument goes like this: Bertha, in personality A, was unconscious of what personality B was up to; therefore, she (personality A) was not simulating. Her hypnoid, unconscious 'bad self,' living the autonomous life of an automaton, was *really* separate from her normal, conscious self. When all is said and done, the hypothesis of the unconscious was (and is) simply an end run around the hypothesis of simulation, by way of arguing that the hysteric's right hand doesn't know (or forgets, or represses) what the left hand is doing" (*R*, 90). On the basis of these and related considerations Borch-Jacobsen concludes that the subject's performance in the hypnotic trance, "amazing though it may be" (*R*, 87), is compatible with the idea of simulation as a "voluntary, deliberate" activity (*R*, 87):

> Hypnotized subjects, even when they are cataleptic and anesthetized, are never just automatons at the mercy of the hypnotist's suggestions. Not only do subjects never lose consciousness, they also always "know all about" what they are pretending to shut out (the pain they are asked not to feel, the muscular fatigue they are asked to ignore, the objects they are asked not to see, the "selves" or "personalities" from which they are asked to dissociate). . . . Bertha, playing her role (her second "personality") on the stage of hysteria, was also watching from the wings, as spectator of her own theatricality. The paradox of the trance (of hysteria) is nothing other than the "paradox of acting," as Diderot put it, which is why Bertha concluded that "the whole business had been simulated." (*R*, 87–89)

Now the first thing to be observed about this statement is that it represents a reversal of Borch-Jacobsen's previous position on the nature of hypnotic imitation. In *The Freudian Subject*, Borch-Jacobsen proposed that the hypnotic-mimetic performance conformed to the *anti*-Diderotian model of acting in that the mimesis involved a complete merger or coalescence between actor and role:

> [T]he "actor" of a fantasy is not the actor described (dreamed?) by playwrights: the playwright bases his actor on an "ideal imagined model," we are told (by Diderot), but he takes care not to enter into the skin of the character whom he merely "quotes," in the last analysis (as Brecht shows). The one who acts the fantasy is mad, however, because he merges with the mimetic model (which

is henceforth no longer a model). He is *caught* in the scene of fantasy, as Laplanche and Pontalis write; by this we are to understand that he takes himself for one or another of the protagonists. A mime through and through, he is thus literally possessed by his role.[22]

The hypnotic trance was understood by Borch-Jacobsen as the very paradigm of that "possession," and the aim of his book was to show the extent to which Freud's writings were implicitly structured by the anti-mimetic ambition to defeat the radical loss of identity, or dedifferentiation between self and other (or role) implied by the hypnotic imitation. But Borch-Jacobsen has changed his mind about the nature of suggestive mimesis and now argues that the hypnotic enactment must be understood on the Diderotian-spectatorial model of acting after all.[23]

Nevertheless, that same Diderotian model of acting poses problems for Borch-Jacobsen. Diderot's *Paradox* argues that the best actor is one who does not get emotionally swept up in his role but is capable of representing emotions that he does not necessarily feel: he can deliberately simulate terror and yet be inwardly calm, can pretend to cry and not feel grief, can feign laughter without being amused, and his feignings are all the more effective as representations for being generated in this way. In short, the Diderotian actor simulates by deliberately producing that divorce between affect and representation that Ferenczi attributed to a traumatically dissociated state. Yet, like others who have adopted the role-theory model of hypnotic acting, Borch-Jacobsen wishes to avoid the implication that the hypnotic subject is lying or malingering—that she is deceiving others, and indeed even herself.[24] As he writes:

22. Mikkel Borch-Jacobsen, *The Freudian Subject* (Stanford, California, 1988), 35.

23. Borch-Jacobson (*R*, 89, n.15) cites William Archer's edition of Diderot's text on acting (Denis Diderot, *The Paradox of Acting*/William Archer, *Masks or Faces* [1888; New York, 1957]).

24. Role-playing theorists such as Sarbin and Spanos believe that although a hypnotized subject may act like, indeed feel like, an automaton, if doing so is part of his role expectations, he never *is* an automaton, never *is* really amnesiac, and never really stops feeling pain. However much he may be absorbed in his role, he is always an actor in Diderot's sense of the term. Such claims make the individual's subjective reports of mimetic absorption and involuntariness, or the failure of some subjects to breach or lift their posthypnotic amnesia even when asked to do so, seem a matter of lying, shamming, or self-deception. An adequate account of how subjects can deceive themselves in this way is thus a key problem for role-playing theorists. Freud's concept of the unconscious attempts to provide a solution to the problem of self-deception by providing a middle term between the dichotomy lying versus truth-telling, but role-playing theorists reject the concept of the unconscious

Simulation is not lying. It is the creation of a new reality. We can't play a role without *incarnating* it. How to play Lady Macbeth without adopting the actual movements of a sleep-walker? How to play Kleist's Penthesileia without actually alternating between two "personalities"? And how to imitate anesthesia without becoming *actually* insensitive to pain? Hysteria and hypnosis are not less real for being simulated. On the contrary, they are *sur*real in the sense of simulation's being pushed to the point where the body goes along with it. How this happens is a mystery (the mystery known as "psychosomatics"), but the mystery in no way negates the fact that the process is a simulative one, and that simulation produces physical effects that are perfectly real . . . . In reality (if we can call it that), simulation is always real—*surreal*—which is why it is so difficult for experimental psychologists to establish a criterion that would allow them to distinguish, with any certainty, a "true" hypnotized subject from a skillful simulator. It's because, as Milton Erickson so neatly puts it, *"The best simulation is an actualization"*: the best way of simulating hypnosis is actually to fall into a hypnotic state. (R, 90–91)[25]

On this basis, Borch-Jacobsen characterizes simulation as "something more, and something else" than lying or fraud, which is how Peter Swales, Fritz Schweighofer, Frederick Crews, and numerous other critics of Freud tend to regard it (R, 85). Yet such a solution to the problem of hypnotic imitation involves a sleight of hand. Wishing to do justice both to the idea that hypnosis is only a form of play-acting under the conscious

---

as unscientific. For an assessment of the theoretical and experimental literature on this topic and of the problems posed by the notion of self-deception see Gauld, *A History of Hypnotism*, 596–608.

25. Borch-Jacobsen refers here to Martin T. Orne's classic article, "The Nature of Hypnosis: Artifact and Essence," *Journal of Abnormal and Social Psychology* 58 (1959): 277–99. Orne demonstrated that it was impossible, based on externally observed behavior, to distinguish between a hypnotic effect due to changes in subjective experience versus one resulting from overt behavioral compliance (or simulation). His findings were taken by skeptics as definitive proof that hypnosis as a special mental state does not exist. However, the fact that behavior alone cannot readily discriminate simulators from "reals" led Orne to conclude rather that the "essence" of hypnosis must be sought in the subjective experience of the individual. Orne reported one such subjective reaction—trance logic, a special mental logic that confuses reality and fantasy and that he linked to Freud's notion of the primary process (295). More recently, Orne has criticized Spanos's version of the social-psychological model of hypnosis because of its premise that the hypnotized subject does not truly experience the suggestions given to him, but merely acts "as if" they were so experienced (reply to Spanos, "Hypnotic Behavior: A Social-Psychological Interpretation of Amnesia, Analgesia, and 'Trance Logic,'" 477).

control of the hypnotic subject *and* to the subjective experience of compulsive absorption in the hypnotic mimesis, Borch-Jacobsen represents the hypnotic subject as simultaneously a spectator of her performance and as incarnating it. But when on that basis he rhetorically asks: "How to play Lady Macbeth without adopting the actual movements of a sleepwalker?" he implies that the best actor is indeed one with her role and hence begs the question at issue: for Diderot the best actor would precisely be someone who could represent those movements and passions with harrowing accuracy without being somnambulistically possessed by, or emotionally merged with, the part. Moreover, Borch-Jacobsen risks anachronism by suggesting that the model of hypnosis that now informs his reinterpretation ought to have been Freud's. During the earlier part of this century the question of the nature of hypnosis was not resolved along the lines Borch-Jacobsen now believes to be correct.[26] Indeed, after decades of theoretical work and laboratory experiments, the question of the nature of hypnosis is still not settled, which suggests that it is unlikely to be resolved on empirical or experimental grounds.[27]

26. Borch-Jacobsen calls Freud's Charcotian ideas about hypnosis "extremely naive" ("N," 28), as if Freud should have recognized the "artifactual" nature of the suggestive effect, should have been aware of his own intervention in the phenomena he observed, and therefore should have sided with the more advanced views of his contemporaries, Bernheim and Delboeuf, who denied that hypnosis constituted a special or altered mental state ("N," 28–29). But he concedes that neither Bernheim nor Delboeuf were completely prepared to "reduce [*diluer*] the scene of suggestion to a pure game of roles (as has been done by experimental psychologists nearer to us in time, such as Sarbin or Spanos)" ("L'Effet Bernheim," 158). For another, brief discussion of Bernheim's and Delboeuf's views on the nature of the hypnotic performance see also Campbell W. Perry et al., "Hypnotic Age Regression Techniques in the Elicitation of Memories: Applied Uses and Abuses," in *Hypnosis and Memory*, ed. Helen M. Pettinati (New York, 1988), 130–34; for the contemporaneous rejection of Bernheim's critique of the notion of hypnosis as an objective state see Gauld, *A History of Hypnotism*, 559–74.

27. To this day a small yet significant group of researchers opposes the position on hypnosis that Borch-Jacobsen now embraces. The post–World War II dispute between those who believe that hypnosis involves a special mental state with specific characteristics, such as amnesia, and those who do not, has lasted for more than thirty years and even some of the participants seem ready to admit that it is not likely to be resolved on strictly empirical grounds. In this regard see especially the discussion in Spanos, "Hypnotic Behavior: A Social-Psychological Interpretation of Amnesia, Analgesia, and 'Trance Logic,'" which contains responses to Spanos's ideas by various critics and his replies (467–97). See also Gauld, *A History of Hypnotism*, 575–630, for a discussion and critique of post–World War II developments.

So that rather than ask: Is what Borch-Jacobsen now says about hypnosis (and by extension about Freud) empirically true? I think it is more profitable to ask: What are the stakes for Borch-Jacobsen in reorienting the debate about hypnosis and trauma along the lines he proposes? Here the answer is clear: what is at stake is the entire validity of the modern discourse of the unconscious or dissocation. Borch-Jacobsen's acknowledgment of the "surreal" nature of the hypnotic acting out leaves unaltered his attack on the claim that hypnosis is an unconscious, involuntary, and "blind" performance; he understands the hysterical mimesis as always controlled, contained, and put in its proper place by the subject's consciousness and will. According to him, the mind is always aware of its own trance confabulations and the mysterious, bodily changes it voluntarily brings about in the hysterical role. He regards the amnesia characteristically observed after trance states as an "artifact" wholly due to hypnotic training and role expectation and hence as easily reversible ("L'Effet Bernheim," 161–63). Not only does he consider the concept of the psychic unconscious a "hypnotic myth" ("N," 43) and psychoanalysis a kind of swindle or fraud,[28] he comes close to characterizing hysteria itself in the same terms. As he puts it, hysteria is not a "real" illness (R, 83) but a "childish" mimetic game of mutual fascination carried out between the patient and the physician (R, 92), a game that presumably could and ought to be abandoned. He even goes so far as to imply that the traumatic neuroses of war are also artifacts of suggestion and that as such they belong to the same category of simulated, hypnotic invention.[29] (It is hardly surprising that Borch-Jacobsen is a strong critic of the multiple personality movement, with its emphasis on the causal role of sexual trauma.) In sum, no repressed or dissociated traumatic event or experience lies behind the dramatic, hysterical-hypnotic scenario.

28. In an interview Borch-Jacobsen has stated that it would not be very far from the truth to say that Freud forged all his case histories ("Basta Così: Mikkel Borch-Jacobsen on Psychoanalysis and Philosophy," interview with Chris Oakley, *Free Associations* 5 [1995]: 434).

29. Thus Borch-Jacobsen criticizes the concept of "incubation"—the idea of a period of latency or unconscious elaboration between the trauma and the onset of hysterical symptoms, a concept which formed the starting-point of Freud's *Nachträglichkeit* and which is seen again in the formulation of post-traumatic stress disorder and dissociation disorder. Borch-Jacoben regards the concept of incubation as a way of avoiding acknowledging the suggested nature of the symptoms in question, and hence as an expedient way of explaining why an injured worker or a soldier wounded in war might develop symptoms of hysteria months or even years later (R, 78). He thus characterizes the phenomenon of the delayed emergence of symptoms after a railway accident as "somewhat suspect" (R, 57).

Suggestion in (after) Analysis

Now it hardly needs saying that Borch-Jacobsen's position is at odds with the general thrust of Ferenczi's ideas. Ferenczi emphatically rejected Babinski's view that the stigmata of hysteria were merely suggested by the doctor, a view that, when applied to the war neuroses, tended to equate them with conscious simulation or malingering. He attributed Babinski's error to the latter's failure to recognize the existence of the unconscious forces that were the true source of the hysterical symptoms.[30] His subsequent emphasis on the sexual traumatic factor in the etiology of hysteria, which coincided with his revival of the Breuer-Freud cathartic method, was represented by him as a return to the "prehistory" of psychoanalysis before Freud lost confidence in the reality of sexual seduction.[31] Throughout his last writings, Ferenczi remained convinced that hysterical fantasies and delusional ideas always contained a grain of historical truth—a traumatic reality that it was the task of analysis to decipher (CD, 58).

More generally, no-one was more sensitive than Ferenczi to human affliction or more committed to the idea that the experience of pain and evil can have profound mental and physical consequences: the modern movement to psychologize suffering finds in Ferenczi its most fervent avatar. For Ferenczi the study of trauma victims was illuminating precisely because, in their ability to mimetically incorporate the thoughts and feelings of the aggressor, they demonstrated the primordial nature of a principle of mental functioning that was opposed to the principle of self-assertion and egoistic resistance that usually governed human behavior. In spite of the vacillations I have emphasized, all of Ferenczi's thought

---

30. Sándor Ferenczi, "Two Types of War Neuroses" (1916–17), *Further Contributions to Psychoanalysis*, 129; idem, "Symposium Held at the Fifth International Psycho-Analytic Congress in Budapest, September 1918," in *Psycho-Analysis and the War Neuroses* (London, 1921), 11; Ferenczi, "An Attempted Explanation of Some Hysterical Stigmata" (1919), *Further Contributions to Psychoanalysis*, 117. Ferenczi's statement, in "Two Types of War Neuroses," that he had obtained case histories from the patients "without their significance for the symptom formation being known either to me or to them, so that suggestive questions on my part were excluded" (129) indicates his desire to refute the charge of suggestion or simulation. For a history of Babinski's dismantling of the hysteria diagnosis and the associated problem of malingering in World War I see especially Marc Oliver Roudebush, "A Battle of Nerves: Hysteria and Its Treatment in France During World War I" (Ph.D. diss., University of California at Berkeley, 1995). See also Joanna Bourke, *Dismembering the Male: Men's Bodies, Britain, and the Great War* (Chicago, 1996), 76–123.

31. Ferenczi, "The Principle of Relaxation and Neocatharsis," 120–21.

pressed towards this insight. The most significant entries in the *Clinical Diary* along these lines are those for 28–30 June, 1932, written as drafts of his "Confusion of Tongues" paper. In those entries Ferenczi interpreted imitation as an elemental principle of adaptation to traumatic disruption that is selfless because it entails a resignation to the other based on the absence of egoistic self-assertion—a resignation he compared to the selflessness of religious renunciation (*CD*, 148). The selfless or "kindness" reaction (*CD*, 146) thus represented the capacity of the traumatized individual to masochistically submit to the violence of the other by mimetically incorporating the unpleasure situation, even as that act of imitation was at the cost of personal unity and autonomy. In the privacy of his *Clinical Diary* (though not in the published version of the "Confusion of Tongues"), Ferenczi gave the principle of mimetic identification an explicitly ethico-political-sexual meaning: the inspired and intelligent ability of the trauma victim to pacify the enemy by imitating or identifying with him represented a general, maternal principle of "appeasement" or "peace" (*CD*, 153). The "wise" child, to whose coping mechanisms the traumatized adult reverted, thus came to stand for a hypersensitive ability to selflessly yield to the selfish other as a means of saving the organism and the world.[32]

Thus if Ferenczi acknowledged the *inescapability* of suffering when he referred to the "unavoidable" traumata (*CD*, 151) associated with the necessary adaptations of infancy (bowel training, training in cleanliness, etc.), he also believed that even such pain was "perhaps partly superfluous and unnecessary" (*CD*, 151). "[M]an becomes passionate and ruthless as a consequence of suffering," Ferenczi wrote (*CD*, 151), thereby attributing the origin of aggression to drives and forces that are originally external and accidental to the innocent child. Ethical behavior did not arise through the sublimation of erotic wishes, as Freud believed, but from an

32. Ferenczi therefore conceptualized the victim's ability to endure pain by mimetically identifying with the traumatic scene or aggressor as a manifestation of the death drive defined as a female principle of masochism or abjection (*CD*, 40–41). This suggests the presence in his thought of a sexual thematization of trauma along conventional gender lines (see for example *CD*, 146, where Ferenczi advocates female "self-denial" to solve the problem of a husband's impotence and infidelity while recognizing that self-abnegation can hardly be expected "with the present trends in the education of women"). Identification with the aggressor was subsequently made famous by Anna Freud as one of the defenses of the ego (Anna Freud, *The Ego and the Mechanisms of Defense* [New York, 1966], 109–21). Like Ferenczi, whom she did not mention, she linked identification to imitation and mimicry, without, however, engaging the problem of mimetic-hypnotic incorporation central to his discussion.

independent mimetic principle of "maternal" altruism, mutuality, and good will. In his more optimistic and utopian moments Ferenczi looked forward to a world cleansed of all destructiveness. He imagined using psychoanalysis to end the "chains of acts of cruelties" (*CD*, 146), which perpetuate ruthlessness and self-assertion, thereby creating a world based on harmony and trust. "Is progress possible to a point where *selfish* (passionate) tendencies are entirely renounced?" he asked, and answered: "Only if the centers of self ceased to exist as such, and, if separate individuals (atoms, etc.) were to come to the 'conviction' that it is better *not* to exist as separate beings. Unification of the universe at an ideal point" (*CD*, 153). In a climax to this line of inquiry, Ferenczi confronted the thought of Descartes and Malebranche on the nature of the passions in order to propose the ethical priority of the "not-I" or feeling of universality represented by the child's or wounded adult's capacity for selfless immersion in the other (*CD*, 150–54).[33]

All this bears on the problem of hypnosis and suggestion because throughout his discussion of traumatic imitation Ferenczi regarded the victim of trauma as precipitated into a special, highly suggestible state, defined as a condition of heightened submissiveness and mimetic dependency on the other. The neocathartic treatment inevitably revived and repeated the traumatico-mimetic origin. And hypnosis was understood by him, as it was by the majority of his contemporaries, as a state of altered consciousness akin to, though not identical with, that of sleep. Ferenczi thus conceived the traumatized person as shocked out of consciousness into a condition of trance-like incorporation or imitation of the violent other. In other words, traumatic mimesis was equivalent to hypnotic possession: the victim of trauma behaved like someone who, shattered and without ego defenses, had no other recourse than to become hysterically and hypnotically possessed by the aggressor (*CD*, 45–

33. The idealism of Ferenczi's philosophy of suffering appears to point once again to the influence of Schopenhauer's thought on the fundamental moral value of the "unegoistic," or instinct of self-sacrifice and pity. In Ferenczi's case that idealism, or utopianism, led to the same simple polarity of good and evil, aggressor and victim, that governed his account of lying, as if the mimetic or "kindness" principle (the "wise" child, or victim of trauma) could be completely purged of the violence that according to Freud necessarily inhered in the identificatory scenario, or as if the negativity and ambivalence constitutive of mimesis could be expelled by attributing it to another subject who is then conceptualized as assaulting the pure and innocent child in the form of a strictly external trauma. In the process, Ferenczi normalized the kindness principle by theorizing it as a principle of female masochism and abjection, just as modern theorists of multiple personality treat the female victim of trauma as a purely passive victim.

46). Moreover, if Ferenczi described the hypnotized victim as capable of beholding the traumatic scene, no ordinary spectating was involved. In his account, the traumatic operation gave the appearance of deploying forces that were capable of brilliantly perceiving how to proceed in order to cope with the trauma—by imitating the aggressor—but the perception involved was not that of a fully integrated or volitional ego. Rather, in line with a durable tradition that stressed the experience of coercion or involuntariness in hypnosis, Ferenczi represented the trauma victim as a sleep-walker (cf. Shakespeare's Lady Macbeth) who performs like a "complicated calculating machine."[34] In her calculated incorporations of the painful external world, the trauma victim thus reverted to that stage of development when, according to Ferenczi's metapsychological speculations, the child in an act of unconscious-hypnotic *reckoning* does not deny or repress unpleasant ideas but instead mimetically incorporates, or introjects, the painful stimuli.[35] The same radicalization of knowing and seeing—of a knowing and seeing at once *machine-like*, because of the automaticity in the absence of emotion and feeling, and *spectral*, because of the capacity to "see" beyond the normal limitations of time and space—was the basis for Ferenczi's equation between the trauma victim and the "medium" or clairvoyant who was able to telepathically read the mind of the aggressor (*CD*, 7–8, 33, 38, 55, 81, 117, 121, 140, 159, 206).[36] The victim's hypnotic-mimetic incorporation of the aggressor was thus simultaneously intelligent and coerced—a stunningly infallible yet compulsory act of "possession" by an aggressor who forced his prey to act against her own immediate interests by mimetically fusing with him in an unconscious, trance-like state of submission. In short, for Ferenczi the

34. Ferenczi, "Notes and Fragments," *Final Contributions to Psychoanalysis*, 230–31; hereafter abbreviated "NF." This might be compared with Georg Simmel's discussion of the city dweller who deals with the unpredictable shocks and disruptions of the modern metropolis with the calculated, rational, and indifferent or objective efficiency of the modern machine (Georg Simmel, "The Metropolis and Mental Life," in *The Sociology of Georg Simmel*, ed. Kurt Wolff [Glencoe, Illinois, 1950], 409–24). Like Ferenczi, Simmel argues that the calculating behavior of the metropolitan under conditions of modernity is due to both the success of defense against stimulation and the failure of defense. On Simmel's inconsistency in this and other regards see Fredric Jameson, "The Theoretical Hesitation: Benjamin's Sociological Predecessor," *Critical Inquiry* 25 (Winter 1999): 267–88.

35. Ferenczi, "The Problem of the Acceptance of Unpleasant Ideas—Advances in Knowledge of the Sense of Reality" (1926), *Further Contributions to Psychoanalysis*, 378–79. On Ferenczi's concept of reckoning, see also Martin Stanton, *Sándor Ferenczi: Reconsidering Active Intervention* (Northvale, New Jersey, 1991), 78–81.

36. Cf. "NF," 254, 256–57.

traumatic-hypnotic mimesis, defined by him as a primordial principle of submission in nature, had exactly the same blindly masochistic, irresistible, and demonic character (*CD*, 76, 139) that Freud associated with the death drive and the repetition compulsion.

It follows that for Ferenczi there was no "hidden observer" during the neocathartic repetition of the trauma, in the sense of a fully conscious observer who deliberately and consciously plays a game of compliance with the hypnotist. Nevertheless, Ferenczi could not escape a version of the same skepticism about the traumatic origin that informs Borch-Jacobsen's recent discussion. This was not just because Ferenczi's introduction of the theme of lying or simulation inevitably tended to cast doubt on the hysteric's testimony. It was also because the difficulty of recovering the memory of the trauma—either because it failed to break into consciousness during treatment or because after the hypnotic-cathartic reconstruction the patient still remained unconvinced about its status qua memory—aroused uncertainty about the reality of the traumatic event. As he confessed in his *Clinical Diary* (though not in print): "Patients tormented in this way start to have doubts about the analysis, accuse us of ignorance, of foolhardiness at their expense, of cruelty, of impotence; try to tear themselves free from us . . . [and] drive us to despair and to doubts about what we are doing" (*CD*, 137). His uncertainty was exacerbated by Freud's misgivings about his renewed emphasis on the cathartic method and the sexual-traumatic factor. Until the end Freud remained highly critical of Ferenczi's reconstructed traumas, seeing in the latter's work nothing but a repetition of his own earlier mistakes about sexual etiology. "He has completely regressed to etiological views I believed in, and gave up, 35 years ago: that the regular cause of neuroses is sexual traumas of childhood, said it in virtually the same words as I had used then," Freud observed to Anna Freud after a crucial last meeting with Ferenczi.[37] In a well-known statement Freud characterized the sexual memories of one of Ferenczi's patients as fairy tales,[38] and in the let-

37. Letter from Sigmund Freud to Anna Freud, 3 September 1932, cited by Peter Gay in *Freud. A Life for Our Time* (New York, 1988), 583–84.

38. Letter from Freud to Ernest Jones, 29 May 1933, *The Complete Correspondence of Sigmund Freud and Ernest Jones 1908–1939*, ed. R. Andrew Paskauskas (London, 1993), 721; Freud used the term *pseudologia phantastica* for the bizarre childhood memories of Ferenczi's patient Elizabeth Severn, a term that was defined at the time as "the invention of experiences that are just fairy tales" (Jeffrey Moussaieff Masson, *The Assault on Truth: Freud's Suppression of the Seduction Theory* [New York, 1984], 182). Severn's memories included the image of having been forced as a young girl to participate in the murder of a

ter to Eitington to which Borch-Jacobsen has drawn attention attributed Ferenczi's errors in this regard to the role of suggestion.[39]

For Borch-Jacobsen, as I have remarked, Freud's letter rebounds on Freud himself, whose early work on sexual seduction can thus be dismissed as a product of the same hypnotic-suggestive methods whose use he blamed for Ferenczi's mistakes. Indeed Borch-Jacobsen characterizes Freud's theory of fantasy as a cynical cover-up designed to "get out of the *cul de sac* he had gotten himself into" ("N," 42). Whatever the truth of such a charge (I return to this issue below), Ferenczi was disturbed by Freud's criticisms—so much so that his *Clinical Diary* and related posthumously published notes can be read as a massive struggle with the problem posed by the role of suggestion in psychoanalysis. On the one hand, Ferenczi reintroduced the question of the relationship between hypnosis and psychoanalysis by claiming that hypnosis was not external to the latter, as Freud wanted to think, but internal to its basic practices; according to Ferenczi the quasi-hallucinatory trance state induced by his relaxation methods was only a more extreme version of the condition of self-forgetful abstraction brought about by the method of free association. Ferenczi thus maintained that mimesis-suggestion was not the antagonist or "other" of psychoanalysis but rather its most intimate interiority; as such, it could not be dogmatically banished from the psychoanalytic project.[40]

At the same time, however, conceiving the technique of neocatharsis as an extension of the psychoanalytic method, Ferenczi wanted to use hypnosis as an objective, contamination-free method for converting neurotic repetition into self-representation and self-recollection. But we have seen that his neocathartic techniques did not appear to work: his patients failed to remember the traumatic events in question. Instead they experienced the (real or fictional) traumatic origin in the mode of an intensely animated miming of the traumatic event that took place in a state of self-forgetfulness that seemed to preclude the requirement of self-observation and self-knowledge. Like William Brown's shell-shocked soldiers, Ferenczi's patients appeared to collapse into such a state of mental confusion that self-representation ceased: they acted out the putative

---

black man, of the latter's genital mutilation, and of having been drugged, poisoned, and prostituted to other men.

39. In another letter to Ernest Jones, 12 September 1932, Freud mentions Ferenczi's unreceptiveness to warnings about his "technical errors," which I take to be an allusion to Ferenczi's cathartic-suggestive methods (*Correspondence of Freud and Jones*, 709).

40. Ferenczi, "The Principle of Relaxation and Neocatharsis," 119.

traumatic scene in a state of complete absence from the self, miming emotions which they could not represent to themselves or others but could only feel in the immediacy of an acting in the present.

But wasn't that the only way traumatic suffering *could* be experienced if, according to the arguments of James, Claparède, and Freud himself discussed in chapter 3, it was the case that the emotions could only be experienced in the immediacy of an acting out or mimetic identification in the present? Ferenczi's work on trauma thus appeared to confirm the relationship between affect, mimesis, and representation that had governed earlier interpretations of the cathartic cure: if the traumatic origin was conceived, not as a violence assaulting the pre-given and innocent subject in the form of an absolute exteriority, as Ferenczi sometimes proposed, but as a primary identification or hypnotic mimesis, as he also argued, then this would explain why traumatic suffering could not be remembered, indeed why it was "relived" in the transferential or hypnotic relationship in the mode of an affective-mimetic identification with another that was characterized by a profound amnesia or absence from the self.

But we have seen that this is a conclusion that Freud and others also resisted, as Borch-Jacobsen does today. For Freud, the merger between self and other that appeared to him to be constitutive of the hypnotic-mimetic state gave way to the requirement that the subject of mimesis *be a subject*, albeit an *unconscious* subject, to whom the scene of identification could be re-presented and therefore recalled. On this principle, if there was suggestion in analysis it did not emanate from the rapport between analyst and patient but from the subject herself in the form of an auto- or self-hypnosis. That was the gist of Freud's claim that Ferenczi's patients suggested to Ferenczi the sexual traumas in whose existence he at first had so naively believed. Mimesis, defined as a blind identification or immersion in the other to the point of forgetfulness of self, was thus denied by way of an antimimetic assertion of distance and difference. As Borch-Jacobsen's *The Freudian Subject* superbly demonstrated, the psychoanalytic project of cure by representation (that is, cure by recollection) depended on that denial, even as Freud also continued to acknowledge the very identity between hypnosis and the transference that made the project of recollection—the problem of the dissolution of the transference—particularly intractable.

The same oscillation between notions of mimesis and antimimesis also governs Borch-Jacobsen's approach to hypnosis. In *The Freudian Subject* and related texts, Borch-Jacobsen theorized hypnosis as a nonspecular mimesis that pervaded the psychoanalytic project. Accepting the classical definition of hypnosis as an immersive identificatory experience, he argued that psychoanalysis was so deeply implicated in the mimetic paradigm as to leave no way out: mimesis emerged in his account as the unmasterable essence of subjectivity, the traumatic origin, and the transferential rapport. As such, it was unrecollectable and untellable. "Is there an end to analysis, after all? Is analysis 'terminable'? Can hypnotic transference be 'dissolved'?" Borch-Jacobsen asked. "Or is it instead the part of analysis that cannot be analyzed, of which the patient cannot be cured, any more than of repetition? . . . In the end, is psychoanalysis, to use Roustang's expression, 'long-drawn-out suggestion'? There is certainly no way to decide on the basis of the facts."[41] Today he thinks otherwise. Rejecting his former position, he describes hypnosis as a specular "game" carried out with the subject's full awareness of her compliant performance. In other words, Borch-Jacobsen now attempts to *resolve* the enigma of hypnosis by deciding in favor of lucid imitation.

This means that for Borch-Jacobsen, too, there is always a subject—albeit, this time, a fully conscious subject—who can master suggestion and put it in its proper place. "The mystery of hypnotic induction disappears the moment one understands that suggestibility is not automatism and that the submission to suggestion is in fact a very voluntary servitude, revocable at any moment," he writes. "In the end, there is no hypnosis, only an auto-hypnosis, or a consent to hypnosis" ("L'Effet Bernheim," 152–53). Precisely because the subject is aware at all times of her trance behavior there is no loss of consciousness in the hypnotic performance and hence no posthypnotic forgetting (indeed, on this account there is no unconscious, only a classical dualism between the conscious mind and the physiological body). For Borch-Jacobsen, all the examples of hysterical and posthypnotic amnesia widely documented in the historical record are simply artifacts, the product of false ideas about the nature of hypnosis on the part of both operator and client. In hypnosis there is no genuine forgetting of the trance enactment, which is merely a suggested scenario undertaken with the patient's voluntary compliance. But there is also no genuine forgetting of the prior traumatic origin, because there is no trauma to forget.

Borch-Jacobsen's antimimetic turn thus involves a radical critique not only of Freud and the concept of the unconscious, his primary target, but also of Ferenczi's entire approach to trauma. I want to make clear that although I am myself not in agreement with Borch-Jacobsen's views, I do not wish to argue that he is simply mistaken, and the opposite

41. Borch-Jacobsen, *The Emotional Tie: Psychoanalysis, Mimesis, and Affect* (Stanford, California, 1992), 57.

of what he now believes to be the truth about hypnosis is in fact the case.[42] Rather, I want to claim that with respect to the episodes and cruxes treated in this book the tension or oscillation between mimesis and antimimesis I have been tracking cannot definitively be resolved by choosing one over the other. Or to put this slightly differently, if one does seek to resolve the problem in that way, the nature of the historical material is such as to raise problems for the consistency of one's position. So for example it would not be difficult to show that Borch-Jacobsen's arguments, for all their forcefulness, are marked by simplifications that beg important questions and by internal contradictions that tend to undermine the coherence of his views.[43] I might state once again that my

42. There are passages in Ferenczi's writings that could be used to support Borch-Jacobsen's position. For example, some of Ferenczi's patients, especially Elizabeth Severn, made suggestions to Ferenczi as to how he ought to conduct the treatment, even to the extent of telling him what questions to ask while in the trance state (Ferenczi, "Child Analysis in the Analysis of Adults," 130; CD, 12). For Ferenczi, such suggestions were evidence of an unconscious intelligence at work in the hypnotic state; Borch-Jacobsen would no doubt take them as evidence that Ferenczi's patients were consciously colluding in the hypnotic "game." Moreover, the florid symptoms observed by Ferenczi often made their first appearance during the treatment itself (Ferenczi, "The Principle of Relaxation and Neocatharsis," 118; idem, "Child Analysis in the Analysis of Adults," 141). Again, Ferenczi and Borch-Jacobsen provide different explanations for this phenomenon. For Ferenczi, the symptoms appear because the neocathartic treatment encourages the expression of hitherto hidden affective experiences for the first time; Borch-Jacobsen would regard the symptoms as simply a product of the hypnotic game.

43. This is not the place for a detailed assessment of Borch-Jacobsen's recent writings, but the following points will at least indicate the direction that such an assessment might take:

1. There is an ambiguity in Borch-Jacobsen's arguments regarding the suggestive scenario. Precisely because there must be a subject, a conscious subject, who is responsible for the suggestive illusion that is hysteria, he oscillates between imagining that illusion as the product of a fully conscious collusion or complicitous game between Freud and his patients, and conceptualizing it is an artifact of Freud's suggestive extortions ("N," 16, 41) or impositions ("N," 38), thereby reducing the patient to the sort of passive automaton that his theory of the subject as conscious agent is designed to forestall.

2. A related question concerns Borch-Jacobsen's attitude towards the problem of testimony. His paper, "Neurotica," is staged as a problem concerning testimony: How can we decide whether a child is telling the truth? Even though he does not wish to equate the problem of simulation with the problem of lying, Borch-Jacobsen (like Ferenczi) approaches the issue in terms of the classical opposition between truth and lies. Thus the patient's testimony is either the source of truth, as when "Anna O.'s" claims about her state of consciousness during the trance are taken by him at face

own approach is to try to do justice to the intractable complexity or doubleness of the historical material with respect to the issues at hand. This has meant, and will continue to mean, acknowledging as fully as possible the extent to which the historical material may seem to *invite* resolution by choosing one account of mimesis over the other, while remaining as

---

value, or it is irredeemably tainted by Freud's suggestion, in which case it amounts to a lie inaugurated by Freud. In short, for Borch-Jacobsen human testimony is simultaneously something we can and must rely on (because human beings tell the truth) and something we cannot rely on (because humans are capable of lying).

3. Borch-Jacobsen places all hypnotic phenomena in the domain of the "artifactual." According to him, we must stop thinking that there is a "psychic reality" that we could study from outside; rather, we must understand that we are always already in the process of manipulating reality and that, above all, there is no error involved. He cites Delboeuf as saying: "'The existence of several schools of hypnotism is thus entirely natural and easily explained. They owe their origin to the reciprocal action of those who are hypnotized on the hypnotist. Only their rivalry has no reason for being: they are all true [elles sont toutes dans le vrai]'" ("L'Effet Bernheim," 168). But this means that the "hidden observer" phenomenon is also artifactual and that Borch-Jacobsen therefore cannot appeal to it for the empirical truth about the hypnotic situation, as he wants to do. In a recent interview Borch-Jacobsen acknowledges this possibility when he volunteers that "I need . . . to preface what I have been saying in a Delboeufian way, and admit that I studied a form of hypnosis propounded by Milton Erickson. Ericksonian hypnosis is based precisely on the idea that hypnosis is not a well-defined state, and that we achieve hypnotic effects in the waking state. Eriksonians are not interested in promoting a deep state of hypnosis. With this in mind, I admit that my views about the unconscious may derive from my own theoretical bias, which [is] based on this particular experience with hypnosis" ("Analysing Freud: An Interview with Mikkel Borch-Jacobsen," *Culture-Art-News* 1 [July 1998]: 10; my thanks to Todd Dufresne, who conducted the interview, for sending me this exchange).

Indeed, there are those who regard the "hidden observer" phenomenon as a function of the demand characteristics of hypnotic experiments. Of special interest in this connection is the use of hypnotic procedures in which witnesses in criminal cases are asked to visualize scenes by "zooming" in on them as if they were a TV camera. This is a technique that yields confabulation and false memories with considerable regularity, yet the testimony is accompanied by a strong sense of subjective conviction as to its veracity on the part of the hypnotized subject and is rarely questioned by the police. For a recent discussion of the contradictory data on the "hidden observer" phenomenon see Campbell W. Perry et al., "Hypnotic Age Regression Techniques in the Elicitation of Memories: Applied Uses and Abuses," in *Hypnosis and Memory*, ed. Helen M. Pettinati (New York, 1988), 128–54.

4. Moreover, if Borch-Jacobsen sometimes appears to imply that we should stop play-

alert as possible to those elements in a given text or situation that resist resolution in those terms and indeed point in exactly the opposite direction. Such an approach is not devoid of anxiety; on the contrary, it almost cultivates it, by which I mean that it underscores the constitutive instability of the mimetic *topos*. But that seems to me preferable to all attempts to quell anxiety by simplifying the material, however invigorating the result.

Fear of Suggestion in Analysis

Ferenczi's late work is played out in the same zone of oscillation and anxiety, and it is not surprising that his thought exhibits tensions and contradictions similar in kind to those found in the work of Freud and Borch-Jacobsen. Throughout Ferenczi's *Clinical Diary* suggestion-hypnosis operates as a switch point between an originary theory of trauma, which imagines the subject as constituted by an unmasterable, immemorial hypnotic identification or mimetic rapport between self and other, and a postoriginary theory of trauma which imagines that the traumatic imitation occurs on the basis of an already existing ego or subject—though not perhaps the fully conscious subject of Borch-Jacobsen's recent analysis. The result is a series of contradictory claims about the function of suggestion in psychoanalysis that cannot be resolved—claims that serve at once to assert and to deny its centrality to the psychoanalytic cure and which, most poignantly, conduce to skeptical conclusions about the traumatic origin that Ferenczi strongly wishes to forestall.

When Ferenczi first reintroduced catharsis into psychoanalysis, he proposed that, if suggestion was indeed internal to the technique of psy-

---

ing the "sterile and desperate game" of hypnosis (*R*, 92), he also believes that we cannot, for suggestion cannot be contained, its limits and boundaries cannot be established with certainty, it overflows. As he concedes, one cannot escape the interaction between hypnotist and subject (which is to say, the influence of one person on another) ("L'Effet Bernheim," 167). The most one can do is to be as explicit as possible about the "artifacts" one is creating ("L'Effet Bernheim," 168). In other words, he simultaneously admits the inescapability of suggestion and implies that it ought to be controlled and mastered. In fact, Borch-Jacobsen himself cannot do without the very concept of the unconscious that he otherwise wishes to discard. For example, the "experimenter expectancy effect" to which he appeals in his discussion of hypnosis ("L'Effet Bernheim," 165), depends on a notion of unconscious communication between subject and experimenter; similarly, the notion of a "culture" of hysteria and hypnosis depends on some notion of unconscious influence, and so on.

choanalysis itself, as appeared to be the case, it emanated from the patient in the form of an auto- or self-suggestion. Unlike the older suggestive techniques, which directly suggested contents and in any case produced only temporary effects, neocatharsis involved a self-induced trance state in which the truth of the traumatic origin was revealed by the halluci-nated images, perceptions, and other symptoms spontaneously produced by the hypnotized subject from her own, authentic material. This was an argument designed to refute the charge of fabrication by assigning the contents of the hypnotic enactments to a fully constituted or postorigi-nary subject. What was important, he wrote in a paper of 1931, was that one should not abuse this phase of helplessness, by urging upon the sub-ject's unresisting mind one's own theories and fantasies: "On the con-trary, we ought to use our undeniably great influence here to deepen the patient's capacity for producing his own material. Putting it in a some-what inelegant way, we might say that in analysis it is not legitimate to suggest or hypnotize things *into* the patient, but it is not only right but ad-visable to suggest them *out.*" But this was to imagine that during the neo-catharsis the patient was simultaneously the author of her own material *and* was reduced to a state of extreme automaticity and suggestibility from which she must be rescued by being "encouraged" to be autono-mous through the analysts's "influence," which is to say, through sugges-tion. And indeed Ferenczi in the same paper of 1931 *also* treated hypnosis as emanating not from the patient but from the hypnotist himself: "Here we get light, of some significance for education, on the course which we ought also to follow in the rational upbringing of children. Their sug-gestibility and their tendency, when they feel themselves helpless, to lean, without any resistance, on a 'grown-up' (that is to say an element of hyp-notism in the relation between children and adults) is an undeniable fact, with which we have to reckon. But instead of doing what is commonly done, that is to say, going on using the great power which grown-ups have over children to stamp upon their plastic minds our own rigid rules as something externally imprinted, we might fashion that power into a means of educating them to greater independence and courage."[44]

In line with the work of William Brown and other practitioners of catharsis, the suggestibility of the victim was thus to be cured by sugges-tion: the bad suggestion of the trauma was to be countermanded, in an act of reparation, by the physician-hypnotist "dehypnotizing" the patient (*CD*, 63)—that is, by the authoritative use of countersuggestion. By strengthening the ego in this way, Ferenczi hoped that traumatic-

44. Ferenczi, "Child-Analysis in the Analysis of Adults," 134; cf.130.

mimetic repetition could be converted into conscious representation: "Help through suggestion, when energy flags: shaking up, encouraging words. Thereupon a sensation of increased strength or decreased weakness of the alloplastic 'capacity for thought and action.' . . . *Recollection* possible only if a sufficiently consolidated ego (integrated, or one that has become so) *resists* external influences" (*CD*, 182). But was recollection achievable by these means? As we have seen, the evidence of the *Clinical Diary* indicated otherwise. Was this because the victim was constitutively unable to recollect the trauma, as Ferenczi's theoretical analysis of traumatic splitting proposed, or was it rather because she was simply simulating or playing a hypnotic-mimetic role, as Borch-Jacobsen's later views imply? Moreover, how could Ferenczi be sure that the process of "gluing" the ego back together again did not degenerate into the direct suggestion of the old mesmeric kind? In short, what guaranteed the historical truth of the analytic-suggestive construction?

The *Clinical Diary* and related posthumously published notes testify to Ferenczi's continued struggle with such questions. On the one hand, he repeatedly presented the traumatic condition as a state of increased suggestibility and loss of will that could only be cured by countersuggestion. "We must stop despising suggestion when faced with the needs of purely infantile neurotics," he wrote. "Kindness alone would not help much either, but only both together" (*CD*, 207). On the other hand, if the traumatic condition was to be cured by "encouragement" or suggestion in this way, then psychoanalysis risked becoming a frankly suggestive technology along the lines of Janet's hypnotic therapy—a therapy in which, as I argue in chapter 3, recollection of the past was subordinated to the general hypnotic-curative task of strengthening the patient's autonomy to the point where the very opposition between remembering and forgetting began to collapse.[45] But Ferenczi also resisted assimilating psychoanalysis to hypnotic-suggestive practices along these lines. He wanted to refute skepticism about the reality of trauma by discovering the truth of the traumatic event, even as he struggled with the extreme difficulty of doing so once it was admitted that suggestion was internal to the practice of psychoanalysis.

At one point he proposed, in an argument reminiscent of Freud's

---

45. Although Ferenczi never mentioned Janet, he described the therapeutic process in terms that recall the latter, as when, reflecting on the function of memory, he observed that instead of an unceasing digging into the past for memories that might never be discovered the patient must be "encouraged" to face the future, thereby implying the same need to forget that Janet sometimes advocated ("NF," 261).

similar effort to resolve the problem of suggestion, that the very absence of memory and conviction on the patient's side testified to the probable truth of the reconstructed infantile experience ("NF," 259–60). But in other passages Ferenczi appeared to recognize that the attempt to pin down the origin was doomed to fail. We have seen that his hypothesis of an ego-memory system which "remembered" the trauma only in auto-plastic or corporeal terms made a certain kind of forgetting inherent to the traumatic experience; according to that hypothesis, trauma could not be cured by conscious recollection but only by interpretive reconstruc-tion and "encouragement"—that is, by suggestion. Yet Ferenczi also wanted to resist the suspicion of origins to which his argument tended to lead. Armed against the charge of fabrication by the claim (or hope) that the hypnotist's influence was limited to a general "encouragement" and that accordingly the patient's fantasies or thoughts were spontaneous and genuine, Ferenczi tried to avoid the extremes of blind credulity and ab-solute skepticism. In an entry in which he explicitly weighed the extremes of credulity and skepticism, he rejected the naively credulous view that the cathartic performances directly represented the traumatic cause ("NF," 263). But he also refused the skeptical position of the kind now represented by Borch-Jacobsen, according to which the patient's trance thoughts and images depicted nothing real at all, or were so distorted as to make the search for origins hopeless.[46] He concluded:

> In fact, *there is* in the end something that cannot, need not, and must not, be in-terpreted—or else analysis becomes an endless substitution of emotions and ideas mostly by their opposites.
>
> On the other hand: the "mental events" of the past (childhood) may have left their memory traces in the language of gestures, un-understandable to our cs (i.e. in the body) as organic-physical 'mnems'; a pcs [preconscious] perhaps did not exist at all, only feeling (pleasure-unpleasure) reactions in the body (subjective memory traces)—so that only *fragments* of the external (traumatic) events can be reproduced. (Perhaps only the first moments of the trauma which could not yet be "repressed" (displaced into the organic) because of the element of surprise (lack, or delay, of countercathexis). If this is so, then some memories of childhood cannot be ever raised as such into the cs, and even in

---

46. "'Psychognostic,' '*Gnosis*' = the view that it is possible to reach by corresponding deep relaxation to the direct experience of the past, which then may be accepted as true without any further interpretation. *Scepticism:* The idea that all thoughts and ideas must be examined first very critically and that they represent: (1) Absolutely nothing of the real happening, or (2) A very distorted version of it ('Telescoping,' Frink)" ("NF," 263).

physical symptoms and hallucinations are always mixed with dreamlike (wish-fulfilling) distortions of defence and turning into the opposite, e.g. as regressions (hallucinations of the moments preceding the trauma). ("NF," 263–64)

On the one hand "there is in the end something that cannot, need not, and must not, be interpreted" (but what is this something, and how is it got at, and what is its relation to the psychoanalytic project?). On the other hand, the insistence on the fragmentariness and distortedness of the traumatic repetition may seem to require, and in that sense to justify, psychoanalytic interpretation, but the extremity of the situation he describes may also be read as presenting a severe challenge to interpretation (to put it mildly). Elsewhere Ferenczi put matters more hopefully: "The dream may therefore be interpreted historically (partly distorted in the direction of *wish-fulfillment*)" ("NF," 266). Some such belief sustained Ferenczi's hope that analysis was not interminable. But it was also a belief that left a nagging doubt which at the end of his life he urgently attempted to defeat with increasingly strained and contradictory arguments. The origin *must* be pinned down somewhere and analysis brought to a halt by the revelation of the traumatic event. But because that appeared all but impossible to accomplish, new rationales had to be devised to secure the certainty his theoretical ideas and therapeutic practices had called into question.

One sign of his growing desperation was the argument he now made to the effect that the very theory of fantasy so central to psychoanalysis was itself a means of refusing or resisting knowledge of the reality of trauma. It is as if Ferenczi thought that analysts (including until recently himself) had traumatically sacrificed their mental integrity on this issue as a result of the very same mechanism of blind submission to or identification with the parental figure (Freud) that protected the traumatized victim from knowledge of the origin. "*Knowledge as a means of doubt* (resistance)," Ferenczi observed. "Trauma having been *told* and not *found out*. Traumatogenesis being *known*; the doubt, whether reality or fantasy, remains or can return (even though everything points to reality). Fantasy-theory = an escape of *realization* (amongst resisting analysts too). They rather accept (and human beings') mind (memory) as unreliable than believe that such things with *those* kinds of persons can *really* have happened. (Self-sacrifice of one's own mind's *integrity* in order to save the parents!)" ("NF," 268).

As a "[c]ure for knowledge-incredulity" ("NF," 268) Ferenczi, in this deeply divided statement in which even the question of who is speaking and to whom is highly unstable, simultaneously and contradictorily demanded belief *and* suspension of belief in the statement of others as the basis for certainty as to the truth of the trauma. In the process he implied, against the evidence of his own findings, that the images, thoughts, or fantasies experienced in the hypnotic-suggestive catharsis could in time become so emotionally charged as to induce the very conviction whose absence had produced the crisis of knowledge he was trying to redress. As if he wished to persuade not only his patients but also himself of the reality of the reconstructed trauma, he imagined himself saying: "'You must not *believe*, you just tell things as they come. Do not force feelings of any kind, least of all the feeling of conviction. You have time to judge things from the reality point of view afterwards.' (In fact, the series of pure images sooner or later turns into highly emotional representations.) 'You have to admit that (exceptionally) even things can have happened of which somebody told you something'" ("NF," 268). But such an argument, based on the acceptance of the statements of others, raised the specter of suggestion once again, which is perhaps why a few days later Ferenczi attempted to secure the historical truth of the trauma by proposing, not that suggestion threatened to waylay the truth, but rather that it was the truth's crucial guardian—for suggestion could suggest only the truth. There was no longer any need for a "Socratic art of suggestion" (as a few entries earlier he had proposed) ("NF," 260), for the truth of the origin or trauma was guaranteed: "Only what is true may be suggested (to children and patients)," he wrote. But "truth cannot be discovered quite spontaneously, it *must* be 'insinuated,' 'suggested.' Children are unable without this help to acquire convictions" ("NF," 269). But that dependence on suggestion left a remainder in the form of a continued attempt to resolve a threatening skepticism. That is why Ferenczi hoped to find in graphology and palm-reading alternative sources of information about the origin of a trauma ("NF," 274). That is why also the question of the termination of the analysis—whether it could be terminated by his neocathartic techniques by helping the patient to separate fantasy from reality and thus establish the traumatic origin—remained unanswered to the end.

From this perspective, of all the statements in Ferenczi's *Clinical Diary* none is more poignant than his admission in the very last entry, only months before his death, that the reality of Thompson's trauma *was still in doubt*. To return to where I began my discussion of Ferenczi's ideas: At the beginning of his *Clinical Diary* Ferenczi reported that Thompson had been grossly sexually abused by her father. But at the close he admitted: "Dm. [Clara Thompson]: Made herself independent—feels hurt because of the absence of mutuality on my part. At the same time, she

becomes convinced that she has overestimated father's (and my) importance. Everything comes from the mother" (CD, 213–14). Doubtful
about the validity of the traumatic recollections ostensibly recovered by
Ferenczi's suggestive techniques, Thompson downplayed the importance of paternal sexual abuse and blamed her mother instead.[47]

47. For Thompson, developing certain themes in Ferenczi's work about the mother's
complicity with the father's abuse (CD, 79, 83, 132–33), the mother's fault was above all her
lack of love and her hypocrisy about the child's developing sexuality.

VI

Splinting the Mind: William Sargant
and Catharsis in World War II

In June 1940 the British psychiatrist William Sargant (1907–1988), who had been assigned to work at an emergency hospital located in the out-skirts of London, confronted cases of acute war neuroses from the retreat from Dunkirk. He later recalled the arrival of soldiers in their tin hats and filthy uniforms, some of them wounded, many in states of total and abject neurotic collapse. What the papers termed a great British achievement, he later remembered, "seemed to us at the time nothing better than a de-feated and defeatist rout." Many of the men were suffering from acute hysteria, reactive depression, functional loss of memory or the use of their limbs, and a variety of other psychiatric symptoms "that one would never see in such abundance except during a war—unless perhaps after an earthquake or railway accident when the most normal people are apt to break down."[1]

Sargant believed that the traumatic neuroses were the result of an in-teraction between the individual's inborn (hysterical) constitution and external circumstances. But as the statement just cited indicates, accord-ing to him under sufficiently stressful or extreme conditions not even the most robust personality was immune to breakdown. This was not the view of the British government, whose Ministry of Pensions' Committee in 1939, in anticipation of hostilities, had adopted the position that the war neuroses only appeared in persons constitutionally disposed to them;

---

1. William Sargant, *The Unquiet Mind: The Autobiography of a Physician in Psychological Medicine* (London, 1967), 86–87; hereafter abbreviated *UM*. For information about Sar-gant's life and career, I thank Ann Dally for allowing me to read her unpublished paper, "Sources for the Life and Work of William Sargant (1907–88)."

accordingly, the Committee had stated as official policy that there would be no pensions for any soldiers suffering from psychoneurotic illness during the war.[2] But when Sargant encountered psychiatric casualties from Dunkirk he recognized in their symptoms the same clinical features of the traumatic neuroses whose prototype was the railway neurosis and that had already been described as shell shock in World War I. Perhaps with the official policy towards war pensions in mind, Sargant in his first paper on the war neuroses observed that such cases proved that "men of reasonably sound personality may break down if the strain is severe enough."[3]

It was while he was using the barbiturate drug, sodium amytal, intravenously for the purposes of sedating Dunkirk casualties that Sargant witnessed abreaction, or catharsis, for the first time. He recalled that one of his Dunkirk cases "accidentally" initiated the treatment. The soldier had arrived at the hospital hysterically mute, hands shaking, and with a nervous paralysis of the bladder. To alleviate his state of terror, Sargant gave him an injection of the sedative. The effect "was startling," Sargant recollected. "His bladder suddenly emptied, his speech returned, his hands stopped trembling, and he became intelligent, articulate and comparatively normal at least until the effects of the injection wore off" (UM, 87). He gave injections of the same drug to other hysterical cases, and again they worked. But he also observed "strange side-effects": a soldier might "suddenly recover the suppressed memories of the gruesome experiences that had caused or hastened his breakdown, and relive them before us." After the discharge of pent-up emotions "soldiers would suddenly improve" (UM, 88). In his autobiography Sargant described an especially dramatic case that "greatly disturbed" him. The patient suffered from gross bodily tremors, total paralysis of the right hand, and an almost complete loss of recent memory. An injection of sodium amytal cured the tremor and restored both the use of his hand and his lost memory, "but only after a frightening emotional release. He described, with dramatic gestures, how during the retreat, he had come across his own brother ly-

2. See Ben Shephard, "'Pitiless Psychology': The Role of Prevention in British Military Psychiatry in the Second World War," *History of Psychiatry*, forthcoming; my thanks to the author for letting me read his valuable paper prior to publication.

3. William Sargant and Eliot Slater, "Acute War Neurosis," *Lancet* 1 (July 6, 1940): 1. Sargant felt that the Pensions Committee's position tended to equate long-term combat disability with malingering. But when the Committee reversed its position in 1944, Sargant claimed that the pendulum had swung back too far in the other direction (UM, 93–95).

ing by the roadside with a severe abdominal wound. At his brother's earnest plea he had dragged him into a field and put him out of his misery with a rifle shot. It was the hand that pulled the trigger that had suddenly become paralysed. After his confession of grief and guilt, this hand worked again" (*UM*, 88).

Using drugs rather than hypnosis to cure the soldier, Sargant—already in 1940 an outspoken critic of psychoanalysis and an advocate instead of physical methods of treatment in psychiatry—rediscovered the method of cathartic abreaction that Breuer and Freud had first introduced in the 1890s to treat hysteria and that during World War I had been revived by William Brown and others to treat the shell-shocked patient. Like the female hysteric and the war neurotic of World War I, the Dunkirk casualty appeared to suffer "mainly from reminiscences"—that is, from the effects of dissociated or repressed experiences that he was unable to dispel. Like them, he was encouraged to abreact or discharge the pent-up experiences held to be at the origin of his neurosis, thereby causing his symptoms to disappear.

Sargant subsequently acknowledged the similarities between his abreactive methods and the cathartic techniques of Breuer and Freud.[4] He also recognized the investigations of several other predecessors who had used drug therapy in various psychiatric contexts during and after World War I, including the work of the British psychoanalyst, Victor Horsley, who in the 1930s had invented "narcoanalysis" by substituting barbiturate abreaction for hypnotic catharsis in certain cases of neurosis (*UM*, 88). In fact, there was in Britain considerably more continuity between the psychiatric experiences of World War I and those of World War II than has generally been realized. The situation was different in the United States where the lessons of the Great War had been largely lost sight of. But in Britain, thanks to the activities of the Ministry of Pensions' Committee major issues concerning diagnosis, treatment, the influence of constitutional factors, and the role of compensation were aired by those who had knowledge of the previous war.[5]

Moreover, the traumatic syndrome was well defined at a descriptive-clinical level, even if nomenclature varied and theories ranged widely. The Pension Committee and the army wanted above all to avoid the

4. William Sargant and H. J. Shorvon, "Acute War Neurosis: Special Reference to Pavlov's Experimental Observations and the Mechanism of Abreaction," *Archives of Neurology and Psychiatry* 54 (1945): 236, hereafter abbreviated "AWN"; H. J. Shorvon and William Sargant, "Excitatory Abreaction: With Special Reference to Its Mechanism and the Use of Ether," *Journal of Mental Science* 93 (1947): 709, hereafter abbreviated "EA."

5. On this point see Shephard, "'Pitiless Psychology.'"

problem of chronicity and the associated costs of pensions for war-induced neuroses. Abram Kardiner's *The Traumatic Neuroses of War* (1941) was important in this regard, not only because it was the first systematic account of the symptoms and psychodynamics of the war neuroses, but more specifically because it explicitly dealt with chronic cases, albeit those from World War I. In that book, today routinely cited as a landmark in the history of posttraumatic disorders,[6] Kardiner continued the dismantling of Freud's libido and repression theory that, as I show in chapter 4, he had begun in 1932 under the influence of Ferenczi's ideas about trauma. Rejecting as inadequate his previous effort at reformulation,[7] he now completely repudiated the value of Freud's libido or instinct theory in favor of a functional-operational emphasis on the traumatized organism's failure of adaptation to the external world.[8] Conceptualizing the traumatic reaction as a response to the breaching of the mind's hypothetical "protective shield" against external stimuli, Kardiner emphasized the production of permanent alterations in the victim's ego consequent on the loss of mastery, especially in cases that had been allowed to become chronic through lack of early treatment or the secondary gain of compensation (*TNW*, 233–39).[9]

Kardiner did not abandon his previous claim that the failure of the

6. Kardiner's text is the source of the symptom list for post-traumatic stress disorder in current psychiatric nosology (Allan Young, *The Harmony of Illusions: Inventing Post-Traumatic Stress Disorder* [Princeton, New Jersey, 1995], 89).

7. Abram Kardiner, *My Analysis with Freud: Reminiscences* (New York, 1977), 111.

8. Abram Kardiner, *The Traumatic Neuroses of War* (Menasha, Wisconsin, 1941), vii, 6, 137–41; hereafter abbreviated *TNW*. In 1947 Kardiner revised this book in collaboration with Herbert Spiegel, who during World War II had treated men at the front and who added recent cases (Abram Kardiner and Herbert Spiegel, *War Stress and Neurotic Illness* [New York, 1947]).

9. See Young, *The Harmony of Illusions*, 90. In his book Kardiner replaced the Freudian concept of "instinct" (or drive) with the concept of "action syndrome," language reminiscent of the functional psychiatry of the psychiatrist Adolf Meyer, with whom Kardiner discussed his manuscript (vii). Kardiner's change of orientation also reflected the influence of anthropological studies he had undertaken during the interwar years that focused on problems of adaptation to "stress" in primitive cultures. In *The Individual and His Society: The Psychodynamics of Primitive Social Organization* (New York, 1939), he compared "voodoo" deaths caused by fright to the traumatic neurosis of war. The implications of Kardiner's functional-sociological effort to convert the political problem of racial oppression into a problem of psychological wound or trauma are discussed in Michael Rogin's important essay, "'Democracy and Burnt Cork': The End of Blackface, the Beginning of Civil Rights," *Representations* 20 (Spring 1994): 1–34.

traumatized ego's mastery involved a kind of fascinated identification or hypnotic imitation with the aggressor, on the model of the child's emotional ties to the mother; nor did he give up a psychodynamic emphasis on conflict and related notions. But those ideas were now in the service of an adaptational approach that posited a stark opposition between the already constituted ego and the external trauma, an opposition that tended to elide or occlude the mimetic-hypnotic dynamic. Moreover, he was now inclined to biologize the traumatic response. The term "physioneurosis," which he introduced for the stabilized, or chronic, core of the traumatic syndrome, helped emphasize the somatic character of the traumatic reaction (*TNW*, 195). Those conceptual changes were accompanied by a related tendency to further accentuate the fixity and asymbolic nature of the traumatic response. Characterizing repetitive traumatic dreams as forms of memory disturbance typical of traumatic neurosis, Kardiner described such dreams in ways that suggested they were almost cinematic replays of the traumatic origin, devoid of fantasy or symbolic meaning (*TNW*, 88–95), a point to which I return.

What of Kardiner's therapeutic observations in *The Traumatic Neuroses of War*? Perhaps most striking, given the widespread use of abreaction during World War II, was his negative assessment of hypnotic abreaction as a therapeutic tool. Kardiner recognized the efficacy of abreaction in acute cases; but he was critical of its value in chronic conditions where structural changes had almost irretrievably calcified the ego (*TNW*, 216–17). Moreover, he argued that even in acute cases no permanent benefits could be expected from abreaction alone, without the analytic measures necessary to give the patient insight into the relations between the trauma and his defensive processes. In short, for Kardiner *reeducation* was the basic goal of therapy; he argued that even the recovery from amnesia ought to be subordinated to the aim of achieving a fundamental alteration in the patient's conscious adaptation to the outer world (*TNW*, 217–19).

Kardiner in 1941 thus raised the same question that had troubled the entire history of catharsis since its invention by Breuer and Freud: How does abreaction cure? (If indeed it does cure; I come back to this.) As shown in chapter 3, in the wake of World War I that question had been posed in the following way: Did abreaction cure the patient primarily because of the emotional discharge brought about by the hypnotic state, as Breuer and Freud seemed to suggest and William Brown also proposed? Or did the relief of symptoms depend not so much on the affective catharsis as on the cognitive dimension of the cure—that is, on the conscious resynthesis and reintegration of the dissociated or repressed memory of the traumatic experience, as William McDougall and C. S. Myers had argued? As I have observed, it is as if two competing models of the role of the patient in treatment opposed one another in the debate over abreaction. One model, which I call the participatory model, imagined that the collaboration of the patient was an inseparable part of the cure. The other model, which I call the surgical model, imagined that, as in the case of drug therapy or surgery, the psychic collaboration of the patient was irrelevant to treatment.

Abreaction: Psychic Integration or Emotional Discharge?

Roy Grinker's and John Spiegel's psychoanalytically oriented abreaction method may be taken as exemplifying the participatory model of cure during the Second World War. As we learn from their influential book, *War Neuroses in North Africa: The Tunisian Campaign* (1943) (republished in 1945 with the title *War Neuroses*), in a method apparently inspired in part by Sargant, the authors used barbiturates to obtain very emotional reenactments of the traumatic scene.[10] They stated that in difficult cases the therapist must himself enter into the hypnotic performance, imitating various roles in order to help the narcotized patient reexperience the traumatic event in its original intensity. In ways that recall the earlier methods of Prince, Brown, Ferenczi and others, the psychiatrist had to join in the hypnotic-abreactive scene. Nevertheless, for Grinker and Spiegel the abreactive reproduction or reliving was only the beginning of the therapeutic process. What was crucial was that the patient should be able to retain and integrate the memory of the reenacted event after the effects of the drug had lifted. Grinker and Spiegel substituted the term "narcosynthesis" for Horsley's earlier term "narcoanalysis" in order to stress the centrality of mental resynthesis or insight in their conception of the cure (*WN*, 157–58).

The authors repeatedly emphasized that psychic reintegration was not

10. Roy Grinker and John Spiegel, *War Neuroses in North Africa: The Tunisian Campaign (January–May 1943)* (Washington, D.C.,1943); hereafter abbreviated *WN*. For Sargant's account of the prehistory of drug abreaction and his claims to priority see Shorvon and Sargant, "Excitatory Abreaction," 709–10; and Sargant, "Eight Years of Psychiatric Work in England," *Journal of Nervous and Mental Disease* 107 (1948): 505; idem, "Some Observations on Abreaction with Drugs," *Digest of Neurology and Psychiatry* 16 (1948): 193–94; idem, *The Unquiet Mind*, 104–5. John Spiegel later acknowledged a debt to Sargant (John Spiegel, personal communication, 1984, cited in Favid R. Jones, "The Macy Reports: Combat Fatigue in World War II Fliers," *Aviation Space and Environmental Medicine* 58 [August 1987]: 808).

an easy task. The soldier resisted remembering, and the therapist might have to compel the patient to recount the traumatic event, as they put it "forcing an eruption of the emotional experience" (*WN*, 290). At the same time, the very intensity of the abreaction risked mimetically immersing the patient so deeply in the nightmare reality of the traumatic scene that he needed the therapist's help in order to face the anxiety and hostility that had been unleashed and reorient himself to the present. The therapist thus played a central role in the confessional process, urging the patient to vividly reexperience the traumatic past, participating in the drama by playing various roles, reassuring him on waking that the traumatic experience was over, and helping him subsequently to consciously integrate and synthesize the uncovered material (*WN*, 161–66, 172–73, 199–205).

The moment of awakening from the drug was especially delicate, for the patient had been so immersed in the atrocious repetition of the battle experience that he was often quite disoriented. Hence the importance of getting the patient to move from the dramatic reliving of the trauma to being able to situate the recovered event in the past where it belonged. Comforting the patient when the emotions became too intense, the physician tried above all to "increase the discriminating and appraising function of the ego, so that the proper distinction can be made between the present safe, protected environment and the hostile atmosphere of the battlefield with which the ego is in such strong contact during the narcosynthesis" (*WN*, 290). In short, for Grinker and Spiegel abreaction was not a mechanical release of bottled up emotion, but an interpretive-confessional practice in which the physician played a key role. Affective discharge as such was not the essence of the cure, but the conscious integration and narration of the historical or psychic truth of the traumatic origin. Manipulating the transferential rapport, the therapist influenced the patient to participate in the cure by urging him into narrative and memory.

In contrast to Grinker's and Spiegel's adoption of the "participatory" model of abreactive treatment, Sargant's conceptualization of abreaction was ambiguous through and through. There were moments when Sargant appeared to accept the participatory model of cure by claiming that the goal of treatment was for the patient to be able to consciously integrate the experiences performed so vividly under the influence of the narcotic. "[I]t may be necessary to use firm and persistent questioning to overcome resistance," Sargant and his colleagues wrote early on in the war. "Sometimes actual memories are falsified by or mixed with fantasies; this may be made use of in subsequent interviews, in giving the patient insight into the deeper causes of his anxiety."[11] But he and his colleagues qualified the demand for knowledge and insight, as when they stated that amnesic patients "can certainly get well without interference," that is, without the full restoration and narration of the dissociated or forgotten material. Indeed, the authors implied that if the emphasis in treatment sometimes fell on restoring the memory of the traumatic event, that was because the men themselves seemed to benefit from it, observing in this regard that the better the personality, "the more is the lack of knowledge of what happened in an amnesic period a source of worry to the patient."[12]

One receives the same impression of the relative unimportance of psychic knowledge and confessional integration in Sargant's approach to abreaction from his comparison of the war neuroses to acute diseases of the abdomen, and from his characterization of treatment as a kind of splinting of the nervous system or "first aid."[13] As in the case of surgery so in the case of acute war neurosis, the first thing the physician had to do was to use the intravenous barbiturates to knock the patient unconscious. Comparing psychic shock to physiological shock, Sargant argued that a speedy response was crucial.[14] Nor was he the only physician to characterize drug treatment in those terms.[15] The use of barbiturates to sedate the war neurotic in this way short-circuited the requirement, stressed by Grinker and Spiegel, of using the preinduction period for therapeutic discussion and rapport. If consent was required in Sargant's approach, it

11. Gilbert Debenham, William Sargant, Denis Hill, and Eliot Slater, "Treatment of War Neurosis," *Lancet* 1 (25 January 1941): 108.

12. William Sargant and Eliot Slater, "Amnesic Syndromes in War," *Proceedings of the Royal Society of Medicine* 34 (1941): 763.

13. Debenham, Sargant, Hill, and Slater, "Treatment of War Neurosis," 109; Sargant, "Eight Years of Psychiatric Work in England," *Journal of Nervous and Mental Disease* 107 (1948): 504.

14. Sargant, "Eight Years of Psychiatric Work in Britain," 504. Cf. Debenham, Sargant, Hill, and Slater, "Treatment of War Neurosis," 109; Sargant, "Physical Treatment of Acute War Neuroses: Some Clinical Observations," *British Medical Journal* 2 (1942): 574. The emphasis on speedy treatment led Sargant to argue that even during the quarter to one-half hour needed for orally administered sodium amytal to take effect lay the seeds of a future phobia, which is why he recommended the use of intravenous barbiturates to induce immediate unconsciousness.

15. See for example John Rawlings Rees, *The Shaping of Psychiatry by War* (New York, 1945), 116, where the sedative drugs are described as a "very effective splint" for treating the war neuroses, though Rees also emphasized the need for subsequent reeducation.

was the kind of consent a patient gives to the surgeon before an operation, not that of a patient who agrees to collaborate in a talking cure.[16]

Indeed, it was largely as a diagnostic-surgical procedure that Victor Horsley characterized Sargant's use of drug abreaction in his book *Narcoanalysis*, published in 1943 at the height of the debate about the value of drug catharsis and its mechanism of cure. There Horsley drew a distinction between what he called the "drug-analytic" method and "narcoanalysis" proper, which he discussed in separate chapters. The distinction served the purpose of simultaneously permitting him to take credit for the ways in which drug abreaction was being used to treat the acute war neuroses while distinguishing that usage from his own more psychoanalytically oriented approach. Thus he presented the drug-analytic method as a biochemical means of inducing deep narcosis in a patient in order to facilitate the physician's knowledge of the case. It was an aid in obtaining data from the patient as quickly and efficiently as possible in emergency conditions where rapid decisions were essential. Although he admitted that the drug-analytic method helped restore amnesias and relieve symptoms, Horsley defined it as essentially a crude and primitive diagnostic measure that could be used by any inexperienced medical officer as an emergency measure in the field.[17]

In contrast, Horsley represented narcoanalysis as a sophisticated treatment procedure by which experienced psychotherapists sought to bring the hidden or repressed causes of illness into the patient's awareness and to integrate them into the latter's consciousness. His aim in narcoanalysis was the psychodynamic one of overcoming resistances efficiently so as to enable the surfacing and airing of repressed experiences and conflicts in a systematic psychotherapeutic technique (*N*, v). Horsley based his title to originality not on the use of barbiturates as such, but on the claim that he was the first to combine a chemical approach with the psychodynamic concepts of conflict, repression, and amnesia. In terms similar to those used in World War I by William Brown to describe hypnotic-cathartic "auto-gnosis," or by J. A. Hadfield to describe "hypno-analysis," Horsley emphasized the goals of self-knowledge and psychic integration (*N*, 10–11, 56–81). From his perspective,

16. It is worth noting in this regard that, as a medical procedure, the patient had the right to refuse the intravenous injection of barbiturates (J. F. Wilde, "Narco-Analysis in the Treatment of War Neurosis," *British Medical Journal* 2 [1942]: 2, 6).

17. J. Stephen Horsley, *Narco-Analysis: A New Technique in Short-Circuit Psychotherapy; A Comparison with Other Methods; And Notes on the Barbiturates* (London, 1943), 34–40; hereafter abbreviated *N*.

all phases of the narco-analytic procedure were important, not only the "analysis" phase, when the patient was questioned while under the influence of the drug, but also the "synthesis" phase, when the crucial task of psychic reintegration in full consciousness was achieved. He recognized that the emotional rapport established between the patient and physician during the administration of the drug was a crucial component of the cure and understood that rapport in classically Freudian terms as a transferential revival of the Oedipus complex, or the patient's earliest ties to the parents (N, 16, 82–98). As such a transferential revival, the emotional rapport ought not to be charmed away (N, 10) but analyzed and understood by the patient. Even as the barbiturates helped shorten the psychotherapeutic procedure, insight and self-knowledge remained the central goal of treatment.

Horsley presented the distinction between the drug-analytic method and narcoanalysis proper as in part a question of drug dosage. He associated the drug-analytic method with the earliest attempts to use the barbiturate sodium amytal in doses of 10–15 grains in order to produce a "profound narcosis" in patients suffering from schizophrenia. It was during the induction of the narcosis and afterwards that the patient became temporarily accessible to the physician, who was then able to learn more about the case.[18] In contrast, his aim in narcoanalysis was to induce a light narcosis with a minimum of confusion or clouding of consciousness (N, 35). For narcoanalysis the drug of choice was not the long-acting barbiturates such as soneryl sodium, or the "fairly short-acting" sodium amytal or nembutal, but the "very short acting" sodium pentothal, which in small doses could be administered intravenously in such a way as to induce the drowsy, semiconscious or hypnotic state of relaxation that seemed to encourage confessional hypermnesia and recall (N, 20–33).

Not surprisingly, Horsley analogized the drug-analytic method to surgery, by comparing the treatment of the acute neuroses by this means

18. For the prehistory of narcoanalysis see W. J. Bleckwenn, "Narcosis as Therapy in Neuropsychiatric Conditions," *Journal of the American Medical Association* 95 (1930): 1168–71; idem, "Sodium Amytal in Certain Nervous and Mental Conditions," *Wisconsin Medical Journal* 29 (1930): 693–96; idem, "The Use of Sodium Amytal in Catatonia," *Association for Research in Nervous and Mental Disease* 10 (1931): 224–29; Carl Phillip Wagner, "Pharmacological Action of Barbiturates," *Journal of the American Medical Association* 101 (1933): 1787–92; F. G. v. Stockert, "Über Evipan-Natrium-Behandlung in der Psychiatrie," *Medizinische Klinik* 31 (1935): 198–201. Cf. A. M. G. Campbell, "Sodium Evipan as an Aid to Psychotherapy," *Guy's Hospital Reports* 88 (1938): 185–98; Morris Herman, "The Use of Sodium Amytal in Psychogenic Amnesic States," *Psychiatric Quarterly* 2 (1938): 738–42.

to the treatment of acute wounds to the abdomen; in his view, the drug-analytic method was simply a first-aid procedure that in emergency circumstances must precede but could not constitute the cure (N, 38–39). He observed that the main advantage of drug abreaction over hypnotic catharsis was precisely that the former did not require specialized skill or experience in hypnotic and analytic techniques (N, 39, 77), a point often mentioned by Sargant and others; this apparent advantage was one of the main reasons why drug abreaction largely replaced hypnotic catharsis during World War II. Horlsey also observed that some physicians preferred the drug-analytic method because it appeared to minimize the extreme dependence on the physician that sometimes followed hypnosis (N, 39).[19] He explicitly associated Sargant with the drug-analytic method described in these medical-surgical terms (N, 39).

We need to question whether Horsley's distinction between the drug-analytic and the narcoanalytic method was valid, and whether cures of the war neuroses, to the extent that they occurred, took place according to the model of remembering and self-recognition that he, like Grinker and Spiegel, advocated. Conversely, we need to question whether, in conceptualizing the drug-abreactive method as an objective surgical-medical procedure, the suggestive-hypnotic relationship to (or as Horsley and others understood it, dependency on) the physician could really be reduced, as Horsley claimed, or even completely eliminated, as others hoped. What is certain, at any rate, is that Sargant's use of abreaction often conformed to Horsley's characterization of it as a kind of surgical or "psychosurgical" procedure designed to help in the anamnesis and to remove symptoms as quickly as possible.

Repetition: Historical Truth or Fictions of Remembering?

Sargant's "surgical" interpretation of abreaction was reinforced by his growing emphasis on the centrality of emotional discharge to the success of the abreactive method and his reliance on a Pavlovian model of conditioning to explain trauma and its cure. The two developments went together. In 1944, following the work of Major Harold Palmer on his experience with ether abreaction in psychiatric casualties in North Africa, Sargant began substituting the inhalation of ether through an open mask for the intravenous injection of sodium amytal in the treatment of the war neuroses. The virtue of ether, in Palmer's view, was that

---

19. On this point see also Margaret Brenman and Merton M. Gill, *Hypnotherapy: A Survey of the Literature* (New York, 1947), 32–34.

it was rapid, cheap, and practical under active service conditions.[20] Back in England, Sargant adopted ether because it produced a very intense emotional excitement of the kind held to be necessary for successful abreaction ("EA," 712–16; *UM*, 68). The shift to ether was an admission that the barbiturates did not always work, especially in the more chronic cases that Sargant was now seeing. Under ether, the kind of patients in whom barbiturate abreaction failed because they continued to report the traumatic scene without emotion and in the past tense, were now able to enter into the cathartic performance with much greater emotional excitement. It was generally easy to produce an emotional release in an individual suffering from a recent acute traumatic neurosis, Sargant and coauthor Shorvon thus observed. But in a long-standing illness, especially one in which reactive depression was a prominent symptom, "the patient may have the utmost difficulty in relaxing and reliving the traumatic situations. He tends to answer in the past tense and in an impersonal manner, and some of the emotion displayed does not ring true, i.e. it is done to satisfy the therapist" ("EA," 714). In these cases, ether showed its superiority over the barbiturates.

We are reminded of Clara Thompson's complaints about the simulated nature of the relivings in Ferenczi's neocathartic methods. Sargant believed that the use of ether solved the problem of simulation precisely by inducing the patient's complete absorption in his role. That is why he began to emphasize the importance of "present-tense" abreaction, in which the patient became so caught up in the performance of the traumatic scene that he reverted to the present tense, emotionally acting out the scene as if it were actually happening all over again. Sargant emphasized that the physician might have to actively enter into the "recalled or created" scene in order to help the patient carry it out in the right way ("EA," 714–16).[21] Like Prince, Brown, and others before him, Sargant

20. H. A. Palmer, "Abreactive Techniques: Ether," *Journal of the Royal Army Medical Corps* 84 (1945): 86–87; idem, "Military Psychiatric Casualties: Experience with 12,000 cases," *Lancet* 2 (20 October 1945): 492–94. The Russians also used ether abreaction in World War II, especially in cases of traumatically induced deaf-mutism (Gregory Zilboorg, "Some Aspects of Psychiatry in the U.S.S.R.," *American Review of Soviet Medicine* 1 [1944]: 575). Sargant observed that Hurst and his collaborators had made use of etherization in the treatment of hysterical conversion symptoms in World War I ("EA," 710). For Hurst's contribution see Arthur Hurst, *Medical Diseases of War* (Baltimore, 1940).

21. Thus in a lecture Sargant observed of the patient in the ether-induced abreaction, "When he is starting to get a little confused, suddenly switch your abreaction over into the present tense. Say 'you are back on the battle field and the tanks are coming down the road.' He will start to violently abreact in the present rather than in the past tense. You have to

thus became an accomplice in the narcotic-hypnotic performance as the soldier's speech was transformed from a past-tense narration into a dramatic, present-tense acting-out.[22]

Moreover, precisely because of the greater emotional release obtained with ether, Sargant in 1944 made what he considered one of his most decisive findings: in abreaction the emotional excitement aroused by the drug was more crucial for the cure than the actual memory of the traumatic event. He thus implicitly resolved the previous World War I debate between Brown, Myers, and McDougall over whether the success of abreaction depended on the recollection and synthesis of the traumatic past or on the affective catharsis, by emphasizing the central role of emotional discharge.[23] It is a sign of the continued relevance yet unsettled nature of that debate that Horsley should have reviewed it in his book, *Narco-Analysis*, with conclusions largely opposed to those of Sargant. For Horsley, as for McDougall, Myers, and other participants in the earlier controversy, the insight and reintegration of the dissociated traumatic past was crucial to the cure (*N*, 69–81). But Sargant held that the emotional excitement outweighed other factors to such an extent that it did not matter if the enacted scene was entirely fictive or suggested: so long as the patient could be encouraged to act out the scene with great emotion and drama the cure would work ("AWN," 235). In fact—and this also deserves emphasis—Sargant claimed that if the reliving of an actual incident did not bring about relief, invented situations could be successfully employed to cure the patient: "one can use fantasy to create excitement,

---

play your part and help him relive the experience. . . . Often at the height of an outburst he will suddenly break off and seem to collapse back on the bed. If you can reach this emotional crescendo and collapse the patient will often wake up and be greatly relieved" (Sargant, Lecture on Narco-Analysis, c. 1940s, D.1/8, Box 7, Sargant Papers, Contemporary Medical Archives Centre, Wellcome Institute for the History of Medicine, London). Palmer also stressed the importance of present-tense abreaction. He encouraged the patient to relive the traumatic scene by visualizing it as if it were being reenacted on a cinema screen, and observed that in many cases the patient spontaneously burst into a flood of tears. If this did not occur, he advocated that every means be used to induce tears, "even to the extent of maudlin suggestion" (Palmer, "Abreactive Techniques: Ether," 87).

22. Sargant credited Grinker and Spiegel with discovering the importance of present-tense abreaction (Sargant, "Some Observations on Abreaction with Drugs," 194). But Grinker and Spiegel did not think that present-tense reenactments were crucial in every case, since they viewed the use of the past tense as a potential sign of the patient's newly-won ability to distance and narrate the trauma *as past* (*WN*, 164).

23. Sargant, "Some Observations on Abreaction with Drugs," 194.

invent false situations or distortions of actual events when the uncovering of a true amnesia or the reliving of an actual experience has not brought about sufficient emotional release to disrupt a deeply ingrained neurotic pattern" ("EA," 717). In short, Sargant claimed that the abreaction of false memories might be more effective than the abreaction of real memories in achieving therapeutic success.[24]

The stakes of Sargant's claim were considerable, for a presumption in favor of the idea that traumatic repetitions were veridical representations of the origin had guided the practice of abreaction from the start. It is true that some World War I physicians had remained cautious. For example, Millais Culpin had doubted whether the terrifying dreams from which shell-shocked patients often suffered were unaltered reproductions of an actual incident, interpreting them instead as disguised and distorted representations of reality.[25] Similarly, Ernst Simmel, who had combined hypnosis with Freudian dream analysis in his treatment of war neurotics, had interpreted traumatic dreams not as literal replays of the past but as sexual-symbolic representations.[26] But the tendency then and in World War II was to view traumatic repetitions, including traumatic dreams and cathartic reenactments, as veridical depictions of the origin. Kardiner lent support to that tendency when, accentuating the fixity and asymbolic nature of the traumatic response, in 1941 he characterized the repetitive and stereotyped traumatic dream as a form of memory disturbance typical of the traumatic neurosis. Echoing Freud's suggestion that the catastrophic nightmare contradicted the wish-fulfillment theory of dreams (WN, 185), he presented such dreams as virtually exact or cinematic replays of the past (TNW, 88–89). This has made Kardiner attractive to recent theorists of trauma such as Bessel van der Kolk and Cathy Caruth, both of whom, radicalizing trends in Kardiner's work, have literalized and materialized the traumatic dream and traumatic memory. Kardiner has thus been inserted retroactively into an account of the history

24. Thus Shorvon and Sargant reported that in one case an ether abreaction, based on the accurate reproduction of the patient's experience of being nearly drowned in his ship, released great emotion and produced subsequent recollection, yet the patient did not get better. In a second abreaction, based on an artificial situation close to the original but falsified in certain ways in order to produce even greater terror, the abreaction ended in the patient's total collapse, with the result that he felt much better ("EA," 718).

25. Millais Culpin, Recent Advances in the Study of the Psychoneuroses (London, 1931), 30.

26. See Ernst Simmel's contribution to "Symposium Held at the Fifth International Psycho-Analytical Congress at Budapest, September 1918," in Psycho-Analysis and the War Neuroses (London, 1921), 30–43.

of mnemotechnics according to which trauma involves the literal registration of the traumatic event in a special traumatic memory system that can never be brought into recollection and self-representation. On this basis, trauma has come to stand for an entire post-Holocaust, post-Vietnam crisis of truth and history, in which not only the actual victim of trauma but everybody in the postcatastrophic condition is trapped (see chapters 7 and 8).

But it is not obvious that the traumatic dreams that Kardiner described can be interpreted in such veridical, or indeed literalist, terms. Although, as Kardiner remarked, the most common content of the dreams he analyzed appeared to be the threat of annihilation, they were intermixed with personal and familial concerns in ways that seemed to make them other than veridical or literal reproductions of the past (*TN*, 87–88). Allan Young has recently observed of the dreams reported by Kardiner: "Of course, the meaning of the patients' dreams—based on the idea that the dreams replay traumatic events—is not as transparent as Kardiner at first suggests. In practice, their signifying power derives not only from their content but also from Kardiner's interpretations of the patient's concurrent symptoms ('action syndrome'), such as symbolically resonant motor dysfunctions."[27] During the Second World War much the same conclusion was reached by Theodore Lidz. "The dream is sometimes the repetition of an actual experience, but a number of patients were certain that the episode had never occurred in reality," Lidz remarked of the repetitive nightmares experienced by veterans of combat in the South Pacific, noting in this regard that similar nightmares were suffered by soldiers who had never experienced combat. He therefore interpreted such dreams in terms of projected death wishes and the influence of precipitating personal conflicts and concerns.[28] Similarly, Lawrence S. Kubie observed that the materials recovered by narcoanalysis or induced hypnagogic reverie were sometimes direct and accurate recollections of the war experience but at other times were subject to fantasmatic and dreamlike condensation with events from earlier years. He warned that

---

27. Young, *The Harmony of Illusions*, 91. Actually, what interested Kardiner was less the claim that the catastrophic dream contradicted the wish-fulfillment of dreams (he noted in this regard that if one examined a dream from the point of view of wish-fulfillment, one could usually find *some* wish), but the manner in which the wish was handled and what this told about the traumatically induced deficits in the structure of the personality or ego (*TN*, 185–86).

28. Theodore Lidz, "Nightmares and the Combat Neuroses" (1946), reprinted in *Essential Papers on Dreams*, ed. Melvin R. Lansky (New York, 1992), 324.

although the material was just as valuable as simpler historical information, its unraveling required experience with dream interpretation.[29]

Grinker and Spiegel, on the other hand, tended to regard traumatic repetition—whether repetitive traumatic dreams or drug-induced traumatic repetitions—as exact reproductions of the traumatic event. They acknowledged that narcotic abreaction did not necessarily reveal veridical memories of the traumatic experience and that abreacted or "uncovered" memories were often mixed with fantasy. Nevertheless, like William Brown they expressed the commonly held view that buried in the unconscious there was an exact or photographic copy of the traumatic event to which the ego must be brought into contact during the narcoanalytic treatment (*WN*, 288).[30] The American physician, Howard Fabing, who conducted mass abreactions at his hospital base in England, was also committed to the notion that the events repeated in abreaction were authentic reproductions of the traumatic past. He acknowledged the possibility of distortion (whether suggested or not is a topic to which I return). But the force of his discussion was to propose that in many cases the recovered memories were completely veridical in nature.[31] He linked the recovery of the historical truth to the participatory model of

29. Lawrence S. Kubie, "Manual of Emergency Treatment for Acute War Neuroses," *War Medicine* 4 (1943): 595.

30. In 1945 Grinker and Spiegel clarified their position by distinguishing between the traumatic event itself and the patient's emotional reaction. "The events which are depicted with the realistic impact of an expert dramatic production are probably always true counter-parts of what actually took place, rather than fantasies such as are produced in dreams or hypnotic states" they observed, adding: "The emotional reactions, however, do not necessarily represent the actual behavior of the flier during the original episode but rather what he repressed and controlled in order to carry on his job" (Roy R. Grinker and John P. Spiegel, *Men Under Stress* [Philadelphia, 1945], 173). For the claim that the narcosynthetic abreaction represented an almost completely exact memory of the original experience see also Herbert Freed, "Narcosynthesis for the Civilian Neurosis," *Psychiatric Quarterly* 20 (January 1945): 39–40.

31. Howard Fabing, "Cerebral Blast Syndrome in Combat Soldiers," *Archives of Neurology and Psychiatry* 57 (1947): 35; hereafter abbreviated "CBS." "Cerebral blast syndrome" was characterized as a condition of temporary unconsciousness and retrograde amnesia, headache, anxiety, noise sensitivity, tremors, and other symptoms that Fabing attributed directly to bomb explosions but that, like hysterical amnesia and other functional disorders, were reversible because they were not accompanied by any discernible lesions of the brain or other physiological changes. The relationship between cerebral blast syndrome and the old notion of "shell shock" was discussed by D. Denny Brown, "'Shell Shock' and Effects of High Explosives," *Journal of Laboratory and Clinical Medicine* 28 (1943): 509–14.

cure by remarking that although success could be achieved by several variations of the drug abreaction method, including one in which the kind of intensely emotional abreaction Sargant advocated was completely missing, any variation that produced an inadequate recovery of the memory for the amnesic material left the patient without therapeutic benefit. From his observations he concluded that "the least common denominator of therapy, the *sine qua non* of therapeutic effectiveness, was the recovery and emotional resynthesis of memory for the amnesic material" ("CBS," 37).

For advocates of chemical catharsis, as for Ferenczi earlier, the question of the truth-status of the abreacted performance was closely tied to the problem of simulation, or malingering—a crucial issue in times of war.[32] From the moment of their invention, the barbiturates had often been regarded as a means of obliging the patient to confess the truth. Horsley himself had originally presented narcoanalysis as a kind of "truth serum," and although on close inspection the various arguments made in support of that idea were frequently contradictory, the general tendency in the literature on drugs and malingering was to treat the barbiturates and related chemicals as a relatively unproblematic method for distinguishing truth from lies.[33] However, some physicians were more circumspect about the possibility of separating truth from fiction by these means.[34] The more skeptical position found expression at the end of World War II in an experimental investigation by Gerson and Victoroff (1948) that queried the historical accuracy of the confessions obtained in interviews conducted under the influence of intravenously injected sodium amytal. Reviewing their own attempts to uncover data in

32. Although Horsley (*N*, 116), Grinker, Spiegel (*WN*, 95), and others regarded the deliberate simulation of symptoms as rare, many were concerned with the topic of malingering. See for example Harold Palmer, "Military Psychiatric Casualties: Experience with 12,000 cases," *Lancet* 2 (13 October 1943): 454–57, and 2 (20 October 1943): 492–94.

33. J. Stephen Horsley, "Narco-Analysis," *Lancet* 1 (4 January 1936): 55–56. See also Robert House, "The Use of Scopolamine in Criminology," *American Journal of Criminal Science* 2 (1931): 328–36; Calvin Goddard, "How Science Solves Crime. III. 'Truth Serum,' or Scopolamine, in the Interrogation of Criminal Suspects," *Hygeia* 10 (1932): 337–40; W. F. Lorenz, "Criminal Confessions Under Narcosis," *Wisconsin Medical Journal* 31 (1932): 245–50; Alfred O. Ludwig, "Clinical Features and Diagnosis of Malingering in Military Personnel," *War Medicine* 5 (1944): 378–82; and Don P. Morris, "Intravenous Barbiturates: An Aid in the Diagnosis and Treatment of Conversion Hysteria and Malingering," *Military Surgeon* 96 (1945): 509–13.

34. See for example J. F. Wilde, "Narco-Analysis in the Treatment of War Neuroses," *British Medical Journal* 2 (1942): 6.

seventeen soldiers known to be guilty of crimes that they nevertheless refused to admit, the authors concluded that the validity of the information
gathered by narcoanalysis was sufficiently ambiguous as to be inadmissi-
ble as legal evidence without further investigation and substantiation.[35]
In a related experiment on students, Redlich, Ravitz, and Dession (1951)
concluded that the contents of a narcoanalytic interrogation were too
imbued with fantasy and other unconscious psychic processes and motives, including in some individuals a neurotic sense of guilt, to be used as
evidence in a court of law. They argued that the subject's temporarily facilitated identification, as manifested by increased suggestibility, played
an important role in the confessional process.[36] The same doubts about
the historical accuracy of narcotically or chemically enhanced recollection, repressed memory, "flashbacks," and related phenomena continue
to haunt the literature on trauma to this day—an especially troubling issue when so much in the literature of PTSD seems to depend on establishing the historical accuracy and indeed literality of the traumatic
repetition. Thus in reviving narcosynthesis in the treatment of Vietnam
veterans with PTSD, Lawrence Kolb has recently cited the earlier work
of Gerson and Victoroff and of Redlich, Ravitz, and Dession in acknowledging that the accuracy of memories uncovered by these means is uncertain, an uncertainty his own procedures do not resolve.[37]

35. "It would seem at the present writing that there is no such thing as a 'truth serum'"
(Martin J. Gerson and Victor M. Victoroff, "Experimental Investigation into the Validity
of Confessions Obtained Under Sodium Amytal Narcosis," *Journal of Clinical Psychopathology and Psychotherapy* 9 [1948]: 374). The authors observed that the patient's loss of
time sense and tendency to confuse reality and fantasy under barbiturate narcosis made
testimony concerning dates, names, and events of questionable veracity; the ambiguities
could be reduced by careful questioning but not eliminated.

36. Fredrick C. Redlich, Leonard J. Ravitz, and George H. Dession, "Narcoanalysis
and Truth," *American Journal of Psychiatry* 107 (1951): 586–93. Sargant later observed that
the lie detector was not a reliable instrument for detecting the truth owing to the role of
suggestion, criticized the evidence obtained by "third degree" police examinations as irredeemably contaminated by the suggestive inmixing of truth and fantasy or falsehood, and
characterized Freud's seduction theory as a suggestive fiction of this kind (William Sargant, *Battle for the Mind* [New York, 1957], 192–96).

37. Lawrence C. Kolb, "Recovery of Memory and Repressed Fantasy in Combat-
Induced Post-Traumatic Stress Disorder of Vietnam Veterans," in *Hypnosis and Memory*,
ed. Helen M. Pettinati (New York, 1988), 265–66. In an official report prepared for the
American Medical Association in 1985, leading experts on hypnosis likewise concluded
that hypnotically induced relivings or "memories" of past experiences were liable to be imbued with confabulation and pseudomemories and hence were not necessarily reliable

Obviously, Sargant's pragmatic-surgical approach was in line with the skeptical position because it made fictive scenarios the basis of the cure. He did not question the authenticity of the patient's suffering, or that it was linked in some way to real experiences. But he maintained that at least in some cases abreaction cured on the basis of entirely suggested or fictional scenarios. For him, the commitment to basing the cure of trauma on the patient's recovery and integration of the historical truth of the origin was therefore misplaced.

Conditioning the Mind

Sargant linked his discovery of the significance of emotional discharge through suggestive-fictional scenarios to a Pavlovian interpretation of the traumatic reaction. Although in his earliest writings on the war neuroses he had described traumatic responses in terms of conditioned reflexes,[38] it was not until 1944, soon after the Normandy invasion, that he read Ivan Pavlov's book, *Conditioned Reflexes and Psychiatry* (1941), and began to apply Pavlov's ideas in a systematic way to the interpretation of the war neuroses. Seeing the parallels between the acute war neuroses and canine neuroses helped reinforce Sargant's anti-Freudian, physiological approach to trauma. The traumatic neuroses were conceptualized by him as neuroses that had less the character of neurotic conflict than of a somatic or biological condition. He thus affirmed the tendency, articulated intermittently in the course of the history of trauma and strongly represented in current neurobiological theories, to interpret the traumatic reaction in physiological or corporeal terms.[39]

Sargant repeatedly singled out one incident in Pavlov's work as crucial to his understanding of the war neuroses. In 1924, many of the dogs on

memories of the past ("Scientific Status of Refreshing Recollection by the Use of Hypnosis," *Journal of the American Medical Association*, 253 [April 5, 1985]: 1918–23).

38. Sargant, "Acute War Neurosis" (1940), 1.

39. Sargant credited Howard Fabing for alerting him in 1944 to the relevance of Pavlov's book ("AWN," 111; *Battle for the Mind*, 21–22). He also acknowledged the work of Harold R. Love, who during the siege of Tobruk conceptualized the war neuroses in Pavlovian terms (Love, "Neurotic Casualties in the Field," *Medical Journal of Australia* 2 [August 22, 1942]: 137–43). Similarly, R. D. Gillespie described the war neuroses as conditioned reflexes (Gillespie, *Psychological Effects of War on Citizen and Soldier* [New York, 1942], 187. For pre– and post–World War II efforts to theorize trauma in terms of conditioning see also Young, *The Harmony of Illusions*, 21–42.

which Pavlov had conducted his experiments on conditioning had nearly drowned in severe floods in Leningrad. They had been rescued at the last minute when they were swimming about, trapped in their cages, with their heads barely above the water. Some of the dogs were in a state of acute excitement that was succeeded by a phase of collapse or stupor that gradually passed. Pavlov then discovered that the terrifying experience, which according to him had produced an "ultraparadoxical" inhibition of the cortex, had destroyed a series of conditioned responses previously implanted in the dogs. More stable dogs were far less affected. Extending Pavlov's arguments to humans, Sargant claimed that the amnesias, paralyses, and other hysterical symptoms characteristic of the acute war neuroses were examples of Pavlovian states of both cortical excitation and inhibition, that is, states in which normal conditioned reflexes were abolished by shock and in which pathological reflexes took their place. He interpreted the excited and purposeless behavior of stunned men in the field, as well as the stupors, amnesias, loss of voice, stammering, exhaustion, and other symptoms of the war neuroses as examples of mixed cerebral inhibition and excitation in this way. By further analogy, he suggested that during cathartic treatment the drug-induced emotional excitement disrupted the pathologically implanted reflexes in turn, thereby curing the patient. Abreaction was thus an artificially created state of brain excitement which repeated the traumatic shock in order to break up morbidly conditioned reflexes and restore normalcy.[40] It was precisely because ether allowed situations of great emotional intensity to be experienced by the patient that it produced the agitation and subsequent stupor necessary to the cure ("AWN," 233–34). In other words, _shock was cured by shock by a process of deconditioning_.

Two aspects of Sargant's approach are worth stressing. First, abreaction on this model was conceptualized as working in a mechanical way, as if the cure depended simply on the automatic release of the repressed or dissociated emotion ("AWN," 237).[41] The fact that fictive scenarios could be employed to produce abreaction was used by Sargant to justify his interpretation ("AWN," 237). Sargant's approach contrasted with that of the psychoanalytically oriented Horsley and others, who decried the tendency to administer narcotic abreaction in a mechanical way and emphasized instead the transferential dimension of the abreactive

40. Sargant and Slater, _An Introduction to Somatic Methods of Treatment in Psychiatry_ (Baltimore, 1944), 9; hereafter abbreviated _ISM_.

41. Cf. Sargant, "Some Observations on Abreaction with Drugs," 205.

process.[42] It is not surprising in this regard that Sargant compared his abreactive methods to the electrical or faradic treatments used in World War I ("AWN," 236). Sargant's comparison is revealing, for one of the main practitioners of the faradic method in Britain, L. R. Yealland, had always insisted on the irrelevance of the subject's participation in the cure. His was a disciplinary-suggestive technique that called for the patient's submission to the doctor's manipulations: discussion was kept to a minimum, and the physician aimed to completely domineer his subject. In some cases Yealland even recommended anesthetizing the patient in order to make the authoritative suggestions more effective.[43] Sargant spoke in similar terms of the need for the doctor to gain complete control of the abreactive process and "quietly dominate the patient" (ISM, 115).[44]

The second aspect of Sargant's Pavlovian approach to abreaction that deserves emphasis is that on the basis of Pavlov's ideas he regarded the moment of emotional collapse as critical to the cure—the moment when the emotional-abreactive excitement had reached such a pitch of intensity that the patient suffered a temporary mental collapse ("ANW," 235). Based on his work with ether, Sargant claimed that what was crucial in abreaction, at least in chronic cases, was not so much the uncovering of the (real or fictive) traumatic experience as the emotional acting out of the traumatic scenario to the point of frenzy and confusion ("EA," 712). He compared the moment of mental collapse to that reached in Pavlov's dogs during the Leningrad floods when the cortex was momentarily incapable of further activity. He repeatedly stressed the importance of that mental collapse, pointing out the frequency of its occurrence in the work

42. Thus Horsley condemned the tendency among physicians, apparent even in the psychoanalytically oriented work of Grinker and Spiegel, to mechanically count backwards to a certain state of narcosis when administering the drug, instead of realizing that the entire process should be considered in psychotherapeutic, transferential terms (N, 55, 59, 68).

43. D. Adrian and L. R. Yealland, "Treatment of Some Common War Neuroses," Lancet 1 (9 June 1917): 867–72.

44. Sargant's approach to the patient recalls the "active therapy" used by the Germans to rationalize the treatment of the war neuroses in World War I, where the emphasis fell on "mesmerizing" or coercing the soldier into compliance in the quickest, most efficient way possible (Paul Lerner, "Rationalizing the Therapeutic Arsenal: German Neuropsychiatry in World War I," in Medicine and Modernity: Public Health and Medical Care in Nineteenth- and Twentieth-Century Germany, ed. Manfred Berg and Geoffrey Cocks [Washington, D.C., 1996], 121–48; and idem, "Hysterical Men: War, Neurosis and German Mental Medicine, 1914–1921" [Ph.D. diss., Columbia University, 1996]).

of others who had used abreaction, such as Brown in World War I. He cited Grinker's and Spiegel's description of the results of that collapse: "'The stuporous became alert, the mute can talk, the deaf can hear, the paralyzed can move, and the terror-stricken psychotics become well organized individuals'" ("AWN," 236, citing Grinker and Spiegel, *WN*, 167).[45] We might put it that for Sargant cure depended on an exorcism of the traumatic experience. Grinker and Spiegel maintained that the effect of trauma was "not like the writing on a slate that can be erased, leaving a slate like it was before. Combat leaves a lasting impression on men's minds, changing them as radically as any crucial experience through which they live."[46] Judith Herman, whose ideas about trauma are discussed in chapter 3, cites approvingly Grinker's and Spiegel's statement to support her contention that abreaction will only succeed if the traumatic memories seemingly restored under the influence of sodium amytal, or hypnosis, are reintegrated into consciousness; the same statement by Grinker and Spiegel has also been cited with approval by van der Kolk and his group.[47] But for Sargant wiping the mind clean was exactly the goal he sought: "[Pavlov] had suddenly found a dramatic way of wiping the 'cortical slate' clean of past implanted habits." On this formulation, Sargant's therapeutics were not so much a therapeutics of remembering as a *therapeutics of erasure or forgetting*.[48]

45. The same moment of mental collapse was the basis for Sargant's comparison between abreaction, sudden religious conversion, and brain washing ("EA," 727). In *Battle for the Mind* Sargant argued that the terrorizing methods of interrogation used by the police in brain washing induced a state of mental collapse that erased the old beliefs and permitted the suggestive implantation of new ones.

46. Grinker and Spiegel, *Men Under Stress*, 371.

47. Judith Herman, *Trauma and Recovery* (New York, 1992), 26; Bessel van der Kolk, Alexander C. McFarlane, and Lars Weisaeth, in *Traumatic Stress: The Effects of Overwhelming Experience on Mind, Body, and Society* (New York, 1996), 59.

48. Sargant, "Some Observations on Phenomena to [sic] Religion and Psychiatry—With Special Reference to Abreactive Techniques" (1947), unpublished talk, PP/WWS, Box 13, F.1/2, 6–7, Sargant Papers, Contemporary Medical Archives Centre, The Wellcome Institute for the History of Medicine. A similar therapeutics of forgetting was advocated by Elmer Klein who, on the basis of his experience with psychiatric casualties from the battle of Iwo Jima, criticized the narcosynthetic emphasis on remembering on the grounds that patients wanted not to be forcibly reminded of their horrifying experiences but wished for the "healing anodyne of merciful forgetfulness" (Elmer Klein, "Acute Psychiatric War Casualties," *Journal of Nervous and Mental Disease* 107 [1949]: 39). In October 1941 Milton H. Erickson and Lawrence S. Kubie reported curing a case of acute hysterical depression in a young woman by hypnotically encouraging her to forget a putative traumatic (sexual) experience—a departure from the technique of psychoanalysis inasmuch as

Forgetting

There is more to be said on this topic. During the war authors began to comment on the fact that if, as was widely claimed, abreaction worked by helping patients to remember forgotten incidents, it also risked failure because patients tended to forget immediately after waking what they had just "remembered"—to the point of doubting the reality of the reenacted traumatic origin. The problem of postabreactive or posthypnotic forgetting was already familiar to physicians from their experiences with hypnotic abreaction in World War I,[49] as well as to Ferenczi, as I have shown. It was also known to physicians who carried out chemical abreaction during World War II.[50] Indeed, years later, when he revived abreactive methods to treat Vietnam veterans, Lawrence Kolb mentioned the problem of postabreactive forgetting as a well-recognized phenomenon.[51]

Psychoanalytically oriented physicians, such as Grinker and Spiegel, attributed the tendency to forget what had just been "remembered" to resistance and to the weakness of the ego's synthetic functions (*WN*, 182).[52]

---

repression was explicitly encouraged by the therapist. "One might well ask for the 'mechanism of cure' where there exists not even a 'rudimentary insight' as in the therapy of direct suggestion of symptom-disappearance, or in the 'cures' of Christian Science or of Our Lady of Lourdes," Brenman and Gill observe of Erickson's and Kubie's report (Brenman and Gill, *Hypnotherapy: A Survey of the Literature*, 77–79).

49. "He was actually in a hypnoidal state, as shown by the fact that some patients, after going through detailed and emotional accounts, forget them soon after they returned to a knowledge of their real surroundings. . . . This lack of recognition is not uncommon" (Millais Culpin, *Recent Advances in the Study of the Psychoneuroses*, 28–29).

50. Thus Louis L. Tureen and Martin Stein observed that although narcoanalysis was valuable as an investigative technique and for clearing up conversion symptoms, it did not furnish insight to the patient himself, "since there is an almost complete amnesia for the session," a claim with which the editors of the work disagreed (Louis L. Tureen and Martin Stein, "The Base Section Psychiatric Hospital," in *Combat Psychiatry: Experiences in the North African and Mediterranean Theaters of Operation, American Ground Forces, World War II*, compiled and edited by Frederick R. Hanson, *Bulletin of the U.S. Army Medical Department*, vol. 9, Supplemental Number [November 1949]: 123).

51. Kolb, "Return of the Repressed: Delayed Stress Reaction to War," 533.

52. Grinker and Spiegel, *Men Under Stress*, 170. In criticizing drug (as opposed to hypnotic) abreaction, Brenman and Gill tended to attribute the tendency of the drug to produce a "split off" experience to the radical physiological changes induced by the drug, changes that made it difficult for physicians such as Grinker and Spiegel to bridge the amnesia, in spite of their efforts to resynthesize the material (Brenman and Gill, *Hypnotherapy: A Survey of the Literature*, 33).

But as we saw in chapter 5, an alternative explanation treats postabreactive or posthypnotic amnesia as an artifact of assumptions about the nature of hypnosis, an artifact that ought to be reversible: all that needs to be done is to suggest to the subject, while he is still hypnotized, that he will remember all that has happened once the hypnosis is over. During World War II this was the position taken by Horsley, who reviewed the problem of postnarcotic forgetting at some length. In some passages, he seems to have thought that the "recurrent amnesia" could be prevented by "waking the patient slowly while the traumatic experiences are recalled again" (*N*, 54) and by subsequently continuing the analysis by free association. But Horsley also observed that during abreaction the sudden return to consciousness of intolerable thoughts might drive the patient to frenzy or suicide, and that it was often an advantage in such cases to follow up the narcoanalytic session with a period of deep narcosis. Recognizing the resulting problem of postnarcotic amnesia he wrote: "Critics have suggested that in such circumstances the patient on waking would have no knowledge of any disclosures confessed or recalled under the influence of the drug: but in my experience this is not so: provided that the patient is given a definite 'posthypnotic' suggestion that on waking he will remember all he has said, he usually does so" (*N*, 80).

Whether the amnesic barrier could always be broken by these means is an important question.[53] Some authors gave as a reason for preferring hypnosis over chemical abreaction the fact that in hypnosis the problem of posthypnotic amnesia could be solved.[54] But for most medical men

---

53. The importance of the question is shown by the fact that in 1942–43 a controversy broke out over the therapeutic efficacy of narcoanalysis. Horsley responded to the problem of postabreactive forgetting with the following, somewhat contradictory, arguments:

1. The problem of postabreactive amnesia might be a function of drug dosage; too large a dose, especially of sodium pentothal, was likely to increase the patient's postabreactive blurring of recall (*N*, 63).

2. In any case, denials by patients of what they had just revealed in treatment were liable to occur in all psychotherapeutic treatments, whether psychoanalysis, hypnotism, or merely sympathetic talks (*N*, 65).

3. The problem might be due to the physician's lack of skill in administering narcoanalysis. Postabreactive amnesia was especially likely to occur if the physician regarded the treatment as merely a mechanical or surgical procedure (*N*, 65).

4. In any case, narcotic postabreactive amnesia could be lifted by posthypnotic suggestion (*N*, 80), an argument that implicitly treated drug abreaction as an essentially hypnotic procedure.

54. J. A. Hadfield, "War Neurosis: A Year in a Neuropathic Hospital," *British Medical Journal* 1 (1942): 320; Charles Fisher, "Hypnosis in Treatment of Neuroses Due to War

untrained in hypnotic procedures or uncomfortable with the idea of treating chemical abreaction as if it were a purely suggestive procedure, alternative solutions to the problem of postabreactive forgetting were necessary. The emphasis fell on using minute doses of the fast-acting barbiturates, such as sodium pentothal, in order to prevent the patient from becoming mentally confused—indeed in order to keep him closer to consciousness throughout the abreactive treatment so that afterwards the memory of the abreactive material could be retained and integrated.[55] This meant in turn that the use of drugs to sedate the patient immediately after he "woke up" from the narcotic was criticized as counter-productive, since sedation contributed to the maintenance of the very amnesia that the abreaction was designed to cure.

These problems placed Sargant in an interesting quandary, and not only because he regularly put his patients into a state of continuous sleep immediately after abreaction. More broadly, he faced the following dilemma: the very intensity of emotional reaction that in certain cases of

---

and Other Causes," *War Medicine* 4 (1943): 565; J. G. Watkins, *Hypnoanalysis* (New York, 1945); and H. S. Alpert, H. A. Carbone, and J. T. Brooks, "Hypnosis as a Therapeutic Technique in the War Neuroses," *Bulletin of the U.S. Army Medical Department* 5 (1946): 315–24. Herbert Spiegel has recalled his effort during World War II to uncover a repressed event with a pentothal interview in a soldier one month after the latter had experienced a combat-induced trauma. Dr. Spiegel failed to find the vein, but nevertheless the soldier abruptly entered into a violent abreaction and relived the terrors of the battle experience. Afterwards, the patient marveled at the effectiveness of a drug that in fact he had never received. On the basis of this and similar experiences, Dr. Spiegel concluded that persons with good hypnotizability were the ones that responded best to the barbiturates. But hypnotizable patients responded even better without the drug because "they were able to use their trance capacity to recover material without the handicap of barbiturate blurring of recall" (Spiegel, "Silver Linings in the Clouds of War: A Five Decade Retrospective," *The History of American Psychiatry Since World War II*, ed. Roy W. Menninger and John Nemiah (American Psychiatric Press, in press). I thank Dr. Spiegel for allowing me to see his article, and for discussing these and related matters with me.

55. That is why Grinker and Spiegel substituted pentothal for the slower-acting sodium amytal, since the former allowed the physician to continue the discussion of "uncovered" material after the effects of the drug injection had worn off and so helped the patient integrate the material revealed in treatment (*WN*, 159). Yet the shift to pentothal did not necessarily solve the problem, as Fabing discovered (see below). Other authors claimed that even with sodium amytal, if used in smaller doses, the patient could be made to retain the memory of the abreaction (see Don P. Morris, "Intravenous Barbiturates: An Aid in the Diagnosis and Treatment of Conversion Hysteria and Malingering," *Military Surgeon* 96 [1945]: 513).

war neurosis he deemed essential for the cure, required a degree of ether-induced excitement that at the limit tended to collapse the patient into mental confusion. But this meant that the patient might be prevented from fully recovering the amnesic material, for "the excitement under ether often interferes with a full recital of events, and there is a tendency for an emotional climax to occur before the whole amnesia has been laid bare" ("EA," 722). Recognizing the difficulty, Sargant suggested that for uncovering a prolonged amnesia barbiturates were more suitable than ether. But that did not necessarily solve the difficulty either, for "one of the drawbacks of the barbiturate method is that if the patient is allowed to fall asleep after his treatment he may deny any knowledge of what has been uncovered whilst under the influence of the drug, i.e. the memories have not been integrated into consciousness" ("EA," 722).

Sargant's various solutions to his dilemma wavered between imagining that drug-induced remembering was constitutive of the cure and the view that forgetting was equally effective. On the one hand, he sometimes attempted to solve the problem of postbarbiturate forgetting by shaming the patient into knowledge and insight. So for example he tried to make the patient conscious of the movement he had just performed during the abreaction with a hitherto paralyzed limb by keeping the soldier conscious afterwards, forcing him to walk back to the ward while still in a groggy state, and obliging him to move the limb in front of the other patients (*ISM*, 116). But he also recommended that patients be sedated immediately after ether abreaction on the grounds that the incidents uncovered were simply too horrible for the patient to cope with ("EA," 715). All this suggests that Sargant's pragmatic approach to narcotic abreaction fell short of the participatory ideal of the cure.

A comparison between Sargant's work and that of Howard Fabing is instructive in this regard. We have seen that Fabing was committed to the therapeutic importance of psychic integration of the trauma. But his work also illustrates the difficulty of achieving cures according to that model. Precisely because of his commitment to obtaining the truth of the traumatic origin Fabing tried but then abandoned the use of ether abreaction, since he felt that the open mask through which the ether was inhaled interfered with the patient's ability to talk and narrate his experience ("CBS," 30). This meant that Fabing rejected the very drug—ether—that Sargant believed was necessary to obtain the intensity of emotional excitement required for the cure. But on selecting the barbiturate sodium pentothal as his drug of choice, Fabing discovered that on recovering from the effects of the drug the patient frequently forgot everything he had just experienced: "Too often he would listen attentively, then shake his head ruefully, saying: 'If you say so, it must be true; but I don't remember a bit of it.' In such cases no therapeutic benefit was obtained" ("CBS," 30).

Fabing was unwilling to consider the idea, implicit in Sargant's more pragmatic-physiological approach to abreaction, that the emotional discharge was in and of itself curative of the traumatic neurosis. Convinced that the patient must be made to recover and integrate the memory of the event, Fabing undertook various experiments to achieve that goal, not all of them successful. For example, he reported inducing the patient during treatment to recite the previously forgotten events while the physician and his staff dramatically simulated the traumatic situation by making battle noises and mimicking the shouts of comrades in order to induce a highly emotional abreaction; the technique produced good results, but still "too many patients continued to fail to recall the amnesic events after full return to consciousness" ("CBS," 30). The attempt to keep the patient awake after treatment in order to make him rehearse the amnesic material also failed—if only because the soldier became too noisy and violent and the demands for personnel to restrain him were too great ("CBS," 31). Fabing finally reported success by following up the sodium pentothal–induced abreaction with a stimulant, Coramine, designed to wake up the patient immediately afterwards and allow a quick review of the amnesic events. Once awake, the soldier was told that while he had been asleep he had related his forgotten experience and that now he was awake he would be able to recall everything. With "a little prodding, and a suggestion here and there," the patient unfolded the whole episode again and was then led back to his bed for further review of the episode. He was also encouraged to write out his experience in order to strengthen its conscious recall. Fabing mentioned the need in some cases for further discussions with the patient in order to help him adopt a "healthy mature attitude toward such emotion-laden material," comparing his approach in this regard to Grinker's and Spiegel's narcosynthesis. He proposed that his methods could be effective in chronic cases of quite long duration ("CBS," 33–34).

The Vicissitudes of Abreaction

Did Fabing succeed by such a combination of drugs in curing his patients? I do not know, and in the absence of follow-up studies it is impossible to be sure. But a reading of his paper leaves one skeptical about the therapeutic effects of narcosynthesis if only because, by documenting the failures of previous efforts at treatment by these means, he renders all optimistic therapeutic claims somewhat suspect.[56] And this leads me to ask: Did abreaction cure? As a preliminary response to that question, three points are especially germane:

1. If Sargant himself succeeded in curing the war neuroses, it was not through the use of chemical abreaction alone. He always insisted on the need for a variety of therapeutic approaches, emphasizing that particular techniques often failed in specific cases.[57] After the war he even repeated Janet's warnings about the limitations of abreaction as a therapeutic procedure.[58] Sargant especially stressed the need for early treatment, noting as others had done that abreaction rarely worked in longer-standing or chronic cases, by which he meant cases of even a few weeks' duration in which the symptoms had become fixed.[59] Thus if his case records and published papers document successes with abreaction, they also reveal many failures. Ever the empiricist, Sargant constantly stated the need to try different drugs or combinations of drugs in order to achieve the desired result, as well as the necessity of tailoring the therapeutic approach to the individual case. He reported favorably on Fabing's technique of using stimulants to wake the patient up after pentothal abreaction.[60] Continuous barbiturate sedation remained for him an important therapy for panic states; he also recommended modified insulin therapy to fatten up the war neurotic, electroconvulsive treatment, group therapy, and even leucotomy in certain cases. His was an eclectic, pragmatic, and essentially symptomatic approach to the war neuroses and a close study of his case records indicates that he did not consider the patient's knowledge or insight into the trauma necessary for the cure.

2. Sargant sometimes wrote as if the significance of his conditioned-reflex approach to trauma was that it undermined the claim that the trau-

56. In any case, Fabing acknowledged several limitations that restricted the applicability of his method. He reported that none of his patients returned to combat duty; he also acknowledged the absence of any follow-up studies ("CBS," 38).

57. Sargant, "Eight Years of Psychiatric Work in England," 506. Sargant and Shorvon stated that ether abreaction did not help hysterical patients who also suffered from depression ("AWN," 235) or cases of chronic obsessional neurosis ("EA," 724).

58. Sargant, "Indications and Mechanisms of Abreaction and Its Relation to Shock Therapies," paper given to the "Premier Congrès Mondial de Psychiatrie," Paris 1950, Box 13, F. 1/5, 201, Sargant Papers, Contemporary Medical Archives Centre, The Wellcome Institute for the History of Medicine.

59. Sargant recorded problems with chronicity as early as July and August of 1940 (Sargant, "Amnesic Syndromes of War," *Proceedings of the Royal Society of Medicine*, 34 [1941]: 761).

60. Sargant, "Some Observations on Abreaction with Drugs," 199.

matic neuroses were cured by psychical suggestion ("EA," 729, 731). But at other times he recognized that if drug abreaction worked it did so by heightening suggestibility.[61] In fact, the role of the barbiturates in inducing hypnoidal states was well recognized, so that drug abreactions were often conceptualized as hypnotic-suggestive in nature.[62] As Sargant appreciated, this meant that the success of abreaction depended on personal factors, including not only the skill and experience of the physician but especially on his relationship to the patient, a relationship in which he and others recognized important suggestive-hypnotic elements. We have seen that Ferenczi had tried to restrict the function of hypnosis in neocatharsis by proposing that suggestion emanated from the patient in the form of an *auto*suggestion. Similarly, psychoanalytically oriented physicians, such as Horsley, conceived the relationship of the patient to the therapist in transferential-libidinal terms, minimizing the role of suggestion by viewing it simply as a means for liberating the patient's authentic memories by releasing inhibitions and censorship. But Sargant had no qualms about conceiving the abreactive rapport between patient and physician in largely suggestive terms.

3. This leads me to my third point, which concerns the problem of cure rates. The efficacy of abreaction was not easy to assess because, as physicians often acknowledged, under conditions of war it was difficult to obtain adequate statistics, especially follow-up statistics. Sargant claimed a return rate of 60 percent in acute cases treated very early, but acknowledged much less success in chronic cases.[63] His experience matched that of physicians in World War I, who started out with considerable therapeutic optimism but ended on a more pessimistic note. Indeed, what is striking about Sargant's case records is that almost every patient was returned not to combat but to civilian life, or at best to his unit for noncombat duty only. Grinker and Spiegel also admitted that less than 2 percent of their patients actually returned to combat (*WN*, 235).

61. Sargant, "Physical Treatment of Acute War Neurosis," 575; idem, "Some Observations on Abreaction with Drugs," 202.

62. For discussions of the relationship between chemical abreaction and hypnosis see Palmer, "Abreactive Techniques: Ether," 86; Brenman and Gill, *Hypnotherapy: Survey of the Literature*, 29–34; C. H. Rogerson, "Narco-Analysis with Nitrous Oxide," *British Medical Journal* 1 (1944): 811; H. J. Eysenck and W. Linford Rees, "States of Heightened Suggestibility: Narcosis," *Journal of Mental Science* 91 (1945): 301–10; Harold Rosen and Henry J. Myers, "Abreaction in a Military Setting," *Archives of Neurology and Psychiatry* 57 (1947): 163.

63. Sargant, lecture given at Nottingham on "the work of neurological unit attached to a general hospital" (circa 1940s),Box 9, E.2/2, Sargant Papers, Contemporary Medical Archives Centre, Wellcome Institute for the History of Medicine.

Like Pavlov's dogs traumatized by the Leningrad floods who continued ever afterwards to be sensitive to the sound of running water, even after treatment the traumatized soldier continued to be so sensitive to the sounds and sights of danger that he remained unfit for duty. The only solution was to discharge him so that he would never again have to face the terrors of battle. Cure by means of emotional discharge? Cure by means of *military* discharge would be closer to the truth.

The problem was that of chronicity. Some authors tried to explain the lack of treatment success by criticizing the selection procedures that had been installed in order to weed out unqualified men. But British and American physicians actively involved in the conduct of the war, such as G. W. B. James, Harold Palmer, Frederick Hanson, Albert Glass, Alfred O. Ludwig, and Stephen W. Ranson, believed that the situation was far more complex. Malcolm Farrell and John Appel reported that in spite of rejecting approximately 12 percent of the men examined at induction sites as mentally or emotionally unfit for service, high rates of breakdown continued in soldiers with no past history of neuropathology.[64] Experienced veterans were especially vulnerable to breakdown when fatigue set in. During the North African campaign the Americans in particular became aware how inadequate their arrangements were for forward psychiatry, with the result that during the subsequent Italian campaign the emphasis began to shift towards the deployment of methods for preventing the production of nonrecoverable cases. The British adopted methods of prevention from the start, but the Americans only belatedly applied such lessons of prevention originally learned from the experience of the First World War. Methods of prevention included the placing of psychiatrists in combat areas, the introduction of incentives in the form of rotating tours of duty, the improvement of group morale and cohesion, better management and training of men, more effective leadership and command, and so on.

One result of the emphasis on prevention was what might be described as a normalization of the war neuroses. Physicians in the military were taught that symptoms that might be considered abnormal in civilian life were to be regarded as normal in the stressful situation of battle; the "normal battle reaction" thus defined was to be the standard against which pathological responses were henceforth evaluated.[65] From this perspec-

64. Malcolm J. Farrell and John W. Appel, "Current Trends in Military Psychiatry," *American Journal of Psychiatry* 101 (1944):12.

65. See for example Stephen W. Ranson, "The Normal Battle Reaction: Its Relation to the Pathological Battle Reaction," in *Combat Psychiatry*, 3–11.

tive, most reactions to battle were considered transient and eminently reversible if treated in a "first aid," firm, and no-nonsense way. Above all, physicians must avoid the mistake of evacuating soldiers with normal battle stress to a hospital in the rear, for this allowed the symptoms to be associated with the "gain" of being removed from combat and hence to become elaborated and fixed.[66]

These developments tended to marginalize the pioneers of war-time chemical abreaction such as Sargant, for although he had always advocated early treatment, except for casualties from Dunkirk and Normandy he was dealing with patients who had been evacuated from front lines hundreds of miles distant from his hospital. Moreover, the merits of narcotic abreaction now came in for reassessment. The cornerstones of forward psychiatric treatment at the battalion level were rest, hot food, showers, sedation, explanation, reassurance, suggestion, and exhortation. There was no place for intravenous therapy of any kind, including chemical abreaction, since intravenous treatment threatened to make a litter patient of an otherwise ambulatory soldier. But at the army neuropsychiatric "exhaustion" centers ten to fifteen miles behind the line, where patients with more severe disorders were sent for further observation and treatment, drug abreaction or narcosynthesis was regarded as

66. See J. R. Rees, "Three Years of Military Psychiatry in the United Kingdom," *British Medical Journal* 1 (1943): 1–6; Herbert Spiegel, "Preventive Psychiatry with Combat Troops," *American Journal of Psychiatry* 101 (1944): 310–15; S. A. MacKeith, "Lasting Lessons of Overseas Military Psychiatry," *Journal of Mental Science* 92 (1946): 542–50; Frederick R. Hanson, ed., *Combat Psychiatry: Experiences in the North Africa and Mediterranean Theaters of Operation, American Ground Forces, World War II, The Bulletin of the U.S. Army Medical Department* 9, Supplemental Number (November 1949); Farrell and Appel, "Current Trends in Military Neuropsychiatry," 12–19; John W. Appel and Gilbert W. Beebe, "Preventive Psychiatry: An Epidemiologic Approach," *The Journal of the American Medical Association* 131 (1946): 1469–75; William C. Menninger, "Lessons from Military Psychiatry for Civilian Psychiatry," *Mental Hygiene* 30 (1946): 571–89; idem, "Development of Psychiatry in the Army in World War II," *War Medicine* 8 (1945): 229–34; idem, "Psychiatric Experience in the War, 1941–46," *American Journal of Psychiatry* 103 (1947): 577–86; R. H. Ahrenfeldt, *Psychiatry in the British Army in the Second World War* (London, 1958); A. J. Glass and R. J. Bernucci, eds., *Neuropsychiatry in World War II* (Washington, D.C., 1966); William Hausman and David McK. Rioch, "Military Psychiatry: A Prototype of Social and Preventive Psychiatry in the United States," *Archives of General Psychiatry* 16 (1967): 727–39; Franklin D. Jones and Robert E. Hales, "Military Combat Psychiatry: A Historical Review," *Psychiatric Annals* 17 (1987): 525–27; G. Belenky, ed., *Contemporary Studies in Combat Psychiatry* (New York, 1987); Gerald N. Grob, "World War II and American Psychiatry," *Psychohistory Review* 19 (1990): 41–69.

one component of an array of abbreviated therapeutic techniques aimed at quickly restoring the patient to duty. Even so, some advocates of barbiturate narcosis stated that its value was limited, of use only in severe anxiety cases accompanied by stupor or amnesia, severe hysterical conversion phenomena, or ticlike disorders; these represented only 5 percent of the total admissions to the centers. Moreover, it was held that very few soldiers treated in this way could be returned to the front; their emotional reactions to battle were deemed to be so overwhelming, or their premilitary personalities so inadequate, that they were unable to resume battle duty.[67]

One influential American figure in the rise of preventive psychiatry, Albert J. Glass, went further in his criticisms of abreaction. The key principles of prevention were immediacy, proximity, and expectancy: the soldier was to be treated as *immediately* as possible, as *proximate* or near to his unit as feasible, and his treatment was to be undertaken with the *expectation* that he would respond favorably and would be returned to duty. Those principles created new disciplinary requirements for patient and physician alike. For the soldier, prevention meant accepting the transient nature and normality of his symptoms and the expectation of returning to duty as soon as possible. For the physician, prevention meant placing the welfare of the group—ultimately, the welfare of the military—above that of the individual patient, in recognition of the paramount obligation to help defeat the enemy. As Glass put it in his reflections on the lessons that had been learned from the experience of forward psychotherapy in World Wars I and II, the physician had to stop being a remote spectator of battle located at a safe distance from the theater of combat, in order to become a forward observer with first-hand knowledge of the battle situation and hence an exponent to the men of the realities of war. From this perspective, the abreactive techniques popularized by Sargant, Grinker, and Spiegel were deemed by Glass to contribute to the very chronicity they were designed to cure. In light of the importance of notions of imitation in the present study, it is interesting that according to Glass the problem of abreaction was precisely the play-acting techniques of *imitative identification* involved. That is, Glass objected to abreaction because

67. See for example Alfred O. Ludwig, "Psychiatry at the Army Level," in *Combat Psychiatry*, 95–96. For a study that also tempered enthusiasm for narcotic abreaction by finding no statistically significant difference in the effectiveness of narcoanalysis as compared to other methods of treatment in combat exhaustion see Leo H. Bartheimer, Lawrence S. Kubie, Karl A. Menninger, John Romano, and John C. Whitehorn, "Combat Exhaustion," *Journal of Nervous and Mental Disease* 104 (1946): 358–89, 489–525.

the procedure required the physician to join in the dramatic performance, thereby encouraging him to identify with the soldier's suffering at
the expense of his identification with the larger goal of winning the war:
"The patients pleaded or insisted that they should not be sent back to
combat. As the therapist participated with his patient in the dramatic reliving of battle scenes, he almost invariably identified with the distress
and needs of the patient and was therefore impelled to promise relief
from future battle trauma."[68] For Glass, the value of forward psychiatry
was that the experience of working in the combat zone hardened the
physician to his own duty to help the soldier fight the war.

In addition, Glass argued that by seeking to uncover neurotic conflicts
and connect the battle symptoms with the patient's previous emotional
difficulties, abreaction readily provided both patient and therapist
with persuasive explanations for the failure in combat ("PCZ," 727). He
therefore deprecated as defeatist and fatalistic the formula "stress +
personality = reaction" that had seemed to explain why everyone had his
breaking-point. He rejected depth analysis, including the recovery of repressed memories, in favor of more authoritarian and suggestive methods contrived to help patients forget their traumatic experiences: "In
essence, the repeated success of brief forward treatment demonstrated
the need for repressive or suppressive therapy rather than for uncovering
depth techniques, for it became clear that the goal of treatment for the
purpose of return to combat duty was the restoration of previous defenses instead of attempts to alter or reorganize the personality" ("PCZ,"
729).[69] In this respect, Sargant received vindication of a sort: for Glass,
good suggestion was to be used to counter bad suggestion by inducing the
patient to erase or forget his sorrows. We might put it, in terms that
would be congenial to Borch-Jacobsen's position, that preventive

68. Albert J. Glass, "Psychotherapy in the Combat Zone," *American Journal of Psychiatry* 110 (1954): 727; hereafter abbreviated "PCZ." Just as in World War I, so in World War
II the theme of the influence of the mollycoddling mother lurked behind the criticism that
in abreaction the physician became identified and hence complicitous with the soldier's
loss of courage and childlike abjection. See for example Ralph Greenson, "Practical Approaches to the War Neuroses," *Menninger Clinics* 9 (1945): 196, where the soldier's breakdown is interpreted as a regression to the oral stage of identification with the mother. The
fear of "momism" as a factor in American psychiatry in World War II and during the Cold
War is the subject of Rebecca Plant's "Philip Wylie's America: Culture and Politics in the
1930s and 1940s" (Ph.D. diss., Johns Hopkins University, in preparation).

69. For a historical review of military psychiatry that criticizes narcosynthesis along the
same lines see Franklin D. Jones and Robert E. Hales, "Military Combat Psychiatry: A
Historical Review," *Psychiatric Annals* 17 (1987): 526.

psychiatry depended for its efficacity on a suggestive expectation of (or insistence on) forgetting that a therapeutics of suggestion was created to achieve.[70]

Advocates of preventive psychiatry claimed great success. The number of reported psychiatric casualties fell dramatically.[71] The assumption was that the psychopathological problems of combat had been overcome, and this assumption also guided the psychiatric approach to the wars in Korea and Vietnam, where similar preventive techniques for the management of combat neuroses were applied.[72] Because combat fatigue, or "gross stress reaction," was considered a transient and reversible condition, there was no recognition of the chronic form of the condition. Yet the optimism that had guided the rise of preventive psychiatry in World War II did not last. In a comparative study of the incidence of war neuroses in combatant nations in the Second World War, Kalinowsky in 1950 argued on the basis of largely anecdotal evidence that countries such as Germany and Russia, which had rejected compensation and had had the toughest attitude towards the patient, had experienced the lowest incidence of traumatic neurosis; he suggested that the continued exis-

70. In using a modified version of narcosynthesis to treat Vietnam veterans with PTSD, Kolb perceives a "set" on the part of the subject to act out emotionally before another whom he trusts to accept and understand—in other words, the abreaction depends on sympathetic imitation or identification. However, unlike Borch-Jacobsen, Kolb does not interpret this "set" or expectation in terms of play-acting or game theory but more psychoanalytically, in terms of the patient's narcissistic and super-ego projections (Lawrence C. Kolb, "The Place of Narcosynthesis in the Treatment of Chronic and Delayed Stress Reactions," in *The Trauma of War: Stress and Recovery in Viet Nam Veterans*, ed. Stephen M. Sonnenberg, Arthur S. Blank, Jr., and John A. Talbott [Washington, D.C., 1985], 215–16).

71. According to one source, 80 percent of the American fighting men who succumbed to acute stress in World War II were returned to some kind of duty, usually within a week. Thirty percent of these were returned to combat units (J. Ellis, *The Sharp End of War: The Fighting Man in World War II* [London, 1980], cited by Judith Herman, *Trauma and Recovery* [New York, 1992], 26).

72. See Arthur J. Glass, "Effectiveness of Forward Neuropsychiatric Treatment," *Bulletin of the U.S. Army Medical Department* 7 (1947): 1034–41; A. J. Glass, K. L. Artiss, and J. J. Gibbs, "The Current Status of Army Psychiatry," *American Journal of Psychiatry* 117 (1961): 673–83; G. W. Beebe and M. E. DeBakey, *Battle Casualties* (Springfield, Illinois, 1952); William J. Tiffany, Jr., "The Mental Health of Army Troops in Viet Nam," *American Journal of Psychiatry* 123 (1967): 1585–86; H. Spencer Bloch, "Army Clinical Psychiatry in the Combat Zone, 1967–68," *American Journal of Psychiatry* 126 (1969): 289–98; Douglas R. Bey, Jr., and Walter E. Smith, "Organizational Consultation in a Combat Unit," *American Journal of Psychiatry* 128 (1971): 401–6.

tence in the United States of an "unexpectedly high percentage of chronic cases" was largely due to the patient's expectation of receiving compensation or the wish for some other form of secondary gain.[73] But his claim, which reflected the values of preventive psychiatry, was challenged when, starting in the 1960s, long-term follow-up studies of World War II and Korean veterans, prisoners of war, and other disaster victims began to show that many individuals retained unaltered their original startle reactions, recurring nightmares, irritability, headaches, and related symptoms 15 or 20 years after the trauma. Since psychiatrists encountered fresh cases that had never sought treatment, it was argued that such cases were not retaining symptoms in order to protect a pension.[74] Studies by William Niederland, Leo Eitinger, Paul Chodoff, and others of "concentration camp syndrome", which also began to appear in the late 1950s and the 1960s, demonstrated a similar tendency to severe chronicity and the delayed onset of symptoms.[75]

73. Lothar B. Kalinowsky, "Problems of War Neuroses in the Light of Experiences in Other Countries," *American Journal of Psychiatry* 107 (1950): 340–46. Kalinowsky acknowledged that after return to the homeland prisoners of war of various nationalities seemed to face greater problems of adjustment than other soldiers (345). Kardiner and Spiegel, among others, also emphasized the role of secondary gain in producing chronicity (Kardiner and Spiegel, *War Stress and Neurotic Illness*, 4). My thanks to Ben Shephard for drawing my attention to Kalinowsky's paper.

74. Norman Q. Brill and Gilbert W. Beebe, *A Follow-Up Study of War Neuroses* (Washington, D.C., 1955); Dobbs and W. P. Wilson, "Observations on Persistence of War Neurosis," *Diseases of the Nervous System* 21 (1960): 1–6; Herbert C. Archibald, Dorothy Long, Christine Miller, and Read D. Tuddenham, "Gross Stress Reaction in Combat—A 15-Year Follow-Up," *American Journal of Psychiatry* 119 (1962): 317–22; Herbert C. Archibald and Read D. Tuddenham, "Persistent Stress Reaction After Combat," *Archives of General Psychiatry* 12 (1965): 475–81; V. A. Kral, L. H. Pazder, and B. T. Wigdor, "Long-Term Effects of a Prolonged Stress Experience," *Canadian Psychiatric Association Journal* 12 (1967): 175–81; Finn Askevold, "War Sailor Syndrome," *Psychotherapy and Psychosomatics* 27 (1976–77): 133–38.

75. See especially J. Bastiaans, *Psychosomatic After Effects of Persecution and Incarceration* (Amsterdam, 1957); Bruno Bettelheim, *The Informed Heart* (New York, 1960); William Niederland, "The Problem of the Survivor, Part I: Some Remarks on the Psychiatric Evaluation of Emotional Disorders in Survivors of Nazi Persecution," *Journal of the Hillside Hospital* 10 (1961): 233–47; idem, "Psychiatric Disorders among Persecution Victims: A Contribution to the Understanding of Concentration Camp Pathology and Its After-Effects," *Journal of Nervous and Mental Disease* 139 (1964): 458–74; idem, "Clinical Observations on the 'Survivor Syndrome,'" *International Journal of Psychoanalysis* 49 (1968): 313–15; idem, "Introductory Notes on Psychic Trauma," in *Psychic Traumatization: After-Effects in Individuals and Communities*, ed. H. Krystal and W. G. Niederland (Boston,

With a sublime disregard for the most vulnerable victims of the war, the concentration camp survivors, whom he did not mention, Kalinowsky in 1950 had emphasized the "almost unlimited endurance of human beings at the time of catastrophes" in terms that made any breakdown appear to be largely the result of either a wish for secondary gain or preexisting personality problems.[76] In keeping with that attitude, there was a decided tendency in the postwar period to downplay the causative role of external trauma in traumatic neurosis in favor of the influence of pretraumatic experiences. In the name of recognizing the "sit-

idem, "The Survivor Syndrome: Further Observations and Dimensions," *Journal of the American Psychoanalytic Association* 29 (1981): 413–25; Henry Krystal, ed., *Massive Psychic Trauma* (New York, 1968); Leo Eitinger, "Pathology of the Concentration Camp Syndrome," *Archives of General Psychiatry* 5 (1961): 371–9; Leo Eitinger, "Concentration Camp Survivors in the Postwar World," *American Journal of Orthopsychiatry* 32 (1962): 367–75; Axel Strom, Sigvald B. Refsum, Leo Eitinger, Odd Gronvik, Arve Lonnum, Arne Engeset, Knut Osvik, and Bjorn Rogan, "Examination of Norwegian Ex-Concentration Camp Prisoners," *Journal of Neuropsychiatry* 4 (1962): 43–62; L. Eitinger, *Concentration Camp Survivors* (London, 1964); Paul Chodoff, "Late Effects of the Concentration Camp Syndrome," *Archives of General Psychiatry* 8 (1963): 323–33; K. R. Eissler, "Perverted Psychiatry?" *American Journal of Psychiatry* 123 (1967): 1352–58; H. Z. Winnik, "Further Comments Concerning Problems of Late Psychopathological Effects of Nazi-Persecution and Their Therapy," *Israel Annals of Psychiatry and Related Disciplines* 5 (1967): 1–16; H. Z. Winnik, "Psychiatric Disturbances of Holocaust ('Shoa') Survivors. Symposium of the Israeli Psychoanalytic Society," *Israel Annals of Psychiatry and Related Disciplines* 5 (1967): 91–100; H. Z. Winnik, "Contribution to Symposium on Psychic Traumatization Through Social Catastrophe," *International Journal of Psycho-Analysis* 49 (1968): 298–301; Ruth Jaffe, "Dissociative Phenomena in Former Concentration Camp Inmates," *International Journal of Psychoanalysis* 49 (1968): 310–12; Erwin K. Koranyi, "Psychodynamic Theories of the 'Survivor Syndrome,'" *Canadian Psychiatric Association Journal* 14 (1969): 165–74; Hector Warnes, "The Traumatic Syndrome," *Canadian Psychiatric Association Journal* 17 (1972): 391–96; Hilel Klein, "Delayed Effects and After-Effects of Severe Traumatization," *Israel Annals of Psychiatry and Related Disciplines* 12 (1974): 293–303.

76. Kalinowsky, "Problems of War Neuroses in the Light of Experiences in Other Countries," 343. Kalinowsky acknowledged that the hysterical symptoms of the First World War might have been manifested in the less flamboyant psychosomatic or "gastric" neuroses of the Second (342). But in favor of his own position he mentioned the absence of breakdowns among the victims of the London blitz, suggesting that such victims had had nothing to gain by emotionally collapsing; he also observed that in Japan, where neurotic reactions among soldiers were rare, psychiatric manifestations had even less importance among civilians in the atomic bomb raids on Hiroshima and Nagasaki (343). That claim would be challenged by Robert Lifton's studies of the psychic effects of Hiroshima, *Death in Life: Survivors of Hiroshima* (Chapel Hill, North Carolina, 1991).

uational" or environmental character of the war neuroses, interest in the actual trauma of combat diminished as the emphasis fell on the soldier's inadequacies and on the potentially transient nature of the traumatic response. Psychoanalysts contributed to the same tendency by stressing the fundamental importance of the patient's infantile-libidinal history. When reparations for concentration camp victims were instituted, the German psychiatrists appointed to the German courts to evaluate cases often refused compensation to survivors on that basis. But starting in the 1960s Niederland and his colleagues condemned German psychiatrists for refusing to acknowledge the connection between the survivor's subsequent mental disorders and the reality of persecution. During the same period we find a comparable emphasis on the reality of the external trauma and its long-term effects in the emerging literature on child abuse.[77]

In a further reaction against the principles of preventive psychiatry, the role of the psychiatrist in the military also came in for criticism.[78] During and after the Vietnam War there began to be a general perception that the methods of preventive psychiatry contributed to problems of chronicity because, by becoming the tool of the military, psychiatrists had ignored the subjective needs and experiences of the individual, especially the possibility of delayed effects, and had downplayed the reality of the external trauma.[79] As an active participant in the movement against

77. For a valuable history of the factors contributing to the conceptualization of child abuse see Ian Hacking, "The Making and Molding of Child Abuse," *Critical Inquiry* 17 (1991): 253–88.

78. Already in 1946 William Needles had strongly criticized preventive psychiatry for its cult of return-to-duty statistics and use of quotas to return men to duty regardless of their psychiatric condition; for failing to back up its therapeutic claims with follow-up studies; for encouraging a hostile and even brutal attitude toward the neurotic patient, treating him as a malingerer rather than as someone who was sick; and for encouraging the psychiatrist to put the needs of the military above those of his patients (William Needles, "The Regression of Psychiatry in the Army," *Psychiatry* 9 [1946]: 167–85).

79. Kolb noted that the unit rotation imposed by DEROS (Date Expected to Return from Overseas) actually undermined group identification and morale, rather than enhanced it, as had been expected (Lawrence C. Kolb, "The Post-Traumatic Stress Disorders of Combat: A Subgroup with a Conditioned Emotional Response," *Military Medicine* 149 [1984]: 239). Kolb, who reported seeing very few chronic cases from World War II during many years of clinical practice, expressed surprise at discovering so many chronic cases among Vietnam veterans. He sought an explanation in the specific character of the Vietnam War, including the public hostility to the war back home and the indifference of the Veterans Administration.

the Vietnam War and a psychiatrist committed to aiding the Vietnam veteran, Robert Lifton strongly criticized physicians for forgetting their duty to serve the patient to the extent of placing the military desire to restore the soldier to service over the needs of the patient. In the United States, concern for the chronic problems of the Vietnam veteran, more than any other factor, precipitated today's consensus about the severity of the effects of external trauma on the human psyche.

Renewed interest in trauma has not only reopened debate over how to conceptualize trauma but also over how to treat the traumatic neuroses, especially chronic cases. With the present revival of hypnotic and chemical abreaction for the treatment of the dissociative disorders and PTSD, remembering has once again replaced forgetting as the therapeutic goal. Of special interest in this regard is Onno van der Hart's and Paul Brown's article, "Abreaction Re-Evaluated" (1992), which critically reviews the literature on abreaction, including the hypnotic-cathartic method of Breuer and Freud, its revival in World War I, the use of chemical abreaction in World War II and the Vietnam War, and present applications. As a commentary on a century of debate, the article is especially arresting because it clearly endorses the participatory model of abreaction and condemns as misconceived and inadequate abreaction on the model of emotional discharge. The authors assess the by now familiar World War I debate between Brown, McDougall, and Myers over abreaction as emotional discharge versus abreaction as psychic integration in order to resolve it in favor of the necessity of integrating and working through the trauma by relivings that are held to uncover accurate memories of the past. Van der Hart and Brown disparage Sargant's use of false scenarios to stimulate abreaction, suggesting that he might well have exposed his patients to retraumatization.[80]

80. Onno van der Hart and Paul Brown, "Abreaction Evaluated," *Dissociation* 5 (1992): 127–40. The authors criticize Sargant's use of invented traumatic scenarios to obtain abreactions by observing: "When a 'cure' was reported, it could well have followed a further dissociation of the traumatic memory, i.e., a retreat into health rather than integration" (135). Cf. Onno van der Hart, Kathy Steele, Suzette Boon, and Paul Brown, "The Treatment of Traumatic Memories: Synthesis, Realization, and Integration," *Dissociation* 6 [1993]: 162–80, where van der Hart and his colleagues advocate a phase-oriented approach to trauma resolution that goes beyond the simple conversion of traumatic memory into narrative memory and that again emphasizes the necessity of personality integration. For other criticisms of the "hydraulic" model of catharsis and abreaction see M. J. Horowitz, *Stress Response Syndromes* (Northvale, New Jersey, 1986); and D. Brown and F. Fromm, *Hypnotherapy and Hypnoanalysis* (Hillsdale, New Jersey, 1986).

When in 1984 Kolb applied narcosynthesis for the first time to Vietnam veterans with chronic symptoms of PTSD, he too rejected the notion of abreaction as a mechanical "derepression," emphasizing instead the need for further "working through" and integration of the trauma. However, noting the continued uncertainty as to whether the memories apparently disclosed by these means were veridical in nature, he concluded with Sargant that what was abreacted and integrated during treatment was not necessarily an accurate memory of what had happened.[81] (Kolb's use of videotaped recordings of the abreactions, to be played back to the patient after treatment, was designed to circumvent postabreactive forgetting or denial—testifying to the persistence of a problem that, as we have seen, had previously troubled other advocates of the participatory or insight model of abreaction.)[82] But the entire force of recent work on trauma is to propose that traumatic relivings are exact and automatic repetitions of traumatic reality. A comprehensive hypothesis about the literal and material nature of traumatic memory is based on that claim. The theoretical and therapeutic problems raised by that hypothesis are the topic of the final two chapters of this book.

81. Kolb, "Recovery of Memory and Repressed Fantasy in Combat-Induced Post-Traumatic Stress Disorder," 273. Kolb's work points up the difficulty of evaluating the therapeutic effects of abreaction. He reported that although abreaction not only gave subjective relief to the patient and also terminated the "flashbacks" and other dissociative events, it did not alter the principal symptoms of the condition. In another report on the same series of patients, he noted that at the time of follow-up two years later, most of the patients were engaged in other forms of treatment so that the specific role of the narcosynthetic treatment in outcomes could not be assessed (Kolb, "The Place of Narcosynthesis," 221).

82. Kolb, "The Place of Narcosynthesis," 218.

VII

The Science of the Literal:
The Neurobiology of Trauma

[What are we to make of the fact that modern neurobiology and certain versions of poststructuralism share virtually the same account of psychic trauma? What does the science of physician Bessel van der Kolk have in common with the literary theory of critic Cathy Caruth?]

I will anticipate my answer to those questions by observing that what van der Kolk and Caruth share is a commitment to two claims: (1) an empirical claim, according to which traumatic symptoms, such as traumatic dreams and flashbacks, are *veridical* memories or representations of the traumatic event; and (2) an epistemological-ontological claim, according to which those same symptoms are *literal* replicas or repetitions of the trauma and that as such they stand outside representation. The two claims are not identical: one might want to assert on the basis of the evidence that traumatic dreams are veridical, in the sense that they accurately reflect or depict or express the traumatic origin, without embracing the view that they have passed beyond representation. But in the work of van der Kolk and Caruth the distinction between those two claims is more or less systematically elided: the veridical becomes the literal, with the result that traumatic repetition is understood as an acting out that completely interrupts the norms of representation. My aim in this and the following chapter is to show that both claims are not only problematic or false but poorly supported by the evidence and arguments adduced in their favor.

The relationship between empiricism and deconstruction is clearly evident in the network of mutual citations by which they are linked. Caruth makes use of the work of van der Kolk and his colleagues to support her arguments, while van der Kolk and his associates return the compliment by appealing to one of Caruth's statements to defend their ideas. In a recent book, *Traumatic Stress: The Effects of Overwhelming Experience on Mind, Body, and Society* (1996), coauthors van der Kolk, McFarlane, and Weisaeth introduce their work by claiming: "Unlike other forms of psychological disorders, the core issue in trauma is reality." They then cite the following assertion by Caruth: "'It is indeed the truth of the traumatic experience that forms the center of its psychopathology; it is not a pathology of falsehood or displacement of meaning, but of history itself.'"[1] The statement by Caruth comes from her introduction to a collection of essays she edited, *Trauma: Explorations in Memory* (1995), a collection which includes an article by van der Kolk and van der Hart and which takes as *its* starting point the modern definition of trauma codified in 1980 by the American Psychiatric Association in the diagnosis of post-traumatic stress disorder, or PTSD. "While the precise definition of post-traumatic stress disorder is contested," Caruth remarks with reference to van der Kolk's work, "most descriptions generally agree that there is a response, sometimes delayed, to an overwhelming event or events, which takes the form of repeated, intrusive hallucinations, dreams, thoughts or behaviors stemming from the event, along with numbing that may have begun during or after the experience, and possibly also increased arousal to (and avoidance of) stimuli recalling the event."[2] According to Caruth, this definition belies a "very peculiar fact" (*TEM*, 4), namely that the pathology cannot be defined either by the event itself or in terms of any distortion of the event due to the personal significance attached to it. Rather, the pathology consists "solely in the *structure of its experience* or reception: the event is not assimilated or experienced fully at the time, but only belatedly, in its repeated *possession* of the one who experiences it" (*TEM*, 4). And this means in turn that the Freudian concepts of repression and unconscious meaning cannot be applied to the traumatic experience. Referring to Freud's discussion in *Beyond the Pleasure Principle* (1920) of the traumatic nightmare, whose tendency to repeat unpleasure posed a challenge to his wish-fulfillment theory of dreams, Caruth asserts:

1. Bessel A. van der Kolk, Alexander C. McFarlane, and Lars Weisath, *Traumatic Stress: The Effects of Overwhelming Experience on Mind, Body, and Society* (New York, 1996), 6; hereafter abbreviated *TS*. In fact, van der Kolk et al. slightly modify Caruth's statement, but without changing the sense.

2. Cathy Caruth, introduction to *Trauma: Explorations in Memory* (Baltimore and London, 1995), 4; hereafter abbreviated *TEM*.

> The returning traumatic dream startles Freud because it cannot be understood in terms of any wish or unconscious meaning, but is, purely and inexplicably, the literal return of the event against the will of the one it inhabits. Indeed, modern analysts as well have remarked on the surprising *literality* and nonsymbolic nature of traumatic dreams and flashbacks, which resist cure to the extent that they remain, precisely, literal. It is this literality and its insistent return which thus constitutes trauma and points toward its enigmatic core: the delay or incompletion in knowing, or even in seeing, an overwhelming occurrence that then remains, in its insistent return, absolutely *true* to the event. . . . The traumatized, we might say, carry an impossible history within them, or they become themselves the symptom of a history that they cannot entirely possess. (*TEM*, 5)

She thus proposes that behind the uncertainty as to the truth of trauma lies a "larger question raised by the fact of trauma, what Shoshana Felman . . . calls the 'larger, more profound, less definable crisis of truth . . . proceeding from contemporary trauma.' Such a crisis of truth extends beyond the question of individual cure and asks how we in this era can have access to our own historical experience, to a history that is in its immediacy a crisis to whose truth there is no simple access" (*TEM*, 6).

In this and the following chapter I examine van der Kolk's and Caruth's commitment to the notion of the literal nature of traumatic dreams, "flashbacks," and other traumatic repetitions, and their literal, belated return. What are the scientific-epistemological origins of the concept of the literal? By what empirical evidence and theoretical-historical arguments is it justified? More generally, what are the stakes involved in conceiving trauma as at once a completely literal record of the truth of trauma and as structured by a temporal gap or aporia—what van der Kolk and his colleagues call the "black hole" of trauma (*TS*, 3)—such that it is always experienced *too late* for knowledge and representation, not only by the actual victims of trauma but by all of us, even those of us who were not there—indeed by history itself?

In this chapter I discuss the work of van der Kolk, a prolific author who in the last decade has emerged as a central figure in the scientific study of trauma. In the next chapter, I focus on Caruth's theories.

Traumatic Dreams, Flashbacks, and the Literal

As I have already observed, the codification of recent scientific ideas about the nature of the traumatic experience may be traced back to the year 1980, when the diagnosis of PTSD officially entered the American

Psychiatric Association's third edition of the *Diagnostic and Statistical Manual of Mental Disorders* (*DSM-III*), the first standardized nosology of American psychiatry and one that became the basis for all subsequent training and research in the field. The *DSM-III*, and the definition of PTSD, have undergone two more revisions since that time. PTSD was originally identified in *DSM-III* as an anxiety disorder with four diagnostic criteria: A. Traumatic event. B. Reexperiences of the event. C. Numbing phenomena. D. Miscellaneous symptoms. The traumatic event was vaguely defined as an event that is "generally outside the range of usual human experience" and as involving a "recognizable stressor that would evoke significant symptoms of distress in almost everyone."[3] From the

3. *Diagnostic and Statistical Manual of Mental Disorders* (Third Edition, *DSM-III*) (American Psychiatric Association, 1980), 236, 238. As Allan Young has shown, the "affect" logic of traumatic stress was expanded in *DSM-III* to include events for which the patient even felt pleasure—for example, the pleasure the American soldier in Vietnam might have experienced in perpetrating unusual violence in an intentional way. Moreover, since *DSM-III* the definition of the traumatic event as lying "outside the range of usual human experience" has been applied in flexible ways in order to accommodate the kinds of events routinely treated as the point of origin of PTSD, which is to say that the traumatic origin depends on contextual and cultural meanings that can be quite local and specific. For instance, the official American condemnation of rape and torture in Vietnam was at odds with the actual morality of some combat units for the majority of whose members, at least, such acts came to be a routine part of daily life and hence did *not* lie outside the range of usual human experience. Such differences of meaning render epidemiological studies of PTSD especially likely to produce widely divergent assessments of the prevalence of the disorder, because meaning variance can be more or less controlled within single studies but is much harder to control across them. Thus a study conducted by the National Vietnam Veterans Readjustment Study (NVVRS) found the prevalence of PTSD among veterans to be *six* times greater than that in a study conducted by the Centers for Disease Control (130). In actual psychiatric practice, however, the problem of meaning variance is resolved into a problem of determining the best technique or instrument for the collection of data, the aim being to produce a plurality of measures which, because of their lack of exact fit, produce gaps or differences that then help stimulate the search for more knowledge (Allan Young, *The Harmony of Illusions: Inventing Post-Traumatic Stress Disorder* [Princeton, New Jersey, 1995], 124–35); hereafter abbreviated *HI*. For a discussion of the failures of the data-gathering methodology in the NVVRS study by a former Vietnam serviceman who believes that the courage of the soldier in Vietnam has been hijacked by phony stories about combat experience and untrue claims about PTSD see B. G. Burkett and Glenna Whitley, *Stolen Valor: How the Vietnam Generation Was Robbed of Its Heroes and Its History* (Dallas, Texas, 1998), 225–30; my thanks to Ben Shephard for the reference to this book.

start reexperiences were linked directly to the trauma in such a way as to imply that traumatic dreams, intrusive images, and other forms of repetition were exact relivings or replicas of the original event or situation. Kardiner's discussion of traumatic nightmares in chronic cases of war neuroses from the First World War was influential in post-Vietnam psychiatry because he seemed to suggest that traumatic nightmares lacked the symbolic transposition, substitutability, displacement, and fantasmatic elaboration characteristic of ordinary dreams and hence could be understood as accurate replays of the traumatic origin—even if, as I argue in chapter 6, his dream descriptions were ambiguous in this regard. But in the early 1980s there were no adequate, controlled studies of people who developed repetitive nightmares after traumatic experiences in adulthood, a topic that was now on the agenda in the wake of the official recognition of PTSD.[4]

In 1984 van der Kolk and his colleagues undertook to make up that lack by examining how the dreams of people suffering persistent traumatic nightmares after combat differed from those of people suffering life-long nightmares without any combat experience. The authors started from the assumption—fairly well accepted in the literature of sleep research, though not securely established—that traumatic nightmares represented a unique kind of dream: like "night-terrors," they tended to occur early in the sleep cycle (that is, in non-Rapid Eye Move-

---

4. The *DSM-III* description of PTSD made the phases of intrusive reexperiences and denial or numbing basic to the diagnosis of post-traumatic stress. More research needs to be done on the period 1960–1980, when studies of long-term psychiatric symptoms in concentration camp survivors slowly converged with follow-up studies of World War II combat troops, investigations of psychiatric problems in the returning Vietnam veteran, and studies of victims of Hiroshima, sexual abuse, and other civilian and natural disasters. From this convergence came the movement in the United States to achieve official recognition of the existence of a single diagnostic entity, post-traumatic stress disorder. It would be interesting to trace the mediating influence of the cognitive-experimental work of the analytically trained Mardi Jon Horowitz, especially his ideas about the role of intrusive imagery in stress disorders, on the formulation of PTSD and the subsequent literalization of traumatic memory. The key text is the first edition of Horowitz's book, *Stress Response Syndromes* (New York, 1976), republished in a revised edition in 1986. Robert Jay Lifton's emphasis on the role of imagery also played an important role in the formulation of PTSD. "We live on images" is the first sentence of his book, *The Broken Connection: On Death and the Continuity of Life* (New York, 1979), cited in Cathy Caruth, "An Interview with Robert Jay Lifton," *TEM*, 133. For an interesting study of the politics of the PTSD movement, see Wilbur Scott, *The Politics of Readjustment* (New York, 1993).

ment, or non-REM, sleep) and were accompanied by body movements and other signs of autonomic arousal; yet unlike night-terrors and like REM-anxiety dreams in this regard, which occur later in the sleep cycle, traumatic nightmares often had elaborate content. In short, traumatic nightmares appeared to have characteristics intermediate between the two most commonly described dream phenomena, night terrors and anxiety dreams. Twenty-five subjects were selected for the experiment, of which fifteen individuals, all diagnosed with PTSD, belonged to the combat nightmare group, and ten belonged to the noncombat life-long nightmare sufferers group. Each subject was given a three- to five-hour semistructured psychiatric interview that covered various routine items plus several additional items pertaining to nightmares, posttraumatic disorder, and aspects of premorbid adjustment. Another interviewer questioned the subjects about their military experiences. The patients were then given a variety of psychiatric tests, including a Rorschach test. Two subjects spent two and three nights respectively in the sleep laboratory for all-night EEG recordings. Apart from the laboratory observations obtained from these two subjects, all the information about the time, nature, and content of the nightmares was obtained from the self-reports given by the patients, presumably during their psychiatric interviews.

On this basis, van der Kolk and his associates drew a contrast between the dreams experienced by the noncombat nightmare sufferers and the dream characteristics of the combat group. They found that the nightmares of the noncombat nightmare sufferers had the typical characteristics of REM-anxiety dreams: they occurred late in the sleep cycle, were rarely repetitive, had "dreamlike content," and were not accompanied by body movements. By contrast, the nightmares of the combat victims were more likely to occur early in the sleep cycle and "when they had content, they were repetitive dreams that were usually exact replicas of actual combat events." The authors also reported that all the subjects in the combat nightmare group experienced body movements accompanying their nightmares. Some subjects reported occasional physical attacks on bed partners. The authors did not specify to which of the groups the two subjects chosen for study in the sleep laboratory belonged; but since both subjects experienced unpleasant nightmare episodes combined with intense bodily arousal (one of them pulled off his EEG wires and walked about the room in a dazed state and later reported having thoughts that he was in an ambush), they presumably belonged to the combat nightmare group. The laboratory findings thus appeared to confirm the validity of the contrast drawn between the kinds of nightmares experienced by the two groups.[5]

Van der Kolk's findings are at odds with those of another researcher in the field of dream studies. In an article on traumatic nightmares in Vietnam veterans published the same year, in a collection of essays edited by van der Kolk, sleep researcher Milton Kramer and his colleagues came to different conclusions about the nature of traumatic dreams. There were disparities between the two studies. Van der Kolk et al. compared a group of combat veterans who had experienced nightmares with a group of nonmilitary persons who nevertheless were life-long nightmare sufferers: for their experiment, the critical difference was therefore between the combat and the noncombat experience. Kramer's was an inter-combat comparison, for he and his associates compared two groups of combat veterans, one of which experienced nightmares and one of which did not. Thus they compared a group of Vietnam veterans with PTSD (seven subjects), all of whom reported experiencing nightmares, with a group of Vietnam veterans (eight subjects), only one of whom suffered from PTSD, and none of whom experienced sleep disturbances. As in the case of van der Kolk's study, each subject in Kramer's study was evaluated by a variety of standard psychiatric assessments. But in addition, and unlike what occurred in van der Kolk's experiment, every subject slept for six consecutive nights in the sleep laboratory. On the first four nights, each time the subject awakened for more than one minute an enquiry was made as to the content of his dreams; on night five, the subject was, in addition, wakened twice from each sleep stage and dream data were collected; night six was a sleep-through night, and no dream information was obtained. Subjects were asked to rate on a one- to five-point scale the fear and anxiety associated with each dream experience at the time the dream was obtained. Each night and each morning additional information about the subject's experience was gathered. This included measures of personal feelings and stress.

Kramer and his colleagues reported that traumatic nightmares occurred in both REM and non-REM sleep, i.e., that traumatic nightmares

5. Bessel van der Kolk, Robert Blitz, Winthrop Burr, Sally Sherry, and Ernest Hartmann, "Nightmares and Trauma: A Comparison of Nightmares After Combat with Lifelong Nightmares in Veterans," *American Journal of Psychiatry* 141 (1984): 187–90. In a subsequent discussion, coauthor Hartmann mentions the existence of a third "combat control" group of eleven men who had had severe combat experience but no nightmares (Hartmann, "Who Develops PTSD Nightmares and Who Doesn't," in *Trauma and Dreams*, ed. Deidre Barrett (Cambridge, Massachusetts, 1996), 103.

were distributed across the sleep periods to which the anxiety dream and the night-terror were usually restricted respectively, thus suggesting that posttraumatic nightmares had specific features that differentiated them from both anxiety dreams and night-terrors. Moreover, many of the dreams of the nightmare sufferers tended to have military content, unlike the dreams of the control group. But from my perspective the Kramer group's most interesting finding was that within the nightmare suffering group there was a discrepancy between the group's impressions about the contents of their dreams and the actual contents as recalled by them under controlled laboratory conditions. Thus the authors observed that although the patients tended to think that their dreams were entirely concerned with their combat experiences, according to the sleep-laboratory evidence military references occurred in only 50 percent of the dreams in this group, so that a large segment of dream life did not revolve around the experience of combat. "It is not that there is no concern about the war," the authors remark, "it is just that it is not the only concern that the patient has. The impression that one might draw is that there are concerns about both areas and that, perhaps, they are interrelated."[6]

In order to illustrate this point, the authors gave a brief synopsis of five dreams of a Vietnam veteran who had volunteered to sleep in the Dream-Sleep Laboratory at the VA Hospital (not a subject from the original study).[7] They reported his dreams as follows:

6. Milton Kramer, Lawrence S. Schoen, and Lois Kinney, "The Dream Experience in Dream-Disturbed Vietnam Veterans," in *Post-Traumatic Stress Disorder: Psychological and Biological Sequelae*, ed. Bessel van der Kolk (Washington, D.C., 1984), 91; hereafter abbreviated "DE."

7. One might speculate that for van der Kolk and his group the fact that the dreams described in detail by Kramer belonged to a subject who was not part of the original dream study, combined with the fact that apparently the patient was not diagnosed with PTSD, undermined the validity and relevance of Kramer's research contribution. On the other hand, Kramer treated those dreams as characteristic of the general nature of traumatic dreams in PTSD patients, dreams he regarded not as literal memories but as symbolic and meaningful in character. Kramer in a later paper described the same dream sequence in order to repeat his conclusions (Milton Kramer, "The Nightmare: A Failure in Dream Function," *Dreaming* 1 [1991]: 284–85). In 1984 also, Peter Ziarnowski and Daniel C. Broida noted that "despite the centrality of nightmares in PTSD's symptom pattern, little research has been done to examine the frequency, content, classification, or imagery of the disturbing dreams" and concluded that the twenty-three Vietnam veterans they studied experienced an equal frequency of dreams about real, possible, and "bizarre" (or unreal) events or experiences ("Therapeutic Implications of the Nightmares of Vietnam Combat Veterans," *VA Practitioner* 1 [1984]: 63).

Dream 1: He was trying to cross a river and a big guy was trying to help him. It seemed that they were in the Mekong Delta. Every time they tried to get to the other side, they were back at the beginning.

Dream 2: He was a young boy at home and was fighting with one of the neighborhood kids. His brother was around. The kid they were fighting was one who later died in Vietnam.

Dream 3: He went with others to a construction site. He was a sapper, like in Vietnam. He blew the site up. The consequence of blowing up the construction site was that his brother's house was flooded.

Dream 4: He remembers being a young child in his parents' home. There was some argument and he left. He wrote them postcards without a forwarding address. Every time they would try to contact him, he had moved somewhere else.

Dream 5: He was driving very fast, and his wife was worried that he would have an accident. ("DE," 91–92)

The authors reported that the patient was surprised by the focus on his family and wife, for prior to his laboratory experience he did not think his family was an issue. They therefore concluded that there was an interaction between the past and the present and between the military and nonmilitary experiences of the patient's life: "[A]ssuming that the dream reflects the current concern of the individual, a great deal of the personal concern of the dream-disturbed patient is on matters not obviously related to the traumatic experience. . . . We suggest that attention to the current and past situation is necessary in the treatment of PTSD victims as their dreams are focused on the traumatic event and other problematic areas in their lives ("DE," 92).

In sum, nothing in the Kramer study indicated that the traumatic dreams of Vietnam veterans were exact replicas of combat experience with no other manifest or latent content. Yet this is precisely what van der Kolk and his associates claimed in their paper of 1984 on the basis of reports by their subjects as to the content of their dreams—reports that, we have just seen, Kramer's study suggested ought not to be taken at face value. Indeed, what is most noteworthy about the paper by van der Kolk and his colleagues in this regard is the flimsiness of the evidence on which its claims are based. The authors seem to have been committed in advance to the idea that, unlike other anxiety dreams, the traumatic nightmare is a unique phenomenon precisely because it is an accurate replica of the traumatic experience. From the table of results presented by them, it seems that the subjects were simply asked to check "Yes" or "No" to the following statements regarding the nightmare characteristics: "Replicates an actual event," and "Repetitive, almost exactly same content." All the PTSD subjects responded "Yes" to the latter item, but only eleven of the PTSD subjects did to the former, which is perhaps why the authors are careful in their wording: "when they [the traumatic nightmares] had content, they were repetitive dreams that were usually exact replicas of actual combat events."[8] But the article contained no reports or descriptions of the actual content of any of the subjects' dreams.[9] Yet from now on, van der Kolk's 1984 paper on traumatic nightmares in PTSD patients (his one publication on the topic, so far as I am aware)[10] will be cited regularly in support of a sweeping neurobiological theory and research program in which the exact, indeed literal, nature of the traumatic experience, as manifested in nightmares, so-called flashbacks, and other reen-

8. Van der Kolk et al., "Nightmares and Trauma," 188.

9. For a discussion of the contents of the dreams of the subjects in the PTSD combat group see Ernest Hartmann, "Who Develops PTSD Nightmares and Who Doesn't," in *Trauma and Dreams*, 100–113. Hartmann begins by saying that "the most common description was something like" the report that follows, suggesting that he may be offering a generalized characterization rather than an exact transcription. He reports one dream in this way: "'I was right back there. Just the way it was. Shells were bursting all over the place. My buddy was hit by a shell right next to me. There was blood and screaming. It was just the way it was. Please don't ask me any more details.'" Though the phrase "just the way it was" captures the veteran's and Hartmann's sense that the dream is a literal "replay" (Hartmann's term) of the traumatic scene, Hartmann reports that there was often "at least one significant alteration" in reported dreams, an "added element": in the dream the subject often saw and experienced himself as suffering the same fate as the actual victim or victims in the scene. For example, one of the dreamers reported that he dreamed that he was having to go through the experience of identifying the dead bodies of his buddies all over again: "'When we went out to the trucks, I identified a couple of guys and in my nightmare then—I identify myself!'" What has "burned itself into memory is the actual scene plus this terrible realization of 'I'm alive; he's dead,'" Hartmann comments. In other words, a theoretically disowned scenario of imitation or identification—a guilty identification, as Hartmann points out, hence an ambivalent one—resurfaces at the center of what is otherwise taken to be a literal replica of the traumatic origin.

10. Van der Kolk and Ducey claimed that Rorschach tests of the same group of fifteen PTSD patients who were the subjects of the 1984 study confirmed their findings (Bessel A. van der Kolk and Charles Ducey, "Clinical Implications of the Rorschach in Post-Traumatic Stress Disorder," in *Post-Traumatic Stress Disorder: Psychological and Biological Sequelae*, 30–42). For the argument that the evidence from such tests may be suspect owing to the role of the subjects' and the examiners' expectations, see Joseph Masling, "The Influence of Situational and Interpersonal Variables in Projective Testing," *Psychological Bulletin* 57 (1960): 65–85; idem, "Differential Indoctrination of Examiners and Rorschach Responses," *Journal of Consulting Psychology* 29 (1965): 198–201, cited by Mikkel Borch-Jacobsen, "L'Effet Bernheim (fragments d'une théorie de l'artefact généralisé)," *Corpus, revue de philosophie* No. 32 (1997): 165–66.

actments, is an absolutely crucial element. Henceforth, traumatic dreams will be treated as *literal memories* of the traumatic event, with the result that the topic of traumatic nightmares does not even appear in the index of van der Kolk's book, *Traumatic Stress*, but is assimilated into the subject of traumatic memory. As a consequence, for at least van der Kolk's influential research group, the traumatic nightmare has disappeared as an independent focus of study.

Moreover, intrinsic to van der Kolk's characterization of the traumatic nightmare as not just an exact or truthful representation of the past but a literal registration of the traumatic event is the claim that, if the trauma is "etched" or "engraved" on the mind and brain with timeless accuracy, this is because it is encoded in the brain in a different way from ordinary memory. According to van der Kolk, the result is that traumatic memory, in its literality, is not integrated into ordinary awareness but is cut off or "dissociated" from consciousness and hence is unavailable for normal recollection. Traumatic disorders are thus simultaneously disorders of remembering and of forgetting: the traumatic "stimulus" seems to be recorded in the brain with unparalleled vividness and accuracy but, precisely because the traumatic event is so shattering, the memory of the trauma is radically dissociated from symbolization, meaning, and the usual processes of integration. Van der Kolk and his colleagues therefore suggest that traumatic memories may be less like what some theorists of memory have called "declarative" or "explicit" memories, involving conscious awareness of events, than like "nondeclarative" or "implicit" or "procedural" memories—"memories of skills and habits, emotional responses, reflexive actions, and classically conditioned responses" (*TS*, 281) involving implicit memory subsystems thought to be associated with particular areas of the brain. When traumatic memories belatedly return to possess the patient they do so in the form of nightmares, flashbacks and other reenactments that are understood to be unchanged over time and, in Caruth's words, "absolutely *true* to the event" (*TEM*, 5).

Van der Kolk and his associates are careful to state their views in conditional terms, implicitly acknowledging their hypothetical-speculative status. But they continually adduce an impressive array of research references and citations that, taken together, lend their views the appearance of solidity. The success of this operation has been so great that their opinion that traumatic dreams are "just about what happened" is now powerfully entrenched.[11] Melvin R. Lansky has observed in this connection

(see note)

---

11. For example, van der Kolk and Greenberg observe: "Although current evidence is far from conclusive, it appears that REM nightmares [anxiety dreams] are dreamlike

that the nightmares characteristic of victims of trauma seem, in the dreamer's experience and when told to others, to be so intimately related to the traumatic event that "questions about the relation of trauma to dream are often not asked, much less answered." He observes with reference to the work of van der Kolk and others that such dreams are "often thought of as more or less like an inflammatory response that follows an injury, that is, simply and unquestionably part of the entire picture of injury and response to it," with the result that in most of the literature on trauma, even the psychoanalytic literature, traumatic nightmares are not conceptualized as "true" dreams subject to the mechanisms of condensation, displacement, symbol formation, and secondary revision, or as subserving, however subtly, the fulfillment of wishes. He remarks that applied to the problem of posttraumatic nightmares, [the wish-fulfillment hypothesis is "particularly counterintuitive: how can one imagine the return to a scene that is, by definition, traumatic to the point of generating overwhelming and unmanageable terror?"]The result is that the dream dynamics specific to posttraumatic nightmares are seldom studied as true dreams by trauma researchers. As he states, "The posttraumatic nightmare is viewed more like an affectively laden memory—more like a nighttime flashback than like a true dream. Such a view of the posttraumatic nightmare, explicit or implicit, is the prevailing one."[12]

Lansky acknowledges that Freud himself can be interpreted as endorsing such a view, a point to which I return when discussing Caruth's interpretation of Freud and the traumatic nightmare. But—continuing a

---

(oneiric), and often have admixtures of other experiences, whereas stage II and III nightmares [traumatic nightmares] are usually exact movielike (eidetic) recreations of the traumatic experience itself" (Bessel A. van der Kolk and Mark S. Greenberg, "The Psychobiology of the Trauma Response: Hyperarousal, Constriction, and Addiction to Traumatic Reexposure," in *Psychological Trauma*, ed. Bessel A. van der Kolk (Washington, D.C., 1987), 70. For similar claims see Ernest Hartmann, *The Nightmare: The Psychology and Biology of Terrifying Dreams* (New York, 1984), 188, 200; idem, "Who Develops PTSD Nightmares and Who Doesn't," 105–9; and Matthew Friedman, "Biological Approaches to the Diagnosis and Treatment of PTSD," in George S. Everly, Jr., and Jeffrey M. Lating, *Psychotraumatology: Key Papers and Core Concepts in Post-Traumatic Stress* (New York and London, 1995), 176–77. But the debate continues. See for example the recent discussion of traumatic dreams in Deirdre Barrett, ed., *Trauma and Dreams* (Cambridge, Massachusetts, 1996) and Ben Shephard's review, "Perchance We Dream?" in the *Times Literary Supplement*, December 12, 1997.

12. Melvin R. Lansky with Carole R. Bley, *Posttraumatic Nightmares: Psychodynamic Explorations* (Hillsdale, New Jersey, 1995), 3–4.

debate that, as I have shown in previous chapters, has a history going all the way back to the First World War—Lansky thinks that matters are considerably more complex, arguing on the basis of detailed analyses of nightmares in Vietnam veterans that the nightmare scenario cannot be reduced to the status of a veridical memory of the traumatic scene, let alone a literal repetition in the sense of van der Kolk and Caruth. He maintains that such dreams contain defensive, screening, and other features that make them, if not identical to other kinds of dreams, at least subject to similar mechanisms of distortion, symbolic substitution, and displacement. He particularly emphasizes the role of family worries and concerns in contributing to the content of such traumatic dreams.

The same debate surrounds the concept of the flashback, to which Lansky refers in one of the passages cited above. In recent years the term flashback has come to be used to describe the daytime reexperiences or reenactments of the traumatic event, reexperiences that are held to be a characteristic symptom of PTSD. The flashback takes the form of recurrent, intrusive images or sensations associated with the traumatic event, or of a sudden feeling that the traumatic event is literally happening all over again. The victim feels as if he has returned to the perceptual reality of the traumatic situation, and it has become orthodox to interpret such flashback experiences as the literal return of dissociated memories of the event. The term flashback implies the cinematic possibility of literally reproducing or cutting back to a scene from the past and hence expresses the idea that the trauma victim's experiences are exact "reruns" or "replays" of the traumatic incident.[13]

The term flashback officially entered American psychiatry in *DSM-IIIR* (1987) when it was used to describe intrusive phenomena in precisely these terms.[14] But the concept of the flashback has a prior history that is at odds with the meaning currently assigned to it. As Frankel has recently shown in a useful review, the term was first employed in the literature on drug-induced hallucinations in the late 1960s and 1970s. Probably bor-

---

13. For a discussion which makes explicit the association between the concept of the flashback and the movies see Arthur S. Blank, Jr., "The Unconscious Flashback to the War in VietNam Veterans: Clinical Mystery, Legal Defense, and Community Problem," in *The Trauma of War: Stress and Recovery in Viet Nam Veterans*, ed. Stephen M. Sonnenberg, Arthur S. Blank, Jr., and John A. Talbott (Washington, D.C., 1985), 297.

14. In *DSM-IIIR* the concept of the flashback was added to the diagnostic criteria for PTSD (250). According to an anecdote, Lawrence C. Kolb used the term "flashback" as early as 1970 or 1971 (Wilbur J. Scott, "PTSD in DSM-III: A Case in the Politics of Diagnosis and Disease," *Social Problems* 37 [1990]: 289).

rowed from the film industry and used widely in the popular media, the flashback label was employed to describe the altered states of awareness, ideation, perception, and emotion experienced by "flashbackers" when they were using hallucinogens. There was no consensus as to whether flashbacks ought to be interpreted as reminiscences or revisualizations of prior drug experiences, themselves understood as dominated by fantasy and imagination, or as fantasies relating to past events; but researchers tended to emphasize the imaginative and role-playing dimension of the flashback experience.

Yet as the concept moved out of the literature on drug-induced hallucinations and into the literature on trauma in the 1980s, including the literature on civilian trauma, the flashback began to be understood as an exact or literal memory of a past event. Indeed, as therapy increasingly focused on the goal of remembering the victimization and hence on retrieving and elaborating traumatic memories, even when such memories had remained dormant for years, there developed a need, or at least a powerful desire, to treat such memories as literal replays of the past. "To be relevant to the theory," Frankel has remarked, "this recall must be viewed as an accurate report or revisualization of the traumatizing past events."[15] Thus Judith Herman and other leading theorists of child abuse interpret flashbacks, dissociative states, intrusions, and other symptoms as exact reiterations of childhood traumas, regarding the recollection and integration of such memories as necessary for effective treatment.[16] (Herman and van der Kolk have jointly authored an article suggesting the same direct links between child abuse and borderline personality disorder.)[17]

15. Fred H. Frankel, "The Concept of Flashbacks in Historical Perspective," *International Journal of Clinical and Experimental Hypnosis* 42 (1994): 328; hereafter abbreviated "CF."

16. See Judith Herman, *Trauma and Recovery: The Aftermath of Violence—From Domestic Abuse to Political Terror* (New York, 1992), and my discussion of Herman's work in chapter 3.

17. Judith L. Herman, J. Christopher Perry, and Bessel A. van der Kolk, "Childhood Trauma in Borderline Personality Disorder," *American Journal of Psychiatry* 146 (1989): 490–95. The authors claim that borderline personality disorders can be traced directly back to childhood trauma, but they rely on the patients' self-reports and make no attempt to verify the accuracy of those self-reports. However, in another much-cited article Herman and Schatzow claim that, in a study of fifty-three victims of incest, three out of four patients were able to validate their memories of abuse by obtaining corroborating evidence of some kind (Judith Herman and Emily Schatzow, "Recovery and Verification of Memories of Childhood Sexual Trauma," *Psychoanalytic Psychology* 4 [1987]: 1–14). For a

In criticizing this development, Frankel comments: ["There are two quite separate but intertwined issues here: Is the remembered trauma historically true, and is the recall of historically true trauma necessary for healing?] Although both of these questions might be answered affirmatively by many clinicians, there is little or no empirical basis for either conclusion" ("CF," 328). He goes on to argue that, although since the early 1980s flashbacks have been managed in the literature and in clinical practice as if they are accurate or literal memories, [most of the studies that have assumed the veridical nature of the flashback have been too poorly designed to evaluate the question at issue; meanwhile, other studies have demonstrated the extent to which the relivings and other reenactments are subject to the effects of distortion and "contagion" from environmental and other cues.] In this context, Frankel emphasizes the potential significance, for the theory of the literal nature of the flashback, of attempts to understand the phenomenon in neurophysiological terms. "If the analogy could be drawn to an epileptogenic focus triggered by specific factors," he writes, "the imagery might be accepted as the reenactment of a fixed memory, and therefore veridical. In other words, the memories could be viewed as pristine rather than affected by psychosocial factors" ("CF," 329–30). Briefly surveying various research efforts along these lines, Frankel notes the absence of convincing evidence and concludes on a skeptical note. According to him, some of the content of a flashback might be accurate, but in many instances there is no evidence to support that belief. "The content might thus be true, false, or confabulated" ("CF," 331). Commenting on the increasing use of imagery-based therapies to elicit long-delayed recall, he argues that dependence on the concept of the flashback to validate such recall is unsupported by evidence in the literature, because "[c]ontextual factors such as expectation, in addition to the suggestibility of patients and the social construct of role-playing, influence in a crucial way the creation and content of flashbacks." He concludes by proposing that the use of the term flashback has misled the field by creating a "false sense of confidence in the accuracy of autobiographical reports" ("CF," 331–32).[18]

---

critique of the evidentiary basis of their study see Richard Ofshe and Ethan Watters, *Making Monsters: False Memories, Psychotherapy, and Sexual Hysteria* (New York, 1994), 305–12.

18. For a similar criticism of the flashback as a depiction of the unvarnished truth see Hacking, *Rewriting the Soul: Multiple Personality*, 253–54. Cf. Fred H. Frankel, "Dissociation in Hysteria and Hypnosis: A Concept Aggrandized," in *Dissociation: Clinical and Theoretical Perspectives*, ed. S. J. Lynn and J. W. Rhue (New York, 1994), 80–93, for a criticism of the current tendency to strip the hysteria concept and PTSD of their psychosocial context

It must be recognized that Frankel's article represents only one side of an intensely polarized debate, pitting the champions of the accuracy of traumatic memories against skeptics, such as Frankel and others, who on a variety of (not always mutually compatible) grounds cast doubt on such claims. In part, the debate over the nature of traumatic nightmares and flashbacks reflects a clash between cognitively oriented scientists, such as van der Kolk, who conceptualize the mind in information-processing, neurophysiological terms and reject the Freudian concept of repression or unconsciously motivated forgetting, and more psychoanalytically oriented clinicians, such as Lansky, who retain a Freudian commitment to notions of psychic conflict and repression and who tend to oppose a literal interpretation of the traumatic dream.[19] But that is not the whole story, for Frankel's criticisms of the theory of the flashback are based not on psychoanalytically derived concerns but problems of hypnotic suggestibility and confabulation. It is a sign that cracks in the edifice of modern trauma theory are beginning to appear that, in its latest edition of the *Diagnostic and Statistical Manual of Mental Disorders* (*DSM-IV*, 1994), the American Psychiatric Association is more cautious than formerly about the veridical or literal nature of traumatic phenomena. It especially notes the vexing question of the role of suggestion and confabulation in the dissociative disorders, although in keeping with its claim to be atheoreti-

---

and to reify instead the simplistic concept of trauma-triggered dissociation, thereby eliminating questions of meaning and ignoring the role of interviewer expectation and imitation-suggestion in the production of the case history. Frankel concludes by noting that "the articulation of the concepts of dissociation, hysteria, and hypnosis continue to elude us, probably largely because the concepts themselves are imprecise" (91).

19. For cognitively oriented critiques of Freud's theory of dreams based on often highly speculative neurobiological ideas that I find unconvincing see *Mind, Psychoanalysis and Science*, ed. Peter Clark and Crispin Wright (Oxford, England, 1988). For the ongoing debate see Nicholas Wade, "Was Freud Wrong? Are Dreams the Brain's Start-Up Test?" (*New York Times*, January 6, 1998) and the ensuing correspondence, "Dream Study Supports Freud" (*New York Times*, January 12, 1998).

20. Thus in its discussion of "Dissociative Identity Disorder" (formerly "Multiple Personality Disorder"), a disorder which is closely allied with PTSD and which is thought to be caused by childhood trauma, the *DSM-IV* observes that controversy surrounds the accuracy of reports of severe physical and sexual abuse in the disorder, because childhood memories may be subject to distortion and also because individuals with the disorder tend to be highly hypnotizable and "especially vulnerable to suggestive influences" (*The Diagnostic and Statistical Manual of Mental Disorders, DSM-IV* [American Psychiatric Association, 1994], 485). In other words, one of the distinguishing marks of the disorder that aids

cal it refuses to adjudicate the dispute.[20] But champions of the literal nature of the traumatic nightmare and flashback are not going to yield their position without a fight.

Defending the Science of the Literal I: Trauma As Icon

Van der Kolk has responded to the controversy over the nature and accuracy of traumatic memories in a variety of ways designed to preserve his position. His responses include, first, a flat denial that traumatic memories can be implanted by suggestion and hence falsified. "The 'false memory' movement claims that thousands of unsuspecting white middle-class women go to therapists who implant false memories of abuse in their minds," he protests. "However, current research shows that . . . there is no evidence that traumatic memories can be simply implanted in people's minds" (TS, 37). Against the charge of the False Memory Syndrome Foundation that an epidemic of false accusations is being created by psychotherapists, van der Kolk writes: "The basis of this organization's existence is the claim that there exists a 'false memory syndrome' that is produced by inappropriate techniques used by therapists working with highly suggestible patients. To date, there is no scientific evidence for the existence of such a syndrome" (TS, 37–38).[21] But he adduces no evidence in support of this claim; nor does he discuss the burgeoning literature on the other side, or the studies made during and after World War II questioning the authenticity of memories recovered by hypnosis or barbiturates or other hypnotic drugs. In fact, he cites with approval studies that use the hypnotic barbiturate, sodium amytal, and hyp-

[handwritten marginal note: false memory syndrom is false]

---

in the diagnosis—high hypnotizability, and consequent tendency to distortion— is the very thing that tends to undermine the credibility of that diagnosis. DSM-IV notes this but is unable to resolve the dilemma.

21. Founded in 1992, the False Memory Syndrome Foundation is a political action group that represents the interests of parents accused by their children of terrible acts of childhood abuse, as well as the interests of skeptical professionals and numerous others, by arguing—in the courts, the media, and elsewhere—that many accusations are due to false memories suggestively produced by ideologically committed therapists in the patient-therapist relationship. For an instructive discussion of the disputes between the multiple personality movement and the False Memory Syndrome Foundation see Hacking, *Rewriting the Soul*, 113–27.

22. Mark S. Greenberg and Bessel A. van der Kolk, "Retrieval and Integration of Traumatic Memories with the 'Painting Cure,'" in *Psychological Trauma*, 193–94; hereafter abbreviated "RI."

nosis to retrieve details of the trauma.[22] Yet in another context he concedes that victims of trauma are likely to be "vulnerable to suggestion and to the construction of explanations for their trauma-related affects that may bear little relationship to the actual realities of their lives"—a statement that acknowledges what he elsewhere denies, namely, the patient's potential for hypnotic confabulation in reconstructing the trauma.[23] On the basis of these contradictions, one might put it that van der Kolk's theory of the literal nature of the traumatic imprint is designed to answer the charge of "false memory" on the grounds that, being automatic, corporeal, and nonautobiographical in character, the imprint cannot be feigned or suggested.

Another of van der Kolk's responses to the controversy over the nature and accuracy of traumatic memories is to reject studies of memory in normal individuals exposed to videotaped stresses in the laboratory—studies associated with researcher Elizabeth Loftus and others that usually emphasize the tendency to memory distortion and decay—on the grounds that laboratory experiments of this kind cannot replicate the eye-witness experience of actual traumas and hence cannot provide meaningful guides to understanding traumatic memories (TS, 279–

23. Bessel van der Kolk and Rita Fisler, "Dissociation and the Fragmentary Nature of Traumatic Memories: Overview and Exploratory Study," *Journal of Traumatic Stress* 8, no. 4 (1995): 510; hereafter abbreviated "DF."

Judith Herman has recently answered the charge of suggestion and confabulation. On the one hand, she acknowledges that the naive use of suggestion can lead to errors in the attribution of sexual abuse, especially errors in the attribution of the trauma of ritual satanic abuse. On the other hand, she recommends the use of hypnotherapy and sodium amytal to fill gaps in the patient's memory, and rejects all skepticism about traumatic recollections: "When these arguments [about fabrication] were first proposed several years ago, I found them almost ludicrously implausible, and thought that their frank appeal to prejudice would be transparent at once. The women's movement had just spent twenty years deconstructing the presumption that women and children are prone to lie, fantasize, or fabricate stories of sexual violation. If any principle has been established, surely it was that victims are competent to testify to their own experience" (Herman, *Trauma and Recovery* [New York, 1997], 180, 245). It is not clear how she proposes to reconcile the contradiction.

24. Loftus has long held that the brain hardly ever represses memories of profoundly important events that it then reproduces accurately or literally at a later time (Elizabeth Loftus and Katherine Kitchum, *The Myth of Repressed Memory: False Memories and Allegations of Sexual Abuse* [New York, 1994]). Hacking reports that in a videotaping of his paper "The Intrusive Past," van der Kolk challenges Loftus by arguing that, although she is right about the sorts of things she studies, such as memories of isolated facts, schoolbook learning, and propositional memory generally, she knows nothing of the kind of memory that

80).[24] A related reaction is to appeal to a small but growing body of research on the study of memory in real-life witnesses of highly emotional events, research that purports to show that personally significant memories are extremely accurate and do not decay over time. For example, in a review cited by van der Kolk (*TS*, 281–82), investigator Sven-Ake Christianson discusses the relation between emotional stress and eye-witness memory. He argues that, contrary to authors earlier this century, the majority of whom questioned the accuracy of eye-witness testimony, and contrary also to the claims of many laboratory studies demonstrating the inaccuracy of memories for highly emotional events, recent studies of real-life events tend to show that strong emotions aid in the accurate recollection and retention of details directly concerned with the emotion-arousing event. Although van der Kolk does not say so, there are problems with such real-life studies, not least the fact that in them "one usually does not know for sure what the original circumstances were, only what the person claims that they were," as Christianson himself puts it.[25] It is precisely in order to circumvent such difficulties of validation that researchers such as Christianson have used a simulation approach in the laboratory, the very approach that van der Kolk rejects. In short, van der Kolk depends in part for proof of the literal nature of traumatic memory on real-life studies that, as one of the leading researchers in the field acknowledges, cannot fully answer the question they are designed to answer.

But the most crucial response by van der Kolk to the controversy over the literal nature of the traumatic nightmare or flashback is to imply that proof of the accuracy and stability of traumatic memories based on the patient's testimony is beside the point. This is because for van der Kolk trauma is not an experience that is subject to the usual "declarative" or "explicit" or "narrative" mechanisms of memory and recall. Rather, what characterizes traumatic memory is precisely that it is "iconic" or "sensorimotor," by which he means that it is dissociated from all verbal-linguistic-semantic representation. Traumatic memories are "mute," because they cannot be expressed in verbal-linguistic terms. Citing the work of certain theorists of cognition and memory, who have suggested that the thought processes of children are different from those of adults in that at first children do not have "symbolic and linguistic" modes of

---

expresses itself not in sentences but in flashback scenes that are constituted by feelings and images (Hacking, *Rewriting the Soul*, 125).

25. Sven-Ake Christianson, "Emotional Stress and Eyewitness Memory: A Critical Review," *Psychological Bulletin* 112 (1992): 288.

representation, van der Kolk proposes that traumatic experiences reduce the adult to the status of a child who can only process experience on a sensorimotor or iconic level (*TS*, 279–302).[26] He also appeals to the work of psychoanalyst Henry Krystal who, on the basis of his work with concentration camp survivors, has identified a condition called *alexithymia* in traumatized patients, defined as a cognitive and affective disturbance involving the inability to differentiate affective states verbally or to grasp their personal meaning. The result is a tendency for such patients to corporealize their feelings by producing psychosomatic symptoms. Krystal has also emphasized the existence of impairments in the patient's dream and fantasy life. "[T]he 'operative thinking' characteristic of alexithymia interferes . . . with the capacity for symbolization, transference elaboration, and achievement of changes and sublimations," he has observed. "These patients do not *recognize* their emotions, because they experience them in an undifferentiated way, poorly verbalized, and because they have poor reflective self-awareness."[27] In an essay cited by Caruth,[28] van der Kolk and coauthor Greenberg build on such proposals in various ways:

> Amnesia can occur when traumatic experiences are encoded in sensorimotor or iconic form and therefore cannot be easily translated into the symbolic language necessary for linguistic retrieval. It is plausible that in situations of terror, the experience does not get processed in symbolic/linguistic forms, but tends to be organized on a sensorimotor or iconic level—as horrific images, visceral sensations, or fight/flight reactions. . . . The essence of the trauma ex-

26. But researcher Daniel Schacter, on whose ideas about implicit and declarative or narrative memory van der Kolk partly relies, disputes the latter's view that memories of emotionally traumatic events are accurately preserved "perhaps for ever" by citing several examples of memory distortion in such cases. He affirms instead that "memories are not simply activated pictures in the mind but complex constructions built from multiple contributions," a principle he believes also applies to traumatic memories (Schacter, *Searching for Memory: The Brain, the Mind, and the Past* [New York, 1996], 5–6, 205–7).

27. Henry Krystal, "Trauma and Aging: A Thirty-Year Follow Up" in Caruth, *TEM*, 83–84, 86. Cf. Krystal, "Trauma: Considerations of Its Intensity and Chronicity," in *Psychic Traumatization: After-Effects in Individuals and Communities*, ed. Henry Krystal and William G. Niederland (Boston, 1971), 11–28; idem, "Trauma and Affects," *Psychoanalytic Study of the Child* 33 (1978): 81–116; idem, "Alexithymia and Psychotherapy," *American Journal of Psychotherapy* 33 (1979): 17–31; idem, "Psychoanalytic Views on Human Emotional Damages," in *Post-Traumatic Stress Disorder: Psychological and Biological Sequelae*, 2–28.

28. Cathy Caruth, *Unclaimed Experience: Trauma, Narrative, and History* (Baltimore, 1996), 132, n. 6; hereafter abbreviated *UE*.

perience is that it leaves people in a state of "unspeakable terror." . . . These various cognitive formulations provide a model for the pathologies of memory associated with psychological trauma without resorting to the psychoanalytic notions of motivated forgetting, censorship, and repression. ("RI," 193–94)

On this basis, van der Kolk and Greenberg suggest that, since the traumatized individual is possessed by images, bodily sensations, and emotions that are dissociated from all semantic-linguistic-verbal representation, such images or memories may not be retrievable by verbal methods of cure but only by therapies of a nonlinguistic or "iconic" kind. In the same essay they therefore advocate the value of a "painting cure" in which a trauma victim's skill in producing "private iconic images" in the form of paintings, drawings, and dream reports is made use of to unlock the dissociated or repressed traumatic memories—the unexamined assumption being that certain elements of the paintings produced in this way are literal reproductions of the traumatic events in question.[29]

It is worth pausing to note that van der Kolk's ideas about the nature of icons and images are mistaken. Throughout his various discussions of the supposed iconic nature of traumatic memories he and his associates treat pictures and visual images as if they were inherently nonsymbolic, which is of course absurd. Worse, they appear to assume that there exists a fundamental opposition between pictures and visual images on the one hand and verbal representations on the other, to the extent that the former are associated with the idea of literal repetition as distinct from representation as such. One way van der Kolk, Caruth, Laub, and others express this idea is by suggesting that trauma creates a structural deficit, wound, or "hole" in the mind where representation ought to be (TS, 3; TEM, 6).[30] For Caruth, van der Kolk, and like-minded theorists, however, the no-

29. Similarly, in an essay also cited by Caruth (UE, 140, n. 6), Brett and Ostroff define trauma in terms of imagery that is held to be unmediated memory (Elizabeth A. Brett and Robert Ostroff, "Imagery and Posttraumatic Stress Disorder: An Overview," American Journal of Psychiatry 142 [1985]: 417–24). For a study that contradicts claims for the literality of traumatic imagery see Richard A. Bryant and Allison G. Harvey, "Traumatic Memories and Pseudomemories in Posttraumatic Stress Disorder," Applied Cognitive Psychology 12 (1998): 81–88.

30. For similar notions of a "gap" or hole in the ego into which are inserted literal memories of the trauma see also Dori Laub, "Bearing Witness or the Vicissitudes of Listening," in Testimony: Crises of Witnessing in Literature, Psychoanalysis, and History, ed. Shoshana Felman and Dori Laub (New York, 1992), 64–65, hereafter abbreviated "BW"; and Jonathan Cohen and Warren Kinston, "Primal Repression: Clinical and Theoretical Aspects," International Journal of Psychoanalysis 67 (1986): 338.

tion of a hole or deficit in representation is compatible with the claim that trauma nevertheless leaves its mark on the mind or brain in the form of a literal, "eidetic,"[31] or iconic "imprint" ("DF," 521). Permanently "etched" or "engraved" in a way that is theorized as standing outside all ordinary cognition, traumatic memory on this hypothesis returns in the form not of recollected representations—the usual understanding of the term "memories"—but of literal icons and sensations. (In Peircian terms, of course, such an imprint, etching, or engraving would be crucially *indexical*; indeed one suspects that the notion of the index comes far closer to the drift of van der Kolk's hypostatization of literalness, about which more is said in the next chapter.)[32]

It is also true that van der Kolk's terminology slips and slides. So for example he writes that "memories of the trauma tend, at least initially, to be experienced as fragments of the sensory components of the event: as visual images; olfactory, auditory, or kinesthetic sensations; or intense waves of feelings that patients usually claim to be representations of elements in the original traumatic event. What is intriguing is that patients consistently claim that their perceptions are exact representations of sensations at the time of the trauma" (*TS*, 287). The word "representations" occurs twice in this passage; but according to van der Kolk's own theories, the word "repetition" comes much closer to what he means. Some measure of the difficulty of tracking his thought stems from inconsistences of this sort, with which his writings are riddled.

Another characteristic slippage concerns the very possibility of verbally reconstructing or narrating the original traumatic experience. Van der Kolk and Fisler have recently invented a "Traumatic Memory Inventory" (TMI) designed to record whether memories of traumatic experiences are retrieved differently from memories of personally significant but nontraumatic events. They report that when asked to respond to a series of questions on the inventory, subjects considered most of the questions relating to nontraumatic memories as nonsensical, because none had olfactory, visual, auditory, or kinesthetic reliving experiences relating to such nontraumatic events, "and they denied having vivid dreams or flashbacks about them." Nor did they have amnesia for these events or "photographic recollections" of any of them. But when asked about their traumatic memories,

31. Bessel van der Kolk, Mark Greenberg, Helene Boyd, and John Krystal, "Inescapable Shock, Neurotransmitters, and Addiction to Trauma: Toward a Psychobiology of Post-Traumatic Stress," *Biological Psychiatry* 20 (1985): 318; hereafter abbreviated "IS."

32. Charles Sanders Peirce, *Collected Papers* (Cambridge, Massachusetts, 1960), 2: 143, 156–73.

all of our subjects reported that they initially had no narrative memories for the events; they could not tell a story about what had happened, regardless of whether they always knew that the traumas had happened, or whether they retrieved memories of the trauma at a later date. All these subjects, regardless of the age at which the trauma occurred, claimed that they initially "remembered" the trauma in the form of somatosensory flashback experiences. These flashbacks occurred in a variety of sensory modalities (visual, olfactory, affective, auditory, and kinesthetic), but initially the different modalities did not occur together. As the trauma came into consciousness with greater intensity, more sensory modalities were activated, and the subjects' capacity to tell themselves and others what actually had happened emerged over time.

This study . . . supports the idea that the very nature of a traumatic memory is to be dissociated, and to be stored initially as sensory fragments that have no linguistic components. All of the subjects in this study claimed that they only came to develop a narrative of their trauma over time. Indeed, five of the subjects who claimed to have been abused as children were, even as adults, unable to tell a complete story of what had happened to them. They merely had fragmentary memories to support other people's accounts and their own intuitive feelings that they had been abused.

Thus, the subjects' traumatic experiences were not initially organized in a narrative form, and they seemed to serve no communicative function. (*TS*, 288–89)

This extract strongly implies that although *initially* the trauma was resistant to narrativization, with the passage of time a narrative of "what actually had happened" became feasible for all but survivors of child abuse.

Elsewhere, however, van der Kolk emphasizes that the endeavor to integrate traumatic memory into consciousness necessarily subjects the former to the control of a "declarative" or "narrative" memory system whose job is precisely to enable the individual to put into words or narrate, for the first time, a past that has never been present to representation. Such a process, he argues, ineluctably distorts the truth, since declarative memory is by definition a construction and misrepresentation of the past. Thus if, as van der Kolk sometimes appears to propose, the narration of the traumatic memory is essential in order to cure PTSD, that narrative treatment is carried out at the cost of falsifying the traumatic origin. ["Merely uncovering memories is not enough; they need to be modified and transformed (i.e., placed in their proper context and reconstructed in a personally meaningful way)," he writes. "Thus, in therapy, memory paradoxically needs to become an act of creation rather than the static recording of events. . . . Like memories of ordinary events, the memory of the trauma needs to become merely a (often distorted) part of the patient's personal past" (*TS*, 19). "People seem to be unable to accept experiences that have no meaning," he also observes; "they will try to make sense of what they are feeling. Once people become conscious of intrusive elements of the trauma, they are likely to try to fill in the blanks and complete the picture." And he adds: "The stories that people tell about their traumas are as vulnerable to distortion as people's stories about anything else" (*TS*, 296–97). Caruth makes a similar point:

> The flashback or traumatic reenactment conveys . . . both *the truth of an event*, and *the truth of its incomprehensibility*. But this creates a dilemma for historical understanding. . . . [T]he capacity to remember is also the capacity to elide or distort . . . The danger of speech, of integration into the narration of memory, may lie not in what it cannot understand, but in that it understands too much. . . . The possibility of integration into memory and the consciousness of history thus raises the question, van der Kolk and van der Hart ultimately observe, "whether it is not a sacrilege of the traumatic experience to play with the reality of the past?" (*TEM*, 153–54)

In other words, fundamentally victims of trauma cannot witness or testify to the trauma in the sense of narrate and represent it to themselves and others: all they can do is perform the experience as if it were literally happening all over again. Caruth's performative theory of traumatic repetition as the literal return of reference finds its alleged scientific validation here. Taken together these passages show that van der Kolk and Caruth are committed to the widespread post-Holocaust assumption according to which any attempt to represent trauma (specifically the trauma of the camps, but here generalized to include any massive trauma) is distortive. As the psychoanalyst Dori Laub has put it: "the very circumstance of *being inside the event* . . . made unthinkable the very notion that a witness could exist. . . . One might say that there was, thus, historically no witness to the Holocaust, either from outside or from inside the event. . . . The historical imperative to bear witness could essentially *not be met during the actual occurrence*."[33] At the same time, the subject's notknowing of the trauma—his inability to speak or represent his experience—is what *guarantees* the return of the truth in the patient's traumatic repetitions. From this perspective, the concept of trauma as literal provides an essentially *ethical* solution to the crisis of representation posed by trauma in our time. As such a solution, trauma in its literality, muteness,

33. Dori Laub, "Truth and Testimony: The Process and the Struggle" (*TEM*, 66–68).

and unavailability for representation becomes a sacred object or "icon" that it would be a "sacrilege" to misappropriate or tamper with in any way.] This is how Laub presents the photograph of a victim of the genocide ("BW," 86).[34] We might put it that the entire theory of trauma proposed by van der Kolk and Caruth is designed to preserve the truth of the trauma as the failure of representation—[thereby permitting it to be passed on to others who can not only imaginatively identify with it but literally share in the communion of suffering.]

[Moreover, if—as van der Kolk and Caruth argue—testimony about the past is necessarily a misrepresentation, then any claim to discover in the traumatic repetition, including the traumatic nightmare and flashback, a content other than that of the literal imprint, has to be viewed as the falsifying effect of a desire to narrate or represent the truth of a traumatic origin that is inherently and constitutively exempt from all such representation—which is also to say that their theory of trauma is *immune to refutation*.[35]] Indeed, there is an important sense in which for

34. This is exactly how Caruth interprets Claude Lanzmann's film *Shoah*, which she understands not as a representation designed to transmit knowledge of the Holocaust but as a means of literally transmitting the trauma in its incomprehensibility to the viewer (*TEM*, 155).

35. See for example the following statement by van der Kolk and Fisler on the results of their "Traumatic Memory Inventory": "The question of whether the sensory perceptions reported by our subjects are accurate representations of the sensory imprints at the time of the trauma is intriguing. The study of flashbulb memories has shown that the relationship between emotionality, vividness and confidence is very complex, and does not necessarily reflect accuracy. While it is possible that these imprints are, in fact, reflections of the sensations experienced at the moment of the trauma, an alternative explanation is that increased activity of the amygdala at the moment of recall may be responsible for the subjective assignment of accuracy and personal significance. Once these sensations are transcribed into a personal narrative, they would presumably be subject to the laws that govern explicit memory: to become a socially communicable story that is subject to condensation, embellishment, and contamination. Thus, while trauma may leave indelible sensory and affective imprints, once these are incorporated into a personal narrative this semantic memory, like all explicit memory, is likely subject to varying degrees of distortion" ("DF," 521). In this slippery passage, the authors begin by suggesting that, like flashbulb memories, traumatic memories may not be accurate; rather, someone's subjective feeling that such memories are veridical may simply be a function of emotional processes aroused by the activities of the amygdala at the time of recall. This seems clear enough. But the authors then go on to imply that traumatic sensations or imprints *are* inherently veridical or literal, it is just that the very act of recall by which they are narrated tends to distort them. [In short, their claim that traumatic memory is literal cannot be refuted by scientific experiment, rational argument, or anything else—not a satisfactory state of affairs.]

van der Kolk the gap between traumatic memory and narrative memory is so radical that it can never be bridged. "Even after considerable periods of time, and even after acquiring a personal narrative for the traumatic experience, most of our subjects reported that these experiences continued to come back as sensory perceptions and as affective states," van der Kolk observes. "The persistence of intrusive sensations related to the trauma, even after the construction of a narrative, contradicts the notion that learning to put the traumatic experience into words will reliably help abolish the occurrence of flashbacks—a notion that seems to be a central assumption in a variety of treatment modalities" (*TS*, 289). The situation appears paradoxical in that without cure the victim's inescapable proclivity to repeat the past has contagious effects—for according to van der Kolk not only do the flashback and other intrusive experiences tend to retraumatize the patient, they also contagiously contaminate the therapist by traumatizing the latter in turn (*TS*, 24–46). On the one hand, for van der Kolk the fear of such contagion helps explain why so many psychiatrists have tried to avoid the sufferings of victims and, more broadly, why psychiatry as a field has been so slow to acknowledge the importance of trauma in our time. Alternatively, van der Kolk explains the avoidance behavior of psychiatry by the very *effects* of traumatic "contagion"—as if psychiatry itself is imagined as a person who has been contagiously traumatized and hence "numbed" or dissociated into forgetting. On the other hand, for Caruth especially, as we shall see, it is precisely the tendency of the victim to contaminate others which ensures that the truth of the past as dislocation and failed witnessing—the truth of history as trauma—will be transmitted to others. The default of representation in trauma certifies that, independent as it is of personal or autobiographical meaning and memory, the truth of the past will be performatively communicated to the collective through the suffering of those who "listen" but were not there.

Defending the Science of the Literal II:
The Neurobiology of Trauma

Now it is obvious in this context that the plausibility of the theory of trauma as literally and timelessly encapsulated in a special memory system would be enormously enhanced if that theory could be supported by neurobiological evidence. This is what van der Kolk has attempted to provide by grounding his theory in assumptions about the neurophysiology of the brain. Given the authority wielded by biological models in contemporary American psychiatry, this is a potentially powerful move.

But how successful has he been? In terms of promoting his research program he has been remarkably effective. As Allan Young has recently observed, the neurohormonal theory of memory advocated by van der Kolk and others is a blueprint for current biological research on PTSD (*HI*, 279). But scrutiny of van der Kolk's publications reveals that here too the scientific evidence is weak.

As is briefly noted in the introduction, in the late nineteenth century the first model for psychical trauma was that of surgical shock involving the sudden prostration of the nervous system. Shock was then linked to the idea of fear and to a corporeal memory system capable of memorializing the experience in ways that suggested not only how the body is ordinarily mobilized to respond to danger but how extreme terror might even induce death.[36] As I show in chapter 6, in a Pavlovian version of these ideas, embodied in the approach of William Sargant and others to the traumatic neuroses of World War II, the traumatic response was conceptualized as a conditioned reflex. Van der Kolk and his colleagues pursue a similar approach by taking as their starting point the organism's ordinary response to the external environment. Under normal conditions human beings react to stimulation by releasing various neurotransmitters and hormones that alert the nervous, circulatory, and other systems in order to ready the organism for "fight or flight." In the brain stem there is a nucleus called the locus coeruleus that projects not only to the cerebral cortex but crucially to the limbic system or "visceral brain," which includes the hypothalamus, hippocampus, and amygdala and regulates the individual's emotional responses to the environment. "The limbic system has the greatest concentration of adrenergic (NOR) cells in the brain," Allan Young has observed in his valuable assessment of the biological theory of traumatic memory:

> A single adrenergic cell, rooted in the locus coeruleus, can innervate the hippocampus, amygdala and cerebral cortex. By making cortical neurons "aware" of changes in the organism's feeling states, the limbic system gives emotional coloring to the individual's perceptions and memories. It infuses them with fear, anger, disgust, pleasure, sadness, and surprise. In this way, the system signals which elements in the environment and memory need attention and demand action.

It is at this point, where the axis linking the locus coeruleus to the limbic sys-

36. For a discussion of these developments see Allan Young, "Bodily Memory and Traumatic Memory," in *Tense Past: Cultural Essays in Trauma and Memory*, ed. Paul Antze and Michael Lambek (New York, 1996), 89–102.

tem intersects the axis connecting the hypothalamus to the adrenal glands, that the stress reaction is produced. (*HI*, 276)

Another component of the biological theory of traumatic memory is connected to the fact that the anterior hypothalamus, amygdala, and hippocampus also have high densities of opioid receptor cells. Exposure to certain stressful events is said to activate these receptors and stimulate the secretion of endorphins (endogenous opioids) in the brain. It is thought that these receptor-endorphin responses explain the numbing or analgesia characteristic of trauma.

The hypothesis behind the biology of traumatic memory is that under conditions of extreme stress—the model is that of an animal's response to inescapable shock—abnormal amounts of neurotransmitters are released, especially the neurotransmitter known as norepinephrine (NOR). According to van der Kolk and his colleagues, this leads to a long-term augmentation of the locus coeruleus pathways leading into the limbic system. "There is no direct evidence that these neuronal changes occur in PTSD, but researchers know how they might occur," Young notes (*HI*, 277). Thus it is known that a neuron's sensitivity to a neurotransmitter is partly determined by the number and molecular properties of its receptors for this neurotransmitter. As Young reports: "These receptor features are known to alter in response to changes in the neuron's biochemical environment, such as the presence of persistently high levels of endogenous opiates. . . . In the case of PTSD, the flood of NOR would, then, precipitate the changes that 'augment' the movement of impulses along (adrenergic) stress response pathways" (*HI*, 277).

Van der Kolk suggests that the long-term potentiation of neural circuits provides the "neurophysiological analogue of memory." He writes: "We hypothesize that such long-term augmentation of the LC [locus coeruleus] pathways following trauma underlies the repetitive intrusive recollections and nightmares that plague patients with PTSD" ("IS," 318). We have seen that traumatic nightmares are thought to share the characteristic of night-terrors in being precipitated during autonomic arousal in stage II sleep. Because autonomic arousal is mediated by the locus coeruleus, van der Kolk suggests that "not only the flashbacks, but also the traumatic nightmares that characterize PTSD, are related to the long-term potentiation of LC pathways to the hippocampus and amygdala. This also could account for the eidetic, instead of the oneiric, quality of traumatic nightmares as described by van der Kolk et al (1984)" ("TS," 518). In short, van der Kolk justifies his claim that traumatic nightmares are literal memories by explaining that traumatic dreams are not processed by the mediated or "declarative" memory system but as "subcortically based emotional memories."[37]

All the symptoms of PTSD are linked to the biological theory of traumatic memory in the same way. For example, according to hypothesis the release of large amounts of NOR during trauma eventually depletes its supply, and this depletion explains the victims's feelings of chronic helplessness, lack of motivation, numbing, and anhedonia, as well as his hypersensitivity to conditioned stimuli. A vicious cycle is created in which NOR depletion produces helplessness, which depletes NOR further, and so on. A related assumption connects the hyperreactivity of PTSD patients, as manifested in startle responses, sleep difficulties, anxiety, and other symptoms, to phases in the production of endogenous opiates, or endorphins, such that the anxiety produced by trauma releases endorphins, which produces a sense of relief, which is followed by endorphin depletion, which induces hyperreactivity and anxiety, with the result that the cycle begins again. The relief associated with endorphins is thought to make the victim of trauma "addicted" to the trauma, as when the victim of sexual abuse repeatedly seeks out similar abusive situations.

Yet in a detailed criticism of a study coauthored by van der Kolk, Young has observed that the predicted outcome failed to occur. In the study eight Vietnam veterans with PTSD were exposed to a 15-minute so-called "neutral" videotape, followed by a 15-minute segment of *Platoon*, Oliver Stone's movie about Vietnam, and then by another 30-minute "neutral" video. The control group consisted of eight Vietnam veterans with no history of PTSD. Thirty minutes before viewing, an intravenous line was inserted into each subject and, immediately after showing the first (neutral) videotape, doses of either the drug naxalone or saline placebo were administered. Booster doses were injected immediately after the second (combat) and third (neutral) videotapes. The hypothesis behind the experiment was that the exposure of the PTSD patients to a combat-related stimulus would provoke a release of endorphins with consequent analgesia, an analgesia that might be reversed by naxalone, a drug known to block interaction between neural receptors and the opiate class of endorphins, but that would not be affected by the saline placebo (*HI*, 281–82).[38]

---

37. Bessel A. van der Kolk, "The Body Keeps the Score: Memory and the Evolving Psychobiology of Posttraumatic Stress," *Harvard Review of Psychiatry* 1 (1994): 253.

38. Young is discussing the paper by Roger Pitman, Bessel A. van der Kolk, and M. S. Greenberg, "Naxalone-Reversible Analgesic Response to Combat-Related Stimuli in Post-Traumatic Stress Disorder," *Archives of General Psychiatry* 47 (1990): 541–44.

The neural-hormonal theory predicted that exposure to combat videotape would produce a marked hormonal response in combat veterans with PTSD. As Young has observed, it is this hormonal response, rather than the release of endorphins, that is the essential component of the theory of traumatic memory. However, the researchers found no significant differences in hormonal responses between the PTSD and the control group, nor did they find any significant differences in endorphin levels. What they found instead was a significant analgesia effect in the PTSD patients receiving the saline placebo: these men reported significant decreases in pain intensity after watching *Platoon*, while the other group reported increases. The anomaly was that the analgesia effect was accompanied by neither the predicted hormonal response nor increases in endorphins (*HI*, 282).

[Van der Kolk and his associates responded in two different, mutually incompatible, ways to this dilemma—ways that according to Young are characteristic of researchers in fields that produce many "findings" but few "facts" and in which it is necessary to "save" the phenomena by various strategies (*HI*, 266–67). The first response was to suggest that the findings were incorrect, that the PTSD group really did experience the predicted increase in hormonal response but that the increase went undetected because of technical difficulties (*HI*, 282). An alternative response was to admit that the findings were correct—there were no significant hormonal differences between the PTSD group and the controls—but that they could be explained by appealing to various, untested, supplementary hypotheses (see *HI*, 282–83 for details).]

Young has observed that by embedding substandard results in a network of facts and citations, connecting PTSD to the neural-hormonal theory of stress, and by invoking ancillary hypotheses and explanations, the appearance of scientific progress has been maintained. He emphasizes the advantage to van der Kolk and his associates of the paradigm of neuroendocrinology in this regard, for it is a well-established field with standardized methods of assessment; as such, it lends its considerable prestige to van der Kolk's enterprise. In other words, although the neurophysiological evidence for van der Kolk's theory of traumatic memory is weak, his ideas continue to carry force because they are associated with successful paradigms, technologies, and practices that conform to the dominant model of what constitutes good psychiatric science. But Young has also remarked that the analogy drawn between the experience of Vietnam veterans and inescapable shock in animals creates inconveniences for van der Kolk and his school. [For the analogy to work, it must either ignore or try to accommodate the heterogeneous nature of the trauma experience in human beings.] "Even among the fraction of diagnosed veterans who traced their present difficulties back to life-threatening combat situations, many (perhaps most) of the events centered around the successful execution of fight or flight impulses," Young remarks. "But it is precisely the impossibility of either fight or flight that defines inescapable stress. (World War I provides a better fit: the image of soldiers huddled in trenches and bunkers over days of unremitting shelling and the image of soldiers buried alive following explosions.) The model of inescapable shock seems likewise problematic in those cases where men were the authors of their etiological events—the perpetrators of violence rather than its victims" (*HI*, 283). Young makes an important point, for although van der Kolk states that trauma is a matter of socially and contextually determined meaning (*TS*, 26–27), his ideas about the literal nature of traumatic memory make such memory in principle a matter of the objective state of the neurons, purged of subjective interpretation.[39]

As I have already noted, van der Kolk and his associates are careful to express their views in hypothetical terms. Indeed, in *Traumatic Stress*, at the end of a chapter reviewing the psychobiology of memory, van der Kolk concludes with the admission that "the question of whether the brain is able to 'take pictures,' and whether some smells, images, sounds, or physical sensations may be etched onto the mind and remain unaltered by subsequent experience and by the passage of time, still remains to be answered" (*TS*, 297). Yet his entire approach to traumatic memory rests on the assumption that the question can be answered in the affirmative.[40]

39. Young points out that van der Kolk attempts to reconcile the question of meaning with the literal, somatic nature of traumatic memory by treating the brain's neocortex as the locus of individuation and a key element in the stress response circuit. "Sited between the environmental stressors (input) and the rest of the central nervous system, the neocortex accounts for interindividual and intergroup variations in both neural activity and endocrine secretions" (*HI*, 284). Narrative coherence is then achieved by imagining that the neocortex can switch on and off at strategic moments. Thus in one study examined by Young, neocortex functioning, such as splitting or denial, is adduced to explain otherwise anomalous results. Similarly, the fact that after watching *Platoon* the veterans with PTSD reported double the amount of "disgust" or "sadness" than the control group, and five times the amount of "guilt," is explained by attributing those subjective feelings to the neocortex without further discussion, leaving the biological story about traumatic memory intact (*HI*, 284–85).

40. For example, van der Kolk has recently observed of the hysterical-traumatic attacks described by Abram Kardiner in patients suffering from battle neuroses: "These patients' attacks were finely conditioned reflexes. Often, patients reproduced the traumatic event

[A recent attempt to prove his theories has involved experiments with brain-imaging, or positron emission tomography (PET), experiments which purport to demonstrate that during the provocation of traumatic memories or flashbacks in subjects with PTSD there is increased activation of the visual area (according to hypothesis, the seat of iconic, traumatic memory) and decreased activation of Broca's area (the part of the central nervous system most centrally involved with speech and hence, according to hypothesis, with narrative memory). The experiments were carried out on a total of eight patients diagnosed with PTSD (six men and two women); there were no controls.]

Van der Kolk has explained that the patients were asked to prepare narratives or "scripts" of their traumatic experiences. The scripts (of 30–40 seconds duration) were recorded by the experimenters in a "neutral" voice on an audiotape and then played back to the patients; when the scripts precipitated marked autonomic responses and triggered flashbacks, a PET scan was made. For comparison, the same subjects wrote and were exposed to "neutral" scripts (about brushing one's teeth, emptying the dishwasher, etc.) of the same short duration. Van der Kolk reports that during exposure to the scripts of their traumatic experiences, the subjects exhibited heightened activity only in the right hemisphere, that is, in the parts of the limbic system connected with the amygdala. Activation of these parts was accompanied by heightened activity in the right visual cortex, according to hypothesis the seat of the patients' visual reexperiencing of their traumas. Most significantly for van der Kolk and his colleagues, Broca's area "turned off." Van der Kolk suggested that the results reflected "the tendency in PTSD to experience emotions as physical states rather than as verbally encoded experiences. Our findings suggest that PTSD patients' difficulties with putting feelings into words are mirrored in actual changes in brain activity" (*TS*, 233).

But it may be objected that the experiment cannot be taken as evidence in support of van der Kolk's theory of traumatic memory, for the good reason that it assumes what has to proved. This is because in the experiment the trauma patient is automatically silenced by being placed in the position of an auditor of his or her own story, while the speaking function is displaced entirely onto the person who makes the audiotaped narration. The silencing of the patient is a function not only of the researchers' presuppositions about the nature of traumatic memory but also of the

---

photographically, without integration into a semantic memory" (Bessel van der Kolk, Nan Herron, and Ann Hostetler, "The History of Trauma in Psychiatry," *Psychiatric Clinics of North America* 17 [September, 1994]: 589).

experimental technique used to investigate traumatic memory. In their PET-scanning experiments the investigators employed radioactively labeled oxygen inhaled into the blood stream of the patient to scan changes in the regional flow of blood in the brain; the changes indirectly reflected alterations in cerebral neuronal activity. The method required the employment of a thermoplastic mask to minimize the patient's head movements during the delicate brain-scanning process, as well as the use of cannulae inserted into the nose for the gas inflow and an overlying face mask attached to a vacuum. All this made it contraindicated if not impossible for the patient to speak during the experiment. No wonder van der Kolk and his associates found that Broca's area was "turned off."[41]

In the past, with different assumptions and methods, different results have been obtained. For example, as shown in the previous chapter, Sargant, Grinker, Spiegel, and Fabing used similar techniques of story-telling in the treatment of war neurotics during the Second World War. They "replayed" or vividly play-acted the soldier's reconstructed story to the patient, while the latter was under the influence of various hypnotic drugs, in order to get him to reexperience the traumatic scenario. Sargant especially emphasized that in order to obtain a therapeutically successful traumatic abreaction or "catharsis," it was necessary to excite the patient

41. For further details see Scott L. Rauch, Bessel A. van der Kolk, Rita E. Fisler, Nathaniel M. Alpert, Scott P. Orr, Cary R. Savage. Alan J. Fischman, Michael A. Jenike, and Roger K. Pitman, "A Symptom Provocation Study of Posttraumatic Stress Disorder Using Positron Emission Tomography and Script-Driven Imagery," *Archives of General Psychiatry* 53 (1996): 380–87. In fairness, it seems to be the case that Broca's area, or rather a part of the brain very close to it, can be "turned on" during the performance of certain cognitive activities even in the absence of verbal expression. But the authors never discuss this issue, focusing on the absence of speech in the traumatized patient in ways that conform to their theory. Another criticism is that the researchers never tested their patients by subjecting them to trauma scripts pertaining to traumas of other people. Finally, a major weakness—one they concede—is the absence of controls, so that there is no way to assess the specificity of the findings. The authors convert this weakness into a virtue in that it allows them to call for more research along the same lines, one way in which the paradigm of the literal is constantly renewed. For the authors' discussion of the technical and other limitations inherent in the procedure of PET-scanning, see 384–85. For a recent assessment of the psychobiology of PTSD, one that frequently cites the experiment of Rauch, van der Kolk et al. while expressing many cautions about the use of neuroimaging in PTSD research, see Rachel Yehuda and Alexander C. McFarlane, *Psychobiology of Posttraumatic Stress Disorder* (New York, 1997). My thanks to Dr. Guy McKhann for discussing the paper by Rauch, van der Kolk, and colleagues with me.

so intensely that the latter would burst into present-tense speech, dramatically acting out the (real or imagined) traumatic scene as if it were happening all over again. For Sargant, as for the entire tradition of cathartic or abreactive therapy, speech was inherent in the traumatic repetition.

Nor did Janet, whose views on the nature of trauma, hysteria, and memory inform much recent work, regard traumatic memory as wordless. Van der Kolk thinks that his neuroimaging experiment confirms Janet's views about the nature of traumatic memory, because Janet made a distinction between "narrative memory" and "traumatic memory" in terms which van der Kolk interprets as anticipating his own distinction between "declarative" (or narrative) and "implicit" (or traumatic) memory ("DF," 520). In chapter 3 I comment on the importance van der Kolk and his school attach to Janet's work in this regard. But Janet would have been surprised by van der Kolk's experiment. For according to Janet, trauma does not entail the loss of all speech but only the loss of the kind of speech associated with what he calls "presentification," involving the victim's capacity for narrative self-reflection and self-knowledge—that is, for self-representation. In describing his famous case of Irène, the young girl who witnessed her mother's death and who repetitively and hysterically acted out the scene of the trauma, Janet reports many of the words the patient uttered during her traumatic reenactments (words that often occurred, as in Sargant's patients, in the present tense, as if the scene were indeed happening all over again).

More generally, the literature on traumatic repetition is full of the victim's speech, as van der Kolk is of course aware (see for example his admiring reference to Lawrence Langer's analysis of the "ruined" (or dissociated) testimony of the Holocaust survivor) ("IP," 177–78).[42] Caruth's emphasis on the necessity of listening to the "crisis of a trauma" also acknowledges that it is the victim's speech that must be attended to. But van der Kolk must "forget" this if he is to cling to the notion of trauma as involving "speechless" terror and if he is to continue to define traumatic memory as a literal impression that is lodged in the brain in strictly "iconic" terms.

*

Van der Kolk's use, or misuse, of Janet is just one example of his general attempt to mobilize the work of his predecessors in ways that allow him

---

42. See Lawrence L. Langer, *Holocaust Testimonies: The Ruins of Memory* (New Haven, 1991) (also cited by Caruth, *UE*, 142, n. 8).

to represent his theory of trauma as the culmination of a lineage that can be made to seem as if it runs from the past to the present in an interrupted yet ultimately progressive way. I do not wish to deny that there are earlier figures who can be used for this purpose, even if, as in the case of Janet, their work has to be distorted in order to be made to fit the story van der Kolk wishes to tell. On the contrary, I think it is important to emphasize how deeply entrenched many of the essentials of van der Kolk's views about memory are in the cultural imaginary of the West, which is why they are so hard to dislodge. In this connection, and as a closing comment, I draw attention to still another, essentially *political*, argument that helps sustain van der Kolk's position. It is a point also made by Frankel. This is the necessity of attempting to tie the signs and symptoms of trauma as directly as possible to the precipitating event in order to prove that there is a direct link between the cause and its consequences. Bennett Simon, in his introduction to Lansky's book on traumatic nightmares in Vietnam veterans, has commented on this aspect of the debate by observing that this is an area where to claim, as Lansky does, that the nightmares and flashbacks of his patients are "not only and not always veridical representations of actual combat trauma, but condensations with prewar and with postwar traumas and humiliations," is "not entirely neutral":

> Those who worked on the task of interviewing Holocaust survivors for purposes of substantiating their claims for reparations from the German government remember that for a period of time some German psychiatrists and courts were arguing (and, to her horror, citing the writings of Anna Freud!) that the personality is formed by age six or that many who suffered in the camps were already damaged personalities before their concentration-camp experience. Any work that points to earlier life traumatic experiences runs the risk of being denounced as "blaming the victim" or as actually or potentially supporting governmental refusal to pay compensation.[43]

Indeed, one has only to read the pioneering work of William G. Niederland on the long-term effects of the concentration-camp experience to realize how the heavy-handed and often heartless response of many German courts and physicians to claims for compensation made it essential for the physicians caring for the victims to insist on the centrality of the horrors of the camps to their patients' subsequent problems. Niederland especially emphasized the necessity of demonstrating that the camp experience could produce delayed effects, so that a relatively symptom-free period after release from the camps could not be taken as

43. Bennett Simon, foreword to Lansky and Bley, *Posttraumatic Nightmares*, xv.

evidence for the absence of traumatic sequelae.[44] In this context, he argued that the nightmares of concentration-camp survivors, which as he observed still occurred with the same intensity and frequency fifteen or sixteen years after liberation, were directly derived from or clearly connected with the harrowing experiences of the persecution years.[45]

The same need to tie pathology directly to the occasioning event by conceptualizing symptoms as if they were the direct, if delayed, reaction

44. William G. Niederland, "The Psychiatric Evaluation of Emotional Disorders in Survivors of Nazi Persecutions," in *Massive Psychic Trauma*, ed. Henry Krystal (New York, 1968), 9–11 (text originally published in 1961). Cf. K. R. Eissler, "Perverted Psychiatry?" *American Journal of Psychiatry* 123 (1967): 1352–58; and Martin S. Bergmann and Milton E. Jucovy, eds., *Generations of the Holocaust* (New York, 1982), 62–79, for a general discussion of the compensation policy.

Relevant to Niederland's work on the concentration-camp survivor are his parallel studies of Freud's famous Schreber case, in which he drew attention to the role of the father's methods of upbringing and ideas in the etiology of Schreber's mental disorder, suggesting that the son's persecutory system contained a nucleus of historical truth (William Niederland, *The Schreber Case: Psychoanalytic Profile of a Paranoid Personality* [New York, 1974]). But unlike Morton Schatzman, who leant on Niederland's research in order to propose a direct, causal link between the father's traumatizing behavior and the son's disorder (*Soul Murder: Persecution in the Family* [New York, 1973]), Niederland remained close to Freud's work on paranoia. He nowhere affirmed the simple causal relationship that Schaztman proclaimed, and emphasized instead the disturbances produced in the son's love for the father. Niederland's and Schatzman's work is critically discussed by Hans Israels, *Schreber: Father and Son* (Madison, Connecticut, 1989). For an interesting attempt to historicize the Schreber case by analyzing Schreber's concern with the Jewish question see Sander L. Gilman, *Freud, Race, and Gender* (Princeton, 1993); for another historicist approach that emphasizes the relationship between Schreber's delusional ideas and the social and political crisis of modernity see Eric L. Santner, *My Own Private Germany: Daniel Paul Schreber's Secret History of Modernity* (Princeton, 1996).

45. Niederland, "The Psychiatric Evaluation of Emotional Disorders in Survivors of Nazi Persecution," 55. Cf. William G. Niederland, "The Survivor Syndrome: Further Observations and Dimensions," *Journal of the American Psychoanalytic Association* 29 (1981): 413–25, in which he identifies anxiety dreams "and certain types of 're-run' nightmares, which reflect the persecution experiences virtually *in situ et concreto*" as among the prominent symptoms of the survivor syndrome (415–16). Niederland also identifies hypermnesia as one of the most tormenting symptoms, by which he means the survivor's "overly sharp, distinct, and virtually indelible memories as far as persecution events are concerned" (416). But Niederland does not endorse a literalist approach to trauma, in the sense of van der Kolk or Caruth, and continues to use the psychoanalytic concepts of defense, including repression, denial, isolation, and projection that van der Kolk and Caruth reject.

to a trauma on the model of an inflammatory response to a foreign body has also been felt by psychiatrists and others in a post-Vietnam era in which the Veterans Administration and other government bureaucracies were initially cautious about accepting a link between the problems of the returning veteran and the war (the VA now provides an important career niche for the study of PTSD). Yet it is worth noting that even at the height of his early commitment to the trauma theory of hysteria, Freud questioned the value of conceptualizing trauma on the model of an infectious foreign agent that automatically produces symptoms of inflammation, on the grounds that the organization of memory is far more complicated than such a causal analogy would imply. But that model has served and continues to serve too important purposes to be easily discarded. The result is that the theory of the literal nature of traumatic memory continues to gain widespread adherence, even as it remains inadequately formulated and weakly supported by the scientific evidence.

VIII

The Pathos of the Literal:
Trauma and the Crisis of Representation

[Cathy Caruth is an exponent of a postmodernist, poststructuralist approach to psychic trauma that has received considerable approbation, not only from humanists in various fields but also from psychiatrists and physicians.]Unlike most commentators on the postmodern condition who regularly invoke themes of trauma, memory, and identity without reference to the large contemporary psychiatric-clinical literature on the topic, Caruth incorporates the neurobiology of trauma into her work. In accordance with the views of the American physician Bessel van der Kolk and his associates, Caruth holds that massive trauma precludes all representation because the ordinary mechanisms of consciousness and memory are temporarily destroyed. Instead, there occurs [an undistorted, material, and—her key term—*literal* registration of the traumatic event that, dissociated from normal mental processes of cognition, cannot be known or represented but returns belatedly in the form of "flashbacks," traumatic nightmares, and other repetitive phenomena.[1]]

[In Caruth's deconstructive version of van der Kolk's neurobiological account of trauma, the gap or aporia in consciousness and representation that is held to characterize the individual traumatic experience comes to stand for the materiality of the signifier in the sense given the concept by the critic Paul de Man, who theorized a "moment" of materiality that on the one hand belongs to language but on the other is aporetically severed from the (speech) act of signification or meaning.[2]More precisely, at the

1. Cathy Caruth, ed., *Trauma: Explorations in Memory* (Baltimore, 1995), 152–53; hereafter abbreviated *TEM*.

2. Paul de Man, *Allegories of Reading: Figural Language in Rousseau, Nietzsche, Rilke, and*

center of Caruth's approach to psychic trauma is a "performative" theory of language derived from de Man. Discussing de Man's examination of the "break within the system" in Kant's formal articulation of the relation between motion and force, Caruth writes:

> The break occurs in Kant's text precisely in the attempt to integrate force into the system of motions. In his analysis of Kant, de Man identifies this break specifically as a disruption in the phenomenal self-representation of language, or in the appearance in language of a performative dimension. . . . Knowing itself as a grammar or a system of tropes, philosophy must, and yet cannot, fully integrate a dimension of language that not only shows, or represents, but acts. Designating this moment as "fatal," de Man associates it . . . with death. It is paradoxically in this deathlike break, or resistance to phenomenal knowledge, that the system will encounter the resistance, de Man suggests, of reference.[3]

For Caruth, an analogous "deathlike break" lies at the heart of trauma; the victim of trauma who cannot symbolize or represent the traumatic event or accident that caused her condition nevertheless obsessively "performs," reenacts, or reexperiences it in the form of flashbacks, dreams, and related symptoms. Empiricism, in the form of an appeal to science, and literary theory are thus appropriately, if oddly, conjoined in Caruth's work.[4] The "flashback," defined as "an interruption . . . that . . . cannot be thought simply as a representation," is regarded by her as the equivalent of the "interruption of a representational mode" that she associates with de Man's deconstruction of language (UE, 115, n. 6). And there is a third element in the mix: a set of widely shared assumptions about the constitutive failure of linguistic representation in the post-

---

Proust (New Haven, 1979); idem, Resistance to Theory (Minneapolis, 1986); idem, "Phenomenality and Materiality in Kant," in Hermeneutics: Questions and Prospects, ed. Gary Shapiro and Alan Sica (Amherst, Massachusetts, 1984). See also Responses: On Paul de Man's War Time Journalism, ed. Werner Hamacher, Neil Hertz, and Thomas Keenan (Lincoln, Nebraska, 1989).

3. Cathy Caruth, Unclaimed Experience: Trauma, Narrative, and History (Baltimore, 1996), 87; hereafter abbreviated UE.

4. Caruth's use of neuroscience is problematic because, as I have shown in the previous chapter, its claims and findings are of questionable validity. But it makes a certain sense, because her de Manian version of deconstruction is empiricist and materialist in nature (as the term has come to be understood in de Manian circles). On de Man's materialism and empiricism see Frances Ferguson, Solitude and the Sublime: Romanticism and the Aesthetics of Individuation (New York, 1992), esp. 1–36. Cf. W. J. T. Mitchell, ed., Against Theory: Literary Studies and the New Pragmatism (Chicago, 1985).

Holocaust, post-Hiroshima, post-Vietnam era, which Caruth never quite puts forward as her own but which is explicit in the writings of other scholars whom she cites with approval, notably literary critic Shoshana Felman and psychoanalyst Dori Laub. For those scholars, the Holocaust in particular is the watershed event of the modern age because, uniquely terrible and unspeakable, it radically exceeds our capacity to grasp and understand it. And since this is so, the Holocaust is held to have precipitated, perhaps caused, an epistemological-ontological crisis of witnessing, a crisis manifested at the level of language itself.[5]

Nevertheless, Caruth and others argue, the testimony of the traumatized subject "reaches a *you*, attains the hearing" through the urgency of an address that contaminates others (*T,* 37). On this view, language succeeds in testifying to the traumatic horror only when the referential function of words begins to break down, with the result that, as Walter Benn Michaels has put it, what is transmitted is "not the normalizing knowledge of the horror but the horror itself."[6] Accordingly, Caruth's book, *Unclaimed Experience,* offers various readings, many of them fairly brief, of exemplary works or passages in works by Freud, Lacan, Duras, Resnais, Kant, Kleist, and de Man, which attempt to demonstrate that language is capable of bearing witness only by a *failure* of witnessing or representation. As for the process by which this occurs, the related notions of speech and listening are crucial to Caruth's account:

> How does one listen to what is impossible? Certainly one challenge of this listening is that it may no longer be simply a choice: to be able to listen to the impossible, that is, is also to have been *chosen* by it, *before* the possibility of mastering it with knowledge. This is its danger—the danger, as some have put it, of the trauma's "contagion," of the traumatization of the ones who listen (Terr, 1988). But it is also its only possibility for transmission. "Sometimes it is better," Dr. Laub suggests, speaking as a clinician, "not to know too much" (Laub, 1991). To listen to the crisis of a trauma, that is, is not only to listen for the event, but to hear in the testimony the survivor's departure from it; the chal-

5. Shoshana Felman and Dori Laub, *Testimony: Crises of Witnessing in Literature, Psychoanalysis, and History* (New York, 1992); hereafter abbreviated *T.* Felman speaks of "the *radical historical crisis in witnessing* the Holocaust has opened up" (201) and characterizes our century as a "post-traumatic century" (1), a notion echoed by Caruth when she refers to our age as "catastrophic" (*TEM,* 11).

6. Walter Benn Michaels, "'You who never was there': Slavery and the New Historicism, Deconstruction and the Holocaust," *Narrative* 4 (1996): 8; hereafter abbreviated "YNT."

lenge of the therapeutic listener, in other words, is *how to listen to departure*. (*TEM*, 10)

[The transmission of the unrepresentable—a transmission imagined by Caruth simultaneously as an ineluctable process of infection and as involving an ethical obligation on the part of the listener—therefore implicates those of us who were not there by making us, as Dori Laub has put it, participants and coowners of the traumatic event (*T*, 57).]

Not that Caruth and Laub are in complete agreement on all issues. When Laub discusses the project to videotape the testimony of Holocaust survivors, he converts the specific technology of the videotape into a dialogic process that he conceptualizes in psychoanalytic terms as a "'brief treatment contract'" (*T*, 70). On this model, survivor and interviewer are brought together in an oral or face-to-face encounter in which the interviewer becomes the listener-analyst who is marked by the trauma but, unlike the victim, has the objectivity and detachment to distance himself from it. In the process, the survivor is encouraged to remember and narrate the past in a procedure that bears some resemblance to the analytic process of "working through."[7] But for Caruth such an act of narration risks betraying the truth of the trauma defined as an incomprehensible event that defies all representation (*TEM*, 153–54). Accordingly, she calls for a mode of responding to trauma that ensures the transmission of the break or gap in meaning that constitutes history as inherently traumatic (*UE*, 18). From this perspective, if history is a symptom of trauma it is a symptom which must not, indeed cannot, be cured but simply transmitted, passed on. "[T]he traumatic nature of history," Caruth sweepingly writes, "means that events are only historical to the extent that they implicate others" (*UE*, 18).

[Caruth attempts to justify her claims by focusing on certain psychoanalytic and literary-theoretical works in order not just to follow each author's explicit arguments about trauma but to trace the "textual itinerary of insistently recurring words or figures" that engender "a literary dimension that cannot be reduced to the thematic content of the text or to what the theory encodes, and that, beyond what we can know or theorize about it, stubbornly persists in bearing witness to some forgotten wound" (*UE*, 5). In this chapter I concentrate on Caruth's treatment of Freud, in

7. Felman conceptualizes the response to Holocaust testimony in the same dialogic-psychoanalytic terms, focusing on the need for the listener—in this case, her students—to recover from the "*loss of language*" and knowledge they themselves suffer from the experience of listening to Celan's poetic "testimony" in order for them to reenter signification and integrate their experience (*T*, 52–56).

order to exhibit what is wrong or problematic about her arguments and to critically assess the politico-ethical stakes involved.

Freud As Theorist of the Literal

In her writings on trauma Caruth insists on the central importance of certain of Freud's texts, which she interprets not only as announcing the performative theory of trauma but also as enacting the supposed failure of representation that performatively constitutes trauma as such. She thereby participates in a general postmodernist tendency to appropriate psychoanalysis for discussions of the trauma of the Holocaust and the post-Holocaust condition. As Caruth acknowledges, her arrogation of Freud depends on a fundamental reorientation of his oeuvre. "Throughout his work," she remarks, "Freud suggests two models of trauma that are often placed side by side":

> the model of castration trauma, which is associated with the theory of repression and return of the repressed, as well as with a system of unconscious symbolic meanings (the basis of the dream theory in its usual interpretation); and the model of traumatic neurosis (or, let us say, accident trauma), which is associated with accident victims and war veterans (and, some would argue, with the earlier work on hysteria; see Herman, *Trauma and Recovery*) and emerges within psychoanalytic theory, as it does within human experience, as an interruption of the symbolic system and is linked, not to repression, unconsciousness, and symbolization, but rather to a temporal delay, repetition, and literal return. Freud generally placed his examples of the two kinds of trauma side by side . . . and admitted . . . that he was not sure how to integrate the two. (*UE*, 135, n. 18)

As do many other theorists of trauma, Caruth rejects Freud's castration model and the associated concepts of repression and unconscious symbolic meaning. What replaces the concepts of castration and repression for Caruth are the notions of the traumatic *accident* and of a *latency* that inheres in the traumatic experience. Specifically, she returns to the railway accident as the archetype for modern theorizations of trauma and shock, and to the connected idea of a temporal delay that intervenes between the fright and the subsequent appearance of the traumatic symptoms. She links the notion of latency both to Janet's concept of "dissociation" (*UE*, 141, n. 8) and to Freud's concept of *Nachträglichkeit* or "deferred action" (*UE*, 133, n. 8), defined by her as a structure of temporal deferral but now stripped of the idea of the retroactive conferral of meaning on past sexual experiences and reduced instead to the idea of a literal if belated repetition of the traumatic event: "If repression, in trauma, is replaced by latency, this is significant in so far as its blankness—the space of unconsciousness—is paradoxically what precisely preserves the event in its literality. For history to be a history of trauma means that it is referential precisely to the extent that it is not fully perceived as it occurs" (*TEM*, 8).

It is important to realize from the outset the extent to which Caruth reconfigures Freud's concept of deferred action. As I observe in chapter 1, for Freud, the temporal logic of *Nachträglichkeit* problematizes the originary status of the traumatic event, because according to that logic trauma involves a dialectic of nonknowledge or "forgetting" and latency through which the sexual past is determined as traumatic by a retroactive conferral of meaning. Caruth attempts to appropriate Freud's concept of *Nachträglichkeit* by emphasizing exclusively its *temporal* aspect—the idea that owing to the individual's lack of preparedness the threat to life that defines trauma "is recognized as such by the mind *one moment too late*. The shock of the mind's relation to the threat of death is thus not the direct experience of the threat, but precisely the *missing* of this experience, the fact that, not being experienced *in time*, it has not yet been fully known" (*UE*, 62). What interests her is the idea that traumatic experience is defined by temporal unlocatability (*UE*, 133, n. 8)—a temporal unlocatability that, as we shall see, is central to her notion of trauma as necessarily implicating others. But when she goes on to define repetition in terms of the belated, literal, and unmediated return of the traumatic event she seems to define trauma in more traditional causal terms, as if trauma involved a linear determinism, or direct action, of the past on the present. She appears to accept this when she observes that her understanding of trauma "corresponds to the deterministic model of the repetition of violence" as exemplified by the traumatic nightmare (*UE*, 136, n. 20).

But this suggests that her model of trauma as defined by latency is much closer to the model of an infectious disease, in which an "incubation period" or period of delay intervenes between the initial infection and the subsequent appearance of the symptoms, than to Freud's concept of *Nachträglichkeit*. In fact, we shall see that when she comes to analyze Freud's discussion of the "accident neurosis" in *Moses and Monotheism* she explicitly accepts the analogy between trauma and infection. That is why Caruth is able to endorse van der Kolk's attempt to link the symptoms of posttraumatic stress to external reality in a causal fashion by conceptualizing traumatic dreams and flashbacks as if they were the direct, if delayed, sequelae of an external trauma, rather like an inflammatory response to a foreign body—an etiological model for psychic trauma that Freud rejected from the start.[8]

Given all this, it is not surprising that Caruth identifies the task of rereading the received understanding of repression theory through trauma theory as "one of the central problems for psychoanalysis today" (*UE*, 136, n. 18) or that, in calling for a basic rethinking of psychoanalysis, she favors interpretations of the traumatic nightmare such as van der Kolk's, which emphasize their literal, memorial character (*UE*, 139, n. 6). In short, Caruth calls for a radical reconfiguration of psychoanalysis in which the traumatic nightmare is defined as an "unclaimed experience"—as a literal, nonsymbolic, and nonrepresentational *memory* of the traumatic event.

A key text for her in this regard is Freud's *Beyond the Pleasure Principle* (1920), in which Freud confronted the problem of the traumatic nightmare for the first time. Freud's text opens with an allusion to the psychic symptoms, including traumatic nightmares, associated with the war neuroses—symptoms that seemed to reflect, Caruth suggests, "nothing but the unmediated occurrence of violent events" (*UE*, 59). In support of this claim she cites a passage in which Freud states:

> "Dreams occurring in traumatic neuroses have the characteristic of repeatedly bringing the patient back into the situation of his accident, a situation from which he wakes up in another fright. This astonishes people far too little. . . .
> . . . Anyone who accepts it as something self-evident that their dreams should put them back at night into the situation that caused them to fall ill has misunderstood the nature of dreams." (*UE*, 59)

Caruth comments on this passage in the following way:

> The returning traumatic dream perplexes Freud because it cannot be understood in terms of any wish or unconscious meaning, but is, purely and inexplicably, the literal return of the event against the will of the one it inhabits. . . . In trauma, that is, the outside has gone inside without any mediation. Taking this literal return of the past as a model for repetitive behavior in general, Freud ultimately argues, in *Beyond the Pleasure Principle*, that it is traumatic repetition, rather than the meaningful distortions of neurosis, that defines the shape of individual lives. (*UE*, 59)

8. See Freud's criticisms of the inflammatory model of trauma in Breuer and Freud, *Studies on Hysteria* (1893–95), in *The Standard Edition of the Complete Psychological Works of Sigmund Freud*, trans. James Strachey (London, 1953–74), 2: 290.

As this passage suggests, Caruth generalizes from *Beyond the Pleasure Principle* in order to claim that for Freud himself traumatic neurosis becomes the paradigm not only of all the neuroses but more broadly for what shapes every individual life—a paradigm, moreover, according to which the traumatic response is defined in terms not of repressed motives, disguised representations, and unconscious symbolic meanings but the literal, unmediated impact of the event. On this interpretation, traumatic dreams are not autobiographically or subjectively mediated or owned by the individual; rather the "self"—which can hardly be characterized as a self any longer (*UE*, 131–32, n. 5)—is possessed by the traumatic dream, which thus bypasses all representation by impersonally memorializing and chronicling the historical truth of the traumatic origin.[9]

Although as an interpretation of the overall meaning of Freud's work this is a controversial assertion, as an interpretation of the traumatic nightmare it is not without precedent. For example, Lansky has observed that when Freud in *Beyond the Pleasure Principle* acknowledged that the traumatic nightmare appeared to prove an exception to the principle that dreams are fulfillments of wishes, he admitted the existence of a "compulsion to repeat" or principle of mental functioning "beyond the pleasure principle" governing the production of traumatic dreams. Since the appearance of Freud's book in 1920 it has seemed to several commentators, especially in recent years, that for Freud the traumatic nightmare

9. Caruth's theory of how the traumatic past is ineluctably registered and transmitted thus comes to seem like an extremely literalist version of history as *chronicle*, conceived as a nonsubjective, nonnarrative, and nonrepresentational method of memorializing the past, a form that Berel Lang considers the most *ethical* form for writing the history of the genocide. On Lang's analysis, the "limit" that is the Holocaust is defined as an object or event that is *literal* because it stands apart from, or before, representation; as such, it requires a historical method that preserves as much as possible the literalness of the past. I call Caruth's version of traumatic history as chronicle "extremely literalist," because although for Lang historical chronicle is a "point zero in historiography," he recognizes that chronicle is never entirely independent of the historian's agency and to some extent also of interpretation and convention, or representation (Berel Lang, "The Representation of Limits," in *Probing the Limits of Representation: Nazism and the "Final Solution,"* ed. Saul Friedlander [Cambridge, Massachusetts, 1992], 300–317). As Amy Hungerford has suggested, it is as if *representation* is imagined as itself a kind of subject who risks murdering or distorting the moral truth of facts concerning the traumatic past, facts that should and do "speak for themselves" (Amy Hungerford, "Surviving Rego Park: Holocaust Theory from Art Spiegelman to Berel Lang," in *The Americanization of the Holocaust*, ed. Hilene Flanzbaum [Baltimore, 1999], 102–24).

must accordingly lack any latent content, wish fulfillment, and symbolic meaning and hence must be understood as a veridical, indeed literal, memory or replay of the traumatic event. As I have shown in the previous chapter, many neuroscientists, notably van der Kolk and his associates, interpret traumatic nightmares in the same literalist terms, even though the evidence for such claims is flimsy and the claim itself badly formulated. Even Lansky interprets Freud's meaning in *Beyond the Pleasure Principle* in this way. He portrays Freud's assumptions about the nature of posttraumatic nightmares as the implicit model for the recent theory of the literal nature of the traumatic dream with which he, Lansky, takes issue.[10]

Caruth leans on this interpretive tradition when she treats Freud in *Beyond the Pleasure Principle* as adhering to the view that the traumatic nightmare is a literal memory of the traumatic event. But is such an interpretation of Freud justified? There is reason to doubt it. It is of course legitimate for Caruth to read Freud's notoriously aporetic texts, such as *Beyond the Pleasure Principle*, against the grain of more orthodox, Oedipal interpretations. Freud's writings on trauma and the mechanisms of defense are disorganized in ways that seem to invite, or necessitate, critical discussion. Moreover, Caruth is right to suggest that the problem of repetition, or the death drive, haunts Freud's oeuvre in ways that pose an internal challenge to his own professed reliance on the theory of libido and repression. She joins forces in this regard with the work of Jonathan Cohen, Warren Kinston, Henry Krystal, and others who have recently returned to the problem of psychic trauma and psychic violence in psychoanalysis and have done so in terms close to hers (*TEM*, 6; *UE*, 131, n. 2, 135–36, n. 18).

Nevertheless, it is far from obvious that Freud's discussion of trauma and the traumatic nightmare in *Beyond the Pleasure Principle* and other texts tends toward the literal in this way, or that he interpreted the "situation" to which the dreamer returns in such literalist terms. A more extensive, historically informed, yet deconstructively inspired appraisal of a much wider range of Freud's writings than those discussed by Caruth, including Freud's discussions of mass psychology in *Group Psychology and the Analysis of the Ego* and *The Ego and the Id*, as well as his key text on anxiety, *Inhibitions, Symptoms and Anxiety*, and emphasizing the enigmatic status of affective or mimetic identification and hypnotic suggestion in Freud's work—a topic Caruth ignores—also raises questions about re-

10. Melvin R. Lansky with Carole R. Bley, *Posttraumatic Nightmares: Psychodynamic Explorations* (Hillsdale, New Jersey, 1995), 7–8. But the authors also go on to emphasize Freud's subsequent changes of mind on this issue.

pression theory without any recourse to the notion of the literal. On the contrary, such an appraisal underscores the fictive-fantasmatic-suggestive dimension of the traumatico-mimetic repetition. Nor does such a critical-genealogical discussion of trauma involve a rejection or critique of the notion of representation as such. Rather, Freud and his contemporaries are understood as having oscillated between two competing notions of representation: one defined in terms of a "representative theatricality" that underscores notions of specular self-distantiation and conscious recollection in trauma; the other defined in terms of an originary and affective "mimesis" or identification that emphasizes notions of nonspecular absorption in a traumatic scenario that is immemorial not because of a constitutive breakdown of all representation but because that scenario is unavailable for a certain *kind* of representation—theatrical self-representation—and hence for conscious memory.[11] As we have seen, hypnotic suggestion has functioned as a switch-point between those two notions of representation.

Another way of putting this is to say that Caruth is attempting not to provide a genealogy of the concept of psychic trauma but to use the notion of trauma as a *critical concept* in order to support her performative theory of language. For her project to succeed it is crucial for her to make the case for the literal in Freud, because the de Manian theory of the performative demands it. This becomes clear in her discussion of Freud's later work, *Moses and Monotheism* (1939). Her decision to focus on that notorious, and notoriously speculative, book is shrewd, because in it Freud attempts to explain the psychology and history of the group (specifically the history of the Jews) by way of the psychology of the individual in terms that might appear to conform to aspects of her performa-

---

11. See Mikkel Borch-Jacobsen, *The Freudian Subject* (Stanford, California, 1988), 39, hereafter abbreviated *FS*; idem, "The Unconscious, Nonetheless," in *The Emotional Tie: Psychoanalysis, Mimesis, and Affect* (Stanford, California, 1992), 123–54. Borch-Jacobsen's new claim that mimesis is always specular, even when most enacted or "surreal" (see chapter 5), does not change the basic point about representation. There are parallels between Caruth's ideas about trauma and Jean-François Lyotard's postmodernist attempt to appropriate Freud's concept of *Nachträglichkeit* in order to define it in terms of an initial moment of traumatic shock involving a pure affect that is experienced in an immediacy and immanence that lies beyond or outside representation and displacement—as if trauma can be conceptualized as an *agent provocateur* leading directly to the symptoms (Caruth cites Lyotard's work on affect in *UE*, 133, n. 8). Elizabeth Bellamy has recently deployed Borch-Jacobsen's discussion of affect to criticize Lyotard's ideas in her *Affective Genealogies: Psychoanalysis, Postmodernism, and the "Jewish Question" After Auschwitz* (Lincoln, Nebraska, 1997), 143–44.

tive theory of trauma. Furthermore, the actual writing of *Moses and Monotheism*, which was drafted during Freud's last years in Vienna and against the backdrop of the rising Nazi persecution of the Jews, is marked by the same compulsive tendency to repetition that Freud is trying to analyze at the level of theory, so that the text may seem to lend itself to Caruth's claim that it performs or "writes" the disaster of the Holocaust itself. Caruth is hardly alone in singling out *Moses and Monotheism* for scrutiny: in the work of Lyotard, Lacoue-Labarthe, Nancy, Derrida, and others the book has recently become a privileged site of investigation into the problematics of trauma, memory, and history so central to discussions of the Holocaust and the postmodern condition.

"Moses [was] an Egyptian."[12] With that apparently outrageous claim Freud launches an argument according to which the religion of monotheism was originally transmitted to the "culturally inferior immigrants," the Israelites, by the non-Jew Moses, an Egyptian prince or priest who had converted to monotheism and who led the Exodus of the Israelites out of Egypt, only to be subsequently murdered by them. The Jewish religion is thus founded on an act of violence, a rebellious slaying of a stranger, and this killing is described by Freud, in conformity with the argument of his earlier speculative-anthropological work, *Totem and Taboo* (1913), as itself a repetition of the killing of the leader or "Father" by the primal horde of "brothers" in the prehistory of the world. But according to Freud, the murder of Moses was destined to be disavowed or repressed and forgotten, with the result that the leaderless Israelites reverted to their primitive, polytheistic idol worship. About one hundred years later the same Israelites who had killed Moses came to a place where there were kindred Semitic tribes who had another leader, a Midianite priest also called Moses, and who worshiped a god called Yahweh. The two groups eventually banded together to form a unified Jewish people. Gradually, the guilt for the atrocious crime against Moses, the memory of which had been repressed in the first group's unconscious, began to increase and make itself felt. In the ensuing struggle between the Egyptian Mosaic and the Yahwistic religions, the Mosaic religion, after a period of several centuries, finally won out. In short, for Freud monotheistic Judaism developed only belatedly, by *Nachträglichkeit*, on the basis of the murder of a stranger whose violent death at the hands of the Israelites

---

12. I follow Caruth in citing two versions of the text: *Moses and Monotheism*, trans. Katherine Jones (New York, 1939), 15; and Freud's original text in *Standard Edition*, 23: 6. Hereafter both sets of page references will be given in the body of the text, using the abbreviations *MM* and *SE*.

was repressed and forgotten, only to return later in the form of a not-fully recognized atonement. According to Freud, the Christian religion is only another disguised repetition of the primal murder, the "redeemer" or Son assuming the tragic guilt of the leader of the primal or brother horde who originally overpowered the "Father" of prehistory and atoning for the sin of murder by the sacrifice of his life.

For Freud, a fundamental problem is how to explain the delayed effects of the murder of Moses on the Jews, that is, the transmission of a "tradition" that makes itself felt in the history of a people in an unconscious and repetitive manner. "How is such a delayed effect to be explained and where do we meet with similar phenomena?" he asks (*MM*, 82; *SE*, 23: 66). Among all the complex questions raised by Freud's work, this is Caruth's central concern also and it is on Freud's answer to this question that her interpretation hinges. After considering and rejecting certain examples of delayed transmission, Freud turns to that of the railway accident as closer to the problem at hand. In *Unclaimed Experience* Caruth argues that the railway accident is not "just any event" but "the exemplary scene of trauma *par excellence*, not only because it depicts what we can know about traumatizing events, but also, and more profoundly, because it tells of what it is, in traumatic events, that is *not* precisely grasped" (*UE*, 6). For Caruth, the railway accident is thus paradigmatic of Freud's treatment of the history of Jews *as* the history of a trauma. When Caruth uses Freud's example of the railway accident in her introduction to her first book *Trauma: Explorations in Memory*, she quotes the following passage:

> "It may happen that someone gets away, apparently unharmed, from the spot where he has suffered a shocking accident, for instance a train collision. In the course of the following weeks, however, he develops a series of grave psychical and motor symptoms, which can be ascribed only to his shock or whatever else happened at the time of the accident. He has developed a 'traumatic neurosis.' This appears quite incomprehensible and is therefore a novel fact. The time that elapsed between the accident and the first appearance of the symptoms is called the 'incubation period,' a transparent allusion to the pathology of infectious disease. . . . It is the feature one might term *latency*." (*TEM*, 7)

Caruth omits two sentences from this quotation. She leaves out the sentence with which the passage begins, in which Freud states that "The next example we turn to seems to have still less in common with our problem" (*MM*, 84; *SE*, 23: 67). And she also omits a sentence which in the original occupies the place indicated by Caruth's ellipses and which reads: "As an afterthought we observe that—in spite of the fundamental difference in the two cases, the problem of traumatic neurosis and that of Jewish monotheism—there is a correspondence in one point" (*MM*, 84; *SE*, 23: 67–68)—namely, <u>the correspondence in regard to latency</u>. This last sentence anticipates the argument Freud will go on to make in the next section to the effect that "[t]he only really satisfactory analogy to the remarkable process which we have recognized in the history of Jewish religion"—an analogy that "is very complete," "approximating to identity" (*MM*, 90; *SE*, 23: 72)—is not the trauma of the railway accident but the <u>etiology of the neuroses in the sexual-aggressive castrative "traumata"</u> of <u>early childhood</u>, their defensive repression, and the delayed return of the repressed.[13] On this basis Freud writes:

> Early trauma—defence—latency—outbreak of the neurosis—partial return of the repressed material: this was the formula we drew up for the development of a neurosis. Now I will invite the reader to take a step forward and assume that in <u>the history of the human species something happened similar to the events in the life of the individual</u>. That is to say, mankind as a whole also passed through conflicts of a sexual-aggressive nature, which left permanent traces, but which were for the most part warded off and forgotten; later, after a long period of latency, they came to life again and created phenomena similar in structure and tendency to neurotic symptoms. (*MM*, 101; *SE*, 23: 80)

Caruth's omissions here are not especially important—when she cites the same passage in her later book, *Unclaimed Experience*, she restores the second of the missing sentences.[14] But they are symptomatic of <u>her general rejection of Freud's Oedipal explanation of the neuroses</u>. Although she concedes that "the analogy with the Oedipal individual constitutes much of his explanation" for the history of the Jews (*UE*, 16), indeed that according to Freud historical memory, or at least Jewish historical mem-

---

13. In making this analogy—or "axiom" (*MM*, 91) or "postulate" (*SE*, 23: 72)—Freud will not attempt to decide whether the <u>etiology of the neuroses in general may be regarded as traumatic</u>. Instead, he appeals to the notion of a "complemental series" in order to avoid any hard-and-fast choice between exogenous or endogenous factors. The notion allows him to treat those two factors as complementary—the weaker the one, the stronger the other—so that any group of cases can in theory be distributed along a scale with the two types of factors varying in inverse ratio. Only at the limits of the series is there any question of a single factor being at work (*MM*, 92; *SE*, 23: 73). See Laplanche and Pontalis, *The Language of Psycho-Analysis* (Baltimore, 1976), s.v. "complemental series."

14. My thanks to Larissa MacFarquhar for drawing my attention to Caruth's omissions here and for a helpful discussion of their significance.

ory "is always a matter of distortion, a filtering of the original event through the fictions of traumatic repression, which makes the event available at best indirectly" (*UE*, 15–16), she finds in the example of the railway accident the "somewhat different understanding" of history that interests her (*UE*, 16):

> In his use of the term *latency*, the period during which the effects of the experience are not apparent, Freud seems to compare the accident to the successive movement in Jewish history from the event to its repression to its return. Yet what is truly striking about the accident victim's experience of the event . . . is not so much the period of forgetting that occurs after the accident, but rather the fact that the victim of the crash was never fully conscious during the accident itself: the person got away, Freud says, "apparently unharmed." The experience of trauma, the fact of latency, would thus seem to consist, not in the forgetting of a reality that can hence never be fully known, but in an inherent latency within the experience itself. . . . And it is this inherent latency of the event that paradoxically explains the peculiar, temporal structure, the belatedness, of the Jews' historical experience: since the murder is not experienced as it occurs, it is fully evident only in connection with another place, and in another time. (*UE*, 17)

As a reading of Freud there is much that is tendentious in these remarks. But they are typical of Caruth's interpretative practices, which involve not so much detailed readings of the texts under consideration as *thematizations* of them in terms of certain privileged figures or tropes. In this instance, by making the accident rather than the child's Oedipal story the model for the history of the Jews, Caruth decisively alters the terms of Freud's analysis. (The concept of the *accident*, defined in terms of contingency, unpredictability, unlocatability, and the unsettling of semantic expectations and meanings, is central to Felman's understanding of the crisis of representation held to characterize our "post-traumatic century"—see especially *T*, 22, where Felman cites Caruth.) Freud describes the experience of the Jews as a history of "what may properly be termed a traumatic experience" (*MM*, 65; *SE*, 23:52) but which he characterizes as a murderous "crime" (*MM*, 109; *SE*, 23, 86) and the guilt-ridden return of the repressed. Caruth rejects Freud's castration model of the trauma in order to thematize the same story as the story of Jewish victimhood—as the history of a murder that, incredibly, "is not experienced [by the perpetrators] as it occurs" (*UE*, 17), of an incomprehensible "missed" trauma that violently "separates" the Jews from Moses (*UE*, 69), of a traumatic "departure" (*UE*, 15), a survival, and a literal return—as if the Jews were victims and survivors of a completely unexpected, unintended, exogenous accident.[15]

Caruth's most glaring alteration in the terms of Freud's explanation of the history of the Jews involves the question of the literal. According to her, the return of the repressed Mosaic religion, like the emergence of symptoms from a train collision or from posttraumatic stress, occurs as the "literal return of the event against the will of the one it inhabits" (*TEM*, 5). She appears to understand the rise of Christianity as the belated return of the trauma of separation from Moses in the same terms. But Freud's account of the transmission of tradition is resistant to Caruth's interpretation. It is true that in *Moses and Monotheism* Freud expresses his desire to know the kernel of "historical truth" in the events he is describing (*MM*, 14, 71, 108; *SE*, 23: 16, 58, 85). But the question posed is: What does Freud mean by the "historical truth"? On the one hand, Freud seems to propose that we can extract the "historical truth" from the legends and other contradictory and distorted materials that are presented to the historian. Indeed, for the events concerning Moses himself he goes to considerable lengths to give concreteness to his arguments, as the historian Yerushalmi has shown, by making selective use of modern critical biblical scholarship.[16] On the other hand, Freud also seems to propose a new concept of historical truth—one that differs from "material truth" (*MM*, 166; *SE*, 23: 129)—when he accepts that knowledge of the past is available retrospectively only through distortion and falsification, a past, moreover, that according to his speculations in *Totem and Taboo* is fundamentally *a*historical, or at least *pre*historical, *primal*.

Thus in the second of two prefatory notes, written in London in June 1938, Freud characterizes his project in terms that may appear to corre-

---

15. Caruth thematizes Freud's famous discussion of the child's "fort-da" game in *Beyond the Pleasure Principle* in the same terms, namely, as "ultimately, and inexplicably, a game, simply, of departure" (*UE*, 66). On this basis, she treats the game as anticipating or bringing into view the concept of trauma as "departure" which she sees at work in *Moses and Monotheism*. She thus displaces the meaning of the mother's departure in the game onto the departure or difference inherent in it, which is to say onto the internal difference that makes trauma always already engage a "notion of history exceeding individual bounds" (*UE*, 66). For a different interpretation of the fort-da game, one that understands it as involving the mimetics of an incorporative identification that is indifferent with respect to pleasure and pain because it lies "beyond" the pleasure principle and "before" unpleasure, see Borch-Jacobsen, *FS*, 32–34.

16. Yosef Hayim Yerushalmi, *Freud's Moses: Judaism Terminable and Interminable* (New Haven and London, 1991).

spond to Caruth's account of the <u>unconscious force of tradition</u> in the history of the Jews: "I have never doubted that <u>religious phenomena</u> are to be understood only on the model of the neurotic symptoms of the individual, which are so familiar to us, as a return of <u>long-forgotten important happenings in the primeval history of the human family</u>, that they owe their obsessive character to that very origin and therefore derive their effect on mankind from the historical truth they contain" (*MM*, 71; *SE*, 23: 58). And in a subsequent discussion of the transmission of knowledge of Moses and his murder—a discussion Caruth ignores—Freud introduces a <u>distinction between written history and oral tradition</u> which also appears to imply that, unlike writing which injures or "murders" the historical record (*MM*, 52; *SE*, 23: 43), oral tradition preserves the historical truth as an undistorted or pristine recollection. But the actual terms of Freud's analysis are such that the distinction between written history and oral tradition cannot be sustained:

> A long time was to elapse . . . before historians came to develop an ideal of objective truth. At first they shaped their accounts according to their needs and tendencies of the moment, with an easy conscience, as if they had not yet understood what falsification signified. In consequence, a difference began to develop between the written version and the oral report—that is, the tradition—of the same subject-matter. [What has been deleted or altered in the written version might quite well have been preserved uninjured in the tradition.] Tradition was the complement and at the same time the contradiction of the written history. It was <u>less subject to distorting influences</u>—perhaps in part entirely free from them—and therefore might be more truthful than the account set down in writing. Its trustworthiness, however, was impaired by being vaguer and more fluid than the written text, being exposed to many changes and distortions as it was passed on from one generation to the next by word of mouth. (*MM*, 85–86; *SE*, 23: 68–69)

At once "in part entirely free" from the injurious distortions to which biblical exegesis or historiography is prey, *and* "vaguer and more fluid" because subject to "many changes and distortions" as it is passed on through the generations, <u>oral testimony or tradition is vulnerable to falsification</u> in ways that collapse the distinction between it and written history which Freud otherwise asserts.

Furthermore if, as Freud argues, [the tradition of monotheism is never completely lost but on the contrary grows more powerful in the course of centuries] and is eventually reproduced with a "faithfulness" that is lacking in imaginative literature (*MM*, 90; *SE*, 23: 72), nevertheless by analogy with childhood neurosis the repressed and forgotten past returns only in a disguised and displaced form. This is what Freud suggests when, in an implicit reference to the argument of his contemporaneous essay, "Constructions in Analysis" (1937), he compares the religion of monotheism or Christianity to the delusions of psychotics—delusions that contain a "piece of forgotten truth" but in an insane and garbled manner (*MM*, 107–8; *SE*, 23: 85). As he writes:

> The psychoanalyses of individuals have taught us that their earliest impressions, received at a time when they were hardly able to talk, manifest themselves later in an obsessive fashion, although those impressions are not consciously remembered. We feel that the same must hold good for the earliest experiences of mankind. One result of this is the emergence of the conception of one great God. It must be recognized as a memory—a distorted one, it is true, but nevertheless a memory. It has an obsessive quality; it simply must be believed. As far as its distortion goes, it may be called a delusion; in so far as it brings to light something from the past, it must be called truth. The psychiatric delusion also contains a particle of truth; the patient's conviction issues from this and extends to the whole delusional fabrication surrounding it. (*MM*, 167; *SE*, 23: 129–30)

It might be maintained that for Freud the force of this argument lies not so much in the declaration that the past returns in a distorted form than in the claim that even the extreme distortions or delusions of the psychotic contain a fragment of historical truth. We might relate this argument to Freud's life-long desire to ground "psychic reality" in an origin antecedent to or independent of that reality, an origin that can therefore be known directly.[17] At the same time—and this is the inherent tension in his thought—since for Freud the commencement is not absolute but is only accessible through the substitutions, displacements and falsifications imposed on it through *Nachträglichkeit* and the motive forces of repression, there can be no simple return to the origin, there is no escaping the uncertainties of analytic interpretation and construction: the origin or trauma does not present itself as a literal or material truth, as Caruth's theory demands, but as a psychical or "historical truth" whose meaning has to be interpreted, reconstructed, and deciphered.[18]

17. The classical discussion of this aspect of Freud's thought is Jean Laplanche and J.-B. Pontalis, "Fantasy and the Origin of Sexuality," *International Journal of Psychoanalysis* 49 (1968): 1–17.

18. Thus in "Constructions in Analysis" Freud is sure that "[a]ll of the essentials of [psychical history] are preserved; even things that seem completely forgotten are present

Moreover, what does Caruth make of the fact that in *Moses and Monotheism*, Freud admits that the archetrauma of history—the hypothesized primal murder of the "father" by the primal horde—might just as well have been a *fantasy*, and that the distinction between fantasy and reality is irrelevant to his argument? His position is congruent with the thesis of the Wolfman case in which, by arguing that the origin of neurosis is either a historical event or a *primal fantasy*, which is itself derived from an archetypical "truth" transcending individual experience, Freud at once elides and displaces the opposition between reality and fantasy in such a way that the status of the event as historical actuality cannot and need not be decided (*MM*, 110–11; *SE*, 23: 87). Indeed for Freud, at the origin there may well be a fundamental or originary dissimulation.[19]

In sum it is only on the basis of an extremely forced reading that Caruth can claim that Freud himself, in *Moses and Monotheism*, proposed a history of Jewish monotheism based on the analogy of the accident and involving a literal engraving of the mind by an incomprehensible reality that "continually returns, in its exactness, at a later time" (*TEM*, 153). I repeat: this is not at all to deny that the concept of trauma has a peculiar place in Freud's thought or that his texts betray manifold hesitations as to how to situate trauma within the psychic economy—hesitations which a psychoanalytically or deconstructively inspired or otherwise motivated commitment to understanding the "latent" content of his ideas might well seek to explore and interpret. But such an undertaking will not find the literal in the sense of Caruth.

---

somehow and somewhere, and have merely been buried and made inaccessible to the subject. Indeed, it may, as we know, be doubted whether any psychical structure can really be the victim of total destruction." Freud extends this argument to psychotic delusionals who, like the hysteric, also suffer from "their own reminiscences" (*SE*, 23: 260, 268) Yet, he adds: "I never intended by this short formula to dispute the complexity of the causation of the illness or to exclude the operation of many other factors" (*SE*, 23: 268). Moreover, owing to the intricacy of psychical structures, for Freud the question of the empirical truth of this "somehow and somewhere" is suspended in favor of attempting to discover the patient's "psychic truth."

19. This is the argument of Borch-Jacobsen who, on the basis of a close reading of Freud's aporetic arguments about the relationship between identification and desire, argues that, for Freud, at the origin of religion is a murderous and "blind" incorporative mimesis that has the structure of an originary dissimulation (*FS*, 48) and that makes the dominating male leader of the Darwinian tribe or horde the "Father" of religion and morals only by *Nachträglichkeit*, that is, by a retroactive act of deferred, identificatory, ambivalent and therefore guilt-inducing obedience to the dead man (Mikkel Borch-Jacobsen, *The Emotional Tie: Psychoanalysis, Mimesis, and Affect*, 15–35).

The Writing of Disaster

In this context, it is worth emphasizing that there is something else that Caruth disregards in Freud's text, an omission that is at first sight surprising in that rather than contradict her position it seems to support it. That omission concerns Freud's adherence to Lamarck's theory of the inheritance of acquired characters as a mechanism for explaining the transmission of traumatic experience across the generations. Inherent in Caruth's theory of trauma is the belief that the trauma experienced by one person can be passed on to others. The basic model for that transmission is the face-to-face encounter between a victim, who enacts or performs his or her traumatic experience, and a witness who listens and is in turn contaminated by the catastrophe. "The final import of the psychoanalytic and historical analysis of trauma is to suggest that the inherent departure, within trauma, from the moment of its first occurrence, is also a means of passing out of the isolation imposed by the event," Caruth writes: "that the history of a trauma, in its inherent belatedness, can only take place through the listening of another" (*TEM*, 10–11). It is a model of transmission that conforms to the claims of physicians such as van der Kolk and Herman that witnesses to the sufferings of others are vulnerable to the same dialectic of trauma as the victims themselves.

But Caruth expands that model to include the transmission of trauma across space and time, so that the trauma of one individual is understood as capable of haunting later generations—as if the ghosts of the past could speak to those living in the present, contagiously contaminating them in turn. As she states in connection with Freud's *Moses and Monotheism* and the "trauma" of the Jews: "Chosenness is thus not simply a fact of the past but the experience of being shot into a future that is not entirely one's own. The belated experience of trauma in Jewish monotheism suggests that history is not only the passing on of a crisis but also the passing on of a survival that can only be possessed within a history larger than any single individual or any single generation" (*UE*, 71). The result is that individuals or groups who never experienced the trauma directly themselves are imagined as "inheriting" the traumatic memories of those who died long ago. Felman has applied the same idea specifically to women by asserting that "every woman's life contains, explicitly or in implicit ways, the story of a trauma," including the story or symptoms of psychic traumas that have occurred "'entirely at second hand, as it were, through the mechanism of insidious trauma'" and that can be transmitted intergenerationally.[20] The group is thus imagined as having the same psychology as the individual, so that history itself can be conceptualized in traumatic terms. In short, for Caruth—as for many others today—history is collapsed into memory by redescribing, as Michaels has recently put it in his critical evaluation of such ideas, "something we have never known as something we have forgotten," thus making the historical past "a part of our own experience" ("YNT," 6).

Michaels is especially interested in the racial-identitarian stakes involved in conceptualizing history as memory in this way; the theory of the intergenerational transmission of trauma (or de Man's theory of the performative), he observes, "makes the Holocaust available as a continuing source of identitarian sustenance" because, by imagining that we who were not there can nevertheless be marked by the trauma of the Jews, it permits the emergence of "an explicitly anti-essentialist" and unreligious Jewish essence based on the possibility of "remembering" someone else's fate ("YNT," 12). Now a version of history as memory is not alien to Freud. For not only does Freud believe in the intergenerational transmission of experience, but he proposes a specific mechanism for such a transmission—a Lamarckian mechanism of inheritance whose notoriety makes it hard to ignore.[21] Thus in *Moses and Monotheism* he states that there is no difficulty in understanding how the tradition of monotheism was passed on by forebears who had been actual participants and eyewitnesses of its murderous origins. "Can we believe the same, however," he asks, "for the later centuries—namely, that the tradition was always based on a knowledge, communicated in a normal way, which had been transmitted from forebear to descendant?" (*MM*, 119; *SE*, 23:93). He does not think so, and argues that it becomes even harder to arrive at this conclu-

20. Shoshana Felman, *What Does a Woman Want? Reading and Sexual Difference* (Baltimore, 1993), 16, and n. 20, citing a paper by clinical psychologist Laura S. Brown, "Not Outside the Range: One Feminist Perspective on Psychic Trauma," published in Caruth's *Trauma: Explorations in Memory* (*TEM*, 100–112).

21. Freud's acceptance of Lamarckianism—or a version of Lamarckianism—has become even harder to ignore since the recent discovery and publication of the previously unknown draft of Freud's twelfth metapsychological paper, "Overview of the Transference Neuroses," dating from the years 1914–15, in which Lamarck's ideas play a prominent role. For a valuable discussion of the history of Lamarckianism in Freud's thought see Ilse Grubrich-Simitis's "Metapsychology and Metabiology: On Sigmund Freud's Draft Overview of the Transference Neuroses," in Sigmund Freud, *A Phylogenetic Fantasy: Overview of the Transference Neuroses*, ed. Ilse Grubrich-Simitis, trans. Axel Hoffer and Peter T. Hoffer (Cambridge, Massachusetts, 1987), 75–107.

sion when we consider the transmission of knowledge of the primal father and his fate in primitive times, in which the assumption of even an oral tradition is hard to justify (*MM*, 120; *SE*, 23: 94). "In what sense, therefore, can there be any question of a tradition? In what form could it have existed?" he again asks (*MM*, 120; *SE*, 23: 94), and gives as his answer: "I hold that the concordance between the individual and the mass in this point is almost complete. The masses, too, retain an impression of the past in unconscious memory traces" (*MM*, 120; *SE*, 23: 94). This is the moment in *Moses and Monotheism* when Freud famously adopts the theory of the inheritance of acquired "qualities" associated with the ideas of Lamarck, in spite of his acknowledgment that modern biology has rejected such a theory (*MM*, 127–29; *SE*, 23: 99–101).

Caruth also believes in the intergenerational transmission of trauma. Yet she does not discuss Freud's theory of the inheritance of acquired characters.[22] The reason for her omission is not hard to detect: she already has a mechanism that is capable of doing all the work that the Lamarckian technology does for Freud. If Caruth does not mention or need Freud's Lamarckian theories, this is because her de Manian version of that technology explains how *texts themselves* performatively achieve the same transformation of history into memory and its transmission across generations that Lamarkianism accomplishes for Freud.

This is evident in the section of *Unclaimed Experience* entitled "The Writing of Disaster"—a deliberate echo of Blanchot's famous book, *The Writing of the Disaster*—in which Caruth tries to demonstrate how texts achieve such a transmission by applying her ideas to *Moses and Monotheism*. Yet when we examine closely how this is supposed to work her argument depends, or appears to depend, on a sleight of hand. Caruth wants to establish that Freud's book is itself the "site of a trauma" (*UE*, 20), a trauma that occurred during its composition between 1934, when Freud first began to write it, and 1938, when he was forced by the threat of Nazi persecution to leave Vienna for exile in England. It is crucial to her argument to show that Freud's trauma is not represented as such in his text but manifests itself only in the aporias or unconsciousness of reference that she believes defines the traumatic experience. *Moses and Monotheism* ap-

22. She mentions only the work of Nicolas Abraham and Maria Torok on the intergenerational transmission of the phantom, because that theory seems to her to "intersect with the notion of intergenerational trauma" (*UE*, 136, n. 22). Her reference is to Nicolas Abraham's "Notes on the Phantom: Complement to Freud's Metapsychology," in Nicolas Abraham and Maria Torok, *The Shell and the Kernel*, ed. Nicholas T. Rand (Chicago, 1991), 1: 171–76.

peared in three separate parts, the last of which Freud withheld from publication until he moved to the safety of London. In two different prefatory notes appended to the beginning of Part III, as well as in a "Summary and Recapitulation" inserted in the middle of the same part, Freud explains the history of the writing and publication of his book. These are the materials around which Caruth's interpretation revolves.

Caruth starts by quoting a long passage from Freud's "Summary and Recapitulation." I cite the passage exactly as she gives it, including her ellipses:

> "The following part of this essay [the second section of part 3] cannot be sent forth into the world without lengthy explanations and apologies. For it is no other than a <u>faithful, often literal repetition of the first part</u>. . . . Why have I not avoided it? The answer to this question is . . . rather hard to admit. I have not been able to efface the traces of the unusual way in which this book came to be written.
>
> In truth it has been written twice over. The first time was a few years ago in Vienna, where I did not believe in the possibility of publishing it. I decided to put it away, but <u>it haunted me like an unlaid ghost</u>, and I compromised by publishing two parts of the book. . . . Then in March 1938 came the unexpected German invasion. It forced me to leave my home, but it also <u>freed me of the</u> fear lest <u>my publishing the book</u> might cause psychoanalysis to be forbidden in a country where its practice was still allowed. No sooner had I arrived in England than I found the temptation of making my withheld knowledge accessible to the world irresistible. . . . I could not make up my mind to relinquish the two former contributions altogether, and that is how the compromise came about of adding unaltered a whole piece of the first version to the second, a device which has the disadvantage of extensive repetition." (*UE*, 19–20)

Caruth omits nothing significant in this passage, but in conformity with the theory of the performative simply claims of it that "in spite of the temptation to lend an immediate referential meaning to Freud's trauma in the German invasion and Nazi persecution, it is not, in fact, precisely the *direct reference* to the German invasion that can be said to locate the actual trauma in Freud's passage"(*UE*, 20). This is because "the invasion is characterized, not in terms of its attendant persecution and threats, of which the Freud family did in fact have their share, but in terms of the somewhat different emphasis of a simple line: 'It forced me to leave my home, but it also freed me' [(Sie) zwang mich, die Heimat zu verlassen, befreite mich aber]. The trauma in Freud's text is first of all a trauma of leaving, the trauma of *verlassen*" (*UE*, 20–21). In short, Caruth focuses on the presence in Freud's "Summary and Recapitulation" of one of the literary figures, "leaving," that she associates with the breakdown of all representation and hence with the performative theory of trauma. She goes on to say that it is the word "leaving" that actually ties the "Summary and Recapitulation" to the "traumatic structuring" of the book, in its "implicit referral" to the two earlier prefatory notes, dated "Before March 1938" (while Freud was still in Vienna) and "In June 1938" (after he had settled in London) (*UE*, 21), in which he describes his reasons for not publishing his book and then his decision to let it appear. She quotes from the second prefatory note, reproduced here exactly as she gives it:

> "The exceptionally great difficulties which have weighed on me during the composition of this essay dealing with Moses . . . are the reason why this third and final part comes to have two different prefaces which contradict—indeed, even cancel—each other. For in the short interval between writing the two prefaces the outer conditions of the author have radically changed. Formerly I lived under the protection of the Catholic Church and feared that by publishing the essay I should lose that protection. . . . . Then, suddenly, the German invasion. . . . In the certainty of persecution . . . I left [*verliess ich*], with many friends, the city which from early childhood, through seventy-eight years, had been a home to me." (*UE*, 21)

On this basis she argues:

> The "interval" between the prefaces, which Freud explicitly notes, and which is also the literal space between "Before March 1938" and "In June 1938," also marks, implicitly, the space of a trauma, a trauma not simply *denoted* by the words "German invasion," but rather *borne* by the words *verliess ich*, "I left." Freud's writing preserves history precisely within this gap in his text; and within the words of his leaving, words that do not simply refer, but, through their repetition in the later "Summary and Recapitulation," convey the impact of a history precisely as what *cannot be grasped* about leaving. (*UE*, 21)

This gloss is characteristic of Caruth's interpretive practices. Her argument depends on the claim that Freud's second prefatory note preserves history only within the gap or aporia produced by words that do not simply refer to his trauma of exile but performatively convey it as something that cannot be grasped or represented. But her analysis depends on a ruse, because it involves the omission of the very words in the passage she cites that would appear to disprove her contention. I quote the passage again, this time with the elided words [in brackets] restored:

"The exceptionally great difficulties which have weighed on me during the composition of this essay dealing with Moses [—inner misgivings as well as external hindrances—] are the reason why this third and final part comes to have two different prefaces which contradict—indeed, even cancel—each other. For in the short interval between writing the two prefaces the outer conditions of the author have radically changed. Formerly I lived under the protection of the Catholic Church and feared that by publishing the essay I should lose that protection [and that the practitioners and students of psychoanalysis in Austria would be forbidden their work]. Then, suddenly, the German invasion [broke in on us and Catholicism proved to be, as the Bible has it, but 'a broken reed.'] In the certainty of persecution [—now not only because of my work, but also because of my 'race'—] I left, with many friends, the city which from early childhood, through seventy-eight years, had been a home to me." (*MM*, 69–70; *SE*, 23: 57)

It is at once apparent that in his allusion to the persecution that threatened him "now not only because of my work but also because of my 'race,'" Freud *does* yield to the temptation to lend a referential meaning to his flight from the Nazis by mentioning explicitly the racial persecutions and threats made against him. Caruth can only disavow this by refusing to quote the words in question. In short, she proceeds as if she cannot afford to quote everything Freud wrote, because the words she omits contradict her argument.

In another sense, what I have just said may not be quite fair—because for Caruth no direct reference, including the direct references of the expunged passages, could ever count as evidence against her position. For the terms in which she frames her argument—"not . . . precisely the direct reference to the German invasion," "a trauma not simply denoted by the words 'German invasion,'" "words that do not simply refer"—are all designed to allow her to maintain her argument about the failure of reference in the face of any counterargument based on the direct references found in the sentences in question. Even if the words in which Freud explicitly states his reasons for leaving Vienna had been left intact, Caruth would still argue that the trauma is "not simply *denoted*" by those words but rather borne by the words "I left" in order to maintain that Freud's trauma of "departure" was opaque to himself and also to the text itself, a text that therefore conveys the historical truth only in the very "unconsciousness of Freud's reference" to his leaving. I note the equation she makes in this connection between Freud's reference to the time interval between the composition of the first and second preface and the "literal space" between the dates of the two prefaces, a literal space that then performatively marks the "space" of the trauma itself. But of course on Caruth's approach *any* point of punctuation, including the absence of punctuation, can serve to prove the literal space of the aporia she is looking for—further evidence of the ideological basis of her claims.

There is more. We have seen that Caruth is committed to demonstrating that Freud's *Moses and Monotheism* performatively communicates the aporia or trauma of history across the generations by demonstrating how it transmits it to another reader living at another time and place—a reader who thereby gains access to the memory of someone else's trauma through the gaps of the linguistic-textual transmission. To this end she draws attention to another moment in Freud's "Summary and Recapitulation" when he says of his departure from Vienna: "It forced me to leave my home, but it also freed me of the fear lest my publishing the book might cause psychoanalysis to be forbidden in a country where its practice was still allowed" (*MM*, 132; *SE*, 23: 103). Caruth interprets this statement to mean not only that Freud now felt free to publish his work, which is what he says, but that he felt free, more specifically, to do so in London, that is, to "bring his voice to another place" (*UE*, 23). She finds confirmation for this interpretation in a letter Freud sent to his son Ernst, who was already in England and to whom he wrote in May 1938 when he was making his final preparations to leave Vienna: "'Two prospects keep me going in these grim times: to rejoin you all and—to die in freedom'" (*UE*, 23). Caruth ignores one of Freud's stated prospects—to rejoin his family—and focuses only on the second, "to die in freedom," which she interprets in terms of the figure of a "departure" and of a voice (or ghost) that in the departure of dying speaks to "us" readers beyond the grave: "Freud's freedom to leave is, paradoxically, the freedom not to live but to die: to bring forth his voice to others in dying. Freud's voice emerges, that is, as a departure. And it is this departure that, moreover, addresses us" (*UE*, 23). She observes in this regard that the last four words of Freud's letter were written in English, and it is in the movement between Freud's native German and English, the language of his new home, that she locates the moment of traumatic transmission:

> I would like to suggest that it is here, in the movement from German to English, in the rewriting of the departure within the languages of Freud's text, that we participate most fully in Freud's central insight, in *Moses and Monotheism*, that history, like trauma, is never simply one's own, that history is precisely the way we are implicated in each other's traumas. For we—whether as German- or as English-speaking readers—cannot read this sentence without,

ourselves, departing. In this departure, in the leave-taking of our hearing, we are first fully addressed by Freud's text, in ways we perhaps cannot yet fully understand. And, I would propose today, as we consider the possibilities of cultural and political analysis, that the impact of this not fully conscious address may be not only a valid but indeed a necessary point of departure. (*UE*, 23–24)

What is the force of this claim? Deploying an "us" and a "we" that refer not to a universal subject but to English- or German-speaking readers exclusively—as if Freud meant to bestow the legacy of psychoanalysis only on an English-speaking or a German-speaking people—Caruth suggests that "we" become the coowners of Freud's trauma because we cannot read his sentence divided between German and English without being split ourselves—without being traumatized or dissociated by the very trauma of "departure" that he himself experienced. Caruth is not interested in offering a psychoanalytic-psychological account of how someone else's trauma might be experienced as our own—for example, through a notion of identification (this would take work)—but in proposing a textual-linguistic explanation based on the idea that the linguistic division or aporia that marks or mars Freud's text is literally reproduced as division or aporia in us, his readers, so that we too necessarily participate in the traumatic dissociation of departure.

Moreover, since for Caruth what is transmitted by Freud's letter to his son is precisely the nonknowledge that constitutes the trauma itself, it is what "we perhaps cannot yet fully understand" in Freud's works that constitutes his legacy as a theorist of trauma. It is this claim that justifies Caruth's entire approach to psychoanalysis. For according to her, the task of the critic or scholar is not so much to try to understand Freud's theory of trauma historically or genealogically by studying the changes in his texts and perhaps by attempting to link those changes to the historical circumstances of their production.[23] It is rather to receive and transmit in turn—to be traumatically *possessed or haunted by*—the aporias of Freud's texts as the enigma of nonknowledge and survival that constitutes trauma as such. If, as Caruth argues, trauma theory is one of the places in which the survival of the Freudian tradition is taking place "in the assuredness of its transformation and appropriation by psychiatry" (*UE*, 72)—presumably a reference to the work of van der Kolk and others—that survival is also occurring in the "creative uncertainties" of this same theory

---

23. Caruth does not deny that a historical approach to Freud is worthy of pursuit (*UE*, 121, n. 17), but that is not her project.

of trauma (*UE*, 72), uncertainties which ensure that the aporia will, to use Caruth's terminology, precisely, survive.

The Parable of the Wound

I am near the end of my analysis, but before I stop I want to comment on a moment in *Unclaimed Experience* that epitomizes the hazards of Caruth's approach to trauma. The moment occurs right at the start of her book when she discusses the parable, or moral, involved in one of the examples Freud uses to exemplify the repetition compulsion. In the third section of *Beyond the Pleasure Principle* Freud turns to literature in order to illustrate the existence in normal persons, not just neurotics, of a "daemonic" compulsion to repeat that appears to lie "beyond" the pleasure principle and hence to necessitate a radical revision of his theory of the psychic economy. What especially interests Freud in the chosen text is that it seems to demonstrate a "'perpetual recurrence of the same thing'" which does not relate to the active behavior on the part of the person concerned, but to a passive experience over which the subject "has no influence, but in which he meets with a repetition of the same fatality." In Freud's summary statement:

> The most moving poetic picture of a fate such as this is given by Tasso in his romantic epic *Gerusalemme Liberata*. Its hero, Tancred, unwittingly kills his beloved Clorinda in a duel while she is disguised in the armour of an enemy knight. After her burial he makes his way into a strange magic forest which strikes the Crusaders' army with terror. He slashes with his sword at a tall tree; but blood streams from the cut and the voice of Clorinda, whose soul is imprisoned in the tree, is heard complaining that he has wounded his beloved once again. (*SE*, 18: 22)

Caruth cites this passage at the beginning of *Unclaimed Experience* because it seems to her that the story of Tancred "can be read . . . as a larger parable, both of the unarticulated implications of the theory of trauma in Freud's writings and, beyond that, of the crucial link between literature and theory that the following pages set out to explore" (*UE*, 3). What exactly is the parable offered by Tasso's poem according to Caruth? In her initial comments on the passage from Freud's text Caruth states that Tancred's wounding of Clorinda represents in Freud's text "the way that the experience of a trauma repeats itself, exactly and unremittingly, through the unknowing acts of the survivor and against his very will." According to her, Tasso's story therefore dramatizes the "repetition at the heart of catastrophe" such that the experience that Freud will call traumatic neurosis "emerges as the unwitting reenactment of an event that one cannot simply leave behind" (*UE*, 2).

For Caruth, in other words, Tancred is the victim of a traumatic neurosis. He is such a victim not only because he experiences a catastrophe for which he is not responsible, but because his murderous act is experienced by him "too soon, too unexpectedly, to be fully known" and is therefore "not available to consciousness until it imposes itself again, repeatedly, in the nightmares and repetitive actions of the survivor" (*UE*, 4). As Caruth puts it: "Just as Tancred does not hear the voice of Clorinda until the second wounding, so trauma is not locatable in the simple violent or original event in an individual's past, but rather in the way that its very unassimilated nature—the way it was precisely *not known* in the first instance—returns to haunt the survivor later on" (*UE*, 4). The claim that Tancred had neither knowledge nor consciousness of the "traumatic event" is crucial to Caruth's interpretation of him as a survivor of traumatic neurosis, just as an incapacity for consciousness and knowledge is a central feature of her performative theory of trauma. Tancred does not know what he has done because the catastrophe of the murder has occurred too soon to be assimilated by him, which is why he cannot represent it but can only repeat it "exactly and unremittingly" in the mode of a traumatic repetition.

Caruth's claims are weird for more than one reason. First, Freud does not cite the story of Tancred as an example of traumatic neurosis but as an example of the general tendency in even normal people to repeat unpleasurable experiences, and hence as an example of the repetition compulsion, or death drive. In the previous section of *Beyond the Pleasure Principle* Freud had discussed the illnesses that occur after railway disasters, other accidents, and the experience of battle as examples of traumatic neurosis, a neurosis whose symptoms, such as traumatic dreams, manifest a tendency to unpleasurable repetition. But he had soon abandoned the "dark and dismal subject of traumatic neurosis" (*SE*, 18: 14) in order to discuss other examples of compulsive repetition, such as the child's "fort-da" game and the story of Tancred and Clorinda—examples not obviously assimilable to the traumatic neuroses. It is only by begging numerous questions about the nature of the death drive that Caruth can incorporate Tancred's experience of being "pursued by a malignant fate or possessed by some 'daemonic power'" (*SE*, 18: 21) into the diagnosis of traumatic neurosis—bearing in mind that for Caruth traumatic neurosis involves the imposition of an unpleasurable "event" or "outside" that has "gone inside without any mediation" (*UE*, 59). What event, with respect to

Freud's summary of the story of Tancred, would count as the literal "out-side" which breaks through the mind's protective shield in this way?[24]

Second, it is not true that Tancred's killing of Clorinda is "unavailable to consciousness" until it imposes itself again in the form of a murderous repetition. Neither Tasso nor Freud makes the claim that Tancred is unconscious or unaware of having killed his beloved the first time. He "unwittingly" murders Clorinda (Freud uses the term *unwissentlich* = "unknowingly") in the sense that he does not intend to kill Clorinda, even if he does intend to kill the enemy she pretends to be. But in the poem he recognizes the dying Clorinda the moment he uncovers her face, and al-though he is described as overcome with such horror at the sight that he becomes delirious and appears to his companions almost dead himself, on coming to his senses he is overwhelmed with grief, self-reproach, and guilt at the knowledge of what he has done and arranges a burial for her.[25] Third, the repetition of the murderous act is not literal and exact, but symbolic, metaphoric: it takes the form of a slashing or wounding of a tree that bleeds, a figure that, as Margaret Ferguson has shown, has a long literary history and whose deployment by Tasso is, Ferguson suggests, charged with symbolic significance, including Oedipal symbolic mean-ing.[26] Caruth must ignore these difficulties if she is to make the claim that Tancred is the unconscious victim of a traumatic neurosis the experience of which remains unavailable to his consciousness except in the form of an exact and unremitting repetition. Just as Caruth converts the Israelites who murdered Moses into passive victims of the trauma of an accidental "separation," so she converts Tancred into the victim of a trauma as well.

These problems point to still another difficulty for Caruth's interpre-tation, one she acknowledges and with which she tries to deal: it is not Tancred but *Clorinda* who is the indisputable victim of a wounding. To re-peat Freud's summary of Tasso's epic, as cited by Caruth: "He [Tancred] slashes with his sword at a tall tree; but blood streams from the cut and the

24. Caruth notes the "curious movement" in Freud's text from the example of combat neurosis to the death drive (*UE*, 66). What interests her is that this movement, thematized by her as a "departure" from the theme of traumatic neurosis, is the same "departure" that according to her is inherent in trauma itself. On this basis she assimilates the fort-da game, and by extension the story of Tancred and Clorinda, to the story of trauma *as* "departure," noting in this regard that by the end of the section in which Freud discusses it the fort-da game is "nevertheless determined as a kind of traumatic play" (*UE*, 134, 13).

25. Torquato Tasso, *Gerusalemme liberata* (Modena, 1991), Canto 12, verses 64–99.

26. Margaret W. Ferguson, *Trials of Desire: Renaissance Defenses of Poetry* (New Haven, 1983), 126–36.

voice of Clorinda, whose soul is imprisoned in the tree, is heard complaining that he has wounded his beloved once again." In short, Tancred is a murderer, albeit an involuntary one, and Clorinda is his victim twice over. Caruth knows and admits this, as when she concedes that "the wound that speaks is not precisely Tancred's own but the wound, the trauma, of another" (*UE*, 8) or refers to "the original wounding of Clorinda" (*UE*, 116, n. 8). Yet she is determined to identify Tancred as a victim of trauma, even though that identification creates problems of yet another kind. For Caruth is interested in a further aspect of Tancred's story, namely, the role Clorinda plays as witness to a trauma. When Caruth first discusses the question of testimony in Tasso's poem she presents Clorinda as the witness of something that Clorinda knows but Tancred does not:

> For what seems to me particularly striking in the example of Tasso is not just the unconscious act of the infliction of the injury and its inadvertent and unwished-for repetition, but the moving and sorrowful *voice* that cries out, a voice that is paradoxically released *through the wound*. Tancred does not only repeat his act but, in repeating it, he for the first time hears a voice that cries out to him to see what he has done. The voice of his beloved addresses him and, in this address, bears witness to the past he has unwittingly repeated. Tancred's story thus represents traumatic experience not only as the enigma of a human agent's repeated and unknowing acts but also as the enigma of the otherness of a human voice that cries out from the wound, a voice that witnesses a truth that Tancred himself cannot fully know. (*UE*, 2–3)

This formulation seems to imply that Clorinda is the witness of *Tancred's* traumatic experience, a witness who knows something—the "truth"—that the victim, Tancred, "cannot fully know." But there is an obvious difficulty with this interpretation in that Tasso and Freud—and even Caruth—all state that Clorinda testifies to her *own* wound. But if Clorinda testifies to her own wound then Tasso's poem cannot serve Caruth's purposes as an exemplary literary example of the performative theory of trauma, because according to that theory the victim of trauma is not capable of witnessing or representing anything of what she has experienced. How can these dilemmas be resolved?

Caruth's highly ingenious answer is to suggest that Clorinda's voice is not exactly her own voice but that of Tancred in the sense that *hers is the voice of the traumatized Tancred's dissociated second self or female "other"*:

> [W]hile the story of Tancred, the repeated thrusts of his unwitting sword and the suffering he recognizes through the voice he hears, represents the experi-

ence of an individual traumatized by his own past—the repetition of his own trauma as it shapes his life—the wound that speaks is not precisely Tancred's own but the wound, the trauma, of another [this is the moment when Caruth acknowledges that *Clorinda* is the victim of a wound]. It is possible, of course, to understand that other voice, the voice of Clorinda, within the parable of the example, to represent the other within the self that retains the memory of the "unwitting" traumatic events of one's past [this is the moment when *Tancred* is reinstated as a trauma victim]. (*UE*, 8)

In other words, Clorinda witnesses Tancred's trauma by virtue of the fact that she is Tancred's internal "alter" who retains the memory of the traumatic experience, a memory that Tancred himself lacks. The advantage of this amazingly resourceful though textually unsupported suggestion is that it not only acknowledges Clorinda's status as victim, as the poem obliges Caruth to do, but it also satisfies Caruth's desire to make Tancred a victim too by incorporating Clorinda *into* Tancred as his second or hidden self who confesses the truth of what Tancred himself does not and cannot know. In short, Caruth resolves the dilemmas posed by her wish to make Tancred the victim of a trauma by presenting him as a *dual personality*, and she does so in terms that conform to both classical (Janetian) and modern theories of traumatic dissociation. (Yet the gender implications of her analysis remain unexplored. It is as if Caruth displaces Clorinda's gender divisions—a warrior maiden who dresses and fights like a man—onto Tancred himself.) Tancred-Clorinda thus comes to represent or figure difference as such, the difference or "other" within the self and language that for Caruth is always implicated in the traumatic scenario and that makes the trauma of one person always appertain or belong to someone else as well.

Caruth's analysis of Tancred-Clorinda as a dual personality expresses her primary commitment to making victimhood unlocatable in any particular person or place, thereby permitting it to migrate or spread contagiously to others. In this case, the primary importance of the contagion is that it allows her to imagine how the reader might be implicated in the trauma of others. That is why she moves so quickly from her proposal that we understand Tancred-Clorinda as a divided self to the suggestion that "we can also read the address of the voice here, not as the story of the individual in relation to the events of his own past, but as the story of the way in which one's own trauma is tied up with the trauma of another, the way in which trauma may lead, therefore, to the encounter with another, through the very possibility and surprise of listening to another's wound" (*UE*, 8). This suggestion epitomizes Caruth's theory of trauma in that it places all of *us* in the position of Tancred-Clorinda and then oscillates, in an unstable yet "exemplary" manner, between imagining oneself as the ineluctable *victim* of a trauma and imagining oneself as the listener to *someone else's* wound—which is to say that the suggestion epitomizes the "unlocatability of any particular traumatic experience" (*UE*, 134, n. 11). In relation to the Holocaust, it is as if she proposes that whether we experienced the trauma of the Holocaust directly or not, each of us, in the post-Holocaust period, is always already a split or dissociated subject, simultaneously victim and witness, and hence always marked by the difference and division that characterizes the traumatized subject.

But her discussion of Tasso's epic has even more chilling implications. For if, according to her analysis, the murderer Tancred can become the victim of the trauma and the voice of Clorinda testimony to *his* wound, then Caruth's logic would turn other perpetrators into victims too—for example, it would turn the executioners of the Jews into victims and the "cries" of the Jews into testimony to the trauma suffered by the Nazis. The *Oxford English Dictionary* defines "parable" as follows: "A fictitious narrative or allegory (usually something that might occur naturally), by which moral or spiritual relations are typically figured or set forth, as the parables of the New Testament." On Caruth's interpretation, what the parable of Tasso's story tells us is that not only can Tancred be considered the victim of a trauma but that even the Nazis are not exempt from the same dispensation.

Conclusion

The guiding thread or central interpretive theme of this book is that from the moment of its invention in the late nineteenth century the concept of trauma has been fundamentally unstable, balancing uneasily—indeed veering uncontrollably—between two ideas, theories, or paradigms:

1. The first or *mimetic* theory holds that trauma, or the experience of the traumatized subject, can be understood as involving a kind of hypnotic imitation or identification in which, precisely because the victim cannot recall the original traumatogenic event, she is fated to act it out or in other ways imitate it. The idea is that the traumatic experience in its sheer extremity, its affront to common norms and expectations, shatters or disables the victim's cognitive and perceptual capacities so that the experience never becomes part of the ordinary memory system. The mimetic hypothesis depends on the traditional account of hypnosis as involving an altered state of consciousness, in that the amnesia held to be characteristic of traumatic shock is understood as analogous to or identical with posthypnotic forgetting. Not surprisingly, therefore, the actual practice of hypnosis has played a major role in attempts to treat victims of trauma, even as different theorizations of hypnosis have been taken to have different implications for the understanding of trauma as well.

An aspect of the mimetic theory that should be stressed is that it leads to doubts about the veracity of the victim's testimony: to the extent that the traumatic occurrence is considered never to have become part of the victim's ordinary memory, it is unclear how she can truthfully testify to what befell her. There is even a sense in which she cannot quite be said to have experienced the trauma in question (another way of putting this is to say that the victim is imagined as having been absent from the traumatic event). Moreover, because the victim is understood as traumatized into a state of imitative-hypnotic suggestibility, the mimetic theory cannot escape worrying about the question of hypnotic suggestion and the "fabrication" of more or less false memories. Finally, since the mimetic theory posits a moment of identification with the aggressor, the victim is imagined as incorporating and therefore sharing the feelings of hostility directed toward herself.

2. The second, or *antimimetic*, theory also tends to make imitation basic to the traumatic experience, but it understands imitation differently. The mimetic notion that the victim is hypnotically immersed in the scene of trauma is repudiated in favor of the antithetical idea that in hypnotic imitation the subject is essentially aloof from the traumatic experience, in the sense that she remains a spectator of the traumatic scene, which she can therefore see and represent to herself and others. The antimimetic theory is compatible with, and often gives way to, the idea that trauma is a purely external event that befalls a fully constituted subject; whatever the damage to the latter's psychical autonomy and integrity, there is in principle no problem of eventually remembering or otherwise recovering the event, though in practice the process of bringing this about may be long and tortuous. And in contrast to the mimetic theory's assumption of an identification with the aggressor, the antimimetic theory depicts violence as purely and simply an assault from without. This has the advantage of portraying the victim of trauma as in no way mimetically complicitous with the violence directed against her, even as the absence of complication as regards the reliability of her testimony shores up the notion of the unproblematic actuality of the traumatic event. No wonder, then, that many women's advocates today, such as Judith Herman, believe in the essential truth of the traumatic memory. The antimimetic hypothesis also lends itself to various positivist or scientistic understandings of trauma.

My goal in this book has been not so much to associate individual interpretations of trauma with one or other of the two theories, though to some extent that has been inevitable, as to show that from the turn of the century to the present there has been a continual oscillation between them, indeed that the interpenetration of one by the other or alternatively the collapse of one into the other has been recurrent and unstoppable. Put slightly differently, my claim is that the concept of trauma has been structured historically in such a way as simultaneously to invite resolution in favor of one pole or the other of the mimetic/antimimetic dichotomy and to resist and ultimately to defeat all such attempts at resolution. This is especially palpable in those instances where attempts have been made to formulate a resolutely antimimetic account of trauma by expunging all hint of mimesis, but it is equally true of mimetic accounts—Ferenczi's, for example—which characteristically find it necessary to imagine a moment in which the play of mimetic imitation is stabilized in and by a relationship of spectatorship between the victim and the aggressor. Each of my eight chapters focuses on various aspects of that fundamental structure. At the same time, I have made an effort not to impose too abstract a grid on my material, in the hope of conveying something of the historical context and precise texture of the different cruxes I consider. For that reason it may be helpful if I briefly review the individual chapters with a view to spelling out their respective relations to the central issue I have just summarized.

Chapter 1 began by engaging directly with Freud, whose writings exemplify both the attractions of mimesis and the perils—conceptual, theoretical, clinical—of yielding to that attraction to the extent of making it fundamental to an account of the human subject. Thus I argued that at the center of Freud's discussion of trauma is a mimetic theory that defines trauma as a situation of unconscious imitation or identification with the traumatic scene. Such identification, which occurs in a state akin to that of trance, renders the subject "blind" to or unaware of the traumatic origin. Consequently, the victim can never be made to remember the traumatic experience, as the cathartic cure would have him do, but can only repeat it in the immediacy of an acting out that is unrepresentable to the patient in the form of a narration of that event as *past*. But I also argued that, made anxious by the loss of identity implicit in the experience of the hypnotic trance, Freud sought to repudiate mimesis, redefining the effects of trauma as the product not of the subject's hypnotic absorption in the other but of the latter's own sexual desire. Freud's theory of the libidinal unconscious, which posits a subject capable of seeing and hence depicting to itself and others the objects of its fantasies and wishes, attempts to solve the problem of mimesis by redefining the relation between the subject and the other in antimimetic, spectatorial terms; the unconscious is understood as a stage on which the subject in its traumatic dreams or fantasies observes himself or herself performing the scene.

But far from giving a unitary account of trauma based on the rejection of mimesis, Freud simultaneously and contradictorily characterized the victim as capable of remembering and testifying to the traumatic experience *and* as prone to an inherent forgetting of it; described hypnosis as basically alien to the psychoanalytic cure *and* as the very model of the relation between subjects, hence as inseparable from the psychoanalytic project; portrayed the transference as a relationship that can in principle be brought to an end by transforming the traumatic origin into self-representation and narration *and* as consisting in a hypnotic rapport between patient and analyst that is inherently unrepresentable to the subject and undissolvable. All these contradictions arise from the same essential source, the structural inability of the two paradigms or theories to rigorously exclude the other.

Similar tensions and contradictions are found in Morton Prince's once-famous book on the Beauchamp case, *The Dissociation of a Personality*, which I examined in chapter 2. In that book Prince often appears to conceptualize the traumatized subject mimetically, as if the effect of trauma were to shatter the victim into a dissociated state of imitative suggestibility and abject submission to the aggressor's hypnotic will. On that mimetic model, the victim of trauma is imagined as so swept up in the traumatic scenario that she is unable to distance herself from what is happening to her and hence incapable subsequently of representing and narrating it to herself and others. But in a gesture designed to repel the mimetic idea, Prince also attributed the effects of trauma to the spontaneity of a subject who, under the impact of a decisive event, splits into component but hitherto latent "selves," each of whom is understood as comprising a distinct personality and at least one of whom turns out to be capable of reporting reliably on the traumatic event. The tension between these two ways of imagining the traumatized subject explains why Prince's text is obsessively preoccupied with scenes of intense, almost cinematic, specularity: more than anything else, it is the ability of his patient's dominant dissociated self, Sally, to *witness* various highly significant scenes that guarantees the independence of a spectatorial subject whose autonomy, on the mimetic paradigm, is otherwise constantly threatened.

The same conflict between mimesis and antimimesis controls the sexual thematics of Prince's text, with the result that his account of the subject's mimetically induced plasticity of gender roles is countered by an antimimetic concept of the self, conceived as an aggregate of preexisting male and female personalities, the distinction between the two being defined in altogether conventional, not to say misogynistic terms. To such an extent does Prince normalize gender roles that the female self is implicitly presented as the very essence of imitative plasticity, which is to say that an excluded mimesis returns to haunt Prince's text in a negative or disavowed form. (At the end of the chapter I noted the similarities between Prince's treatment of sexual difference and the accounts of gender found in recent interpretations of multiple personality.) Finally, even as Prince tried to deploy hypnosis as a scientifically neutral or objective mnemotechnology for uncovering the traumatic event, he showed him-

self troubled by the threat of confabulation posed by his suggestive practices. That is, his text betrays considerable skepticism about the reality of
a trauma that his suggestive therapy was supposed to uncover and validate.

In chapter 3, I turned to the problem of trauma in men at war so as to
examine a question that had already been central to the work of Freud
and Prince but that took on new urgency in World War I: How does hypnosis or abreaction cure? In a 1920 debate, the key issues of which have
remained alive to this day, British physicians who had revived Breuer's
and Freud's hypnotic-cathartic method to treat the shell-shocked soldier
asked whether the putative relief of symptoms depended on the patient
being made to recall and "synthesize" the traumatic experience, or
whether on the contrary what produced the beneficial effects was the discharge of emotions brought about by the hypnotic-suggestive therapy.
Whereas the first position tended to regard hypnosis antimimetically as
an objective instrument for helping the patient uncover and narrate a
memory that was held to be an accurate representation of the traumatic
origin, the second position was compatible with the mimetic view that
hypnosis was not an instrument of self-representation but worked instead to induce an emotionally charged, imitative acting-out that could
only be felt and experienced in the present.

Once again, my aim was less to identify various participants in the debate with one or other version of cure than to show the difficulties they
encountered in trying to sustain one or other version in its pure form. In
this connection I discussed key writings by the long-overlooked French
psychologist and hypnotist Pierre Janet, who has recently been praised as
an advocate of the view that the goal of therapy is to convert traumatic
memory into narration by hypnosis and other means; in actual fact Janet
not only recognized the therapeutic value of forgetting or altering the
past but also called into question the entire opposition between remembering and forgetting on which the psychotherapy of the trauma victim is
now largely thought to depend.

I pursued related themes in chapters 4 and 5 on the work of Freud's
brilliant (and brilliantly wayward) disciple Sándor Ferenczi, often cited
today by trauma theorists as a heroic, because wrongly slighted, figure
who was concerned with the reality of trauma at a time, the interwar
years, when physicians in general and psychoanalysts in particular had
turned their backs on the problem. In chapter 4 I also considered the
work of Abram Kardiner, who drew inspiration from Ferenczi in his own
work on chronic cases of war neurosis. In these chapters—in some re-

spects the most difficult in the book, hence also the most difficult to summarize adequately—I showed how Ferenczi's ideas about trauma, which were informed by his experience with war neurosis cases in World War I, embodied both mimetic and antimimetic accounts of imitation; the writings that result, especially the late *Clinical Diary*, demonstrate the extent to which those differing accounts complicated his efforts to master trauma by bringing the traumatic event into consciousness and narration. I showed too that Ferenczi was continually troubled by his therapeutic dependence on the trance state, which he recognized—in the teeth of a psychoanalytical orthodoxy that wished to banish suggestion once and for all—as an inescapable component of the psychoanalytic relationship between patient and analyst, yet sought to conceptualize as an objective instrument for the recovery of traumatic memories uncontaminated by suggestion and confabulation. Among the issues that arose in the course of my discussion of Ferenczi was the question of simulation or malingering; accordingly, I addressed the topic head on by examining critically certain recent writings by Mikkel Borch-Jacobsen, a contemporary philosopher whose previous work on Freud and mimesis, above all his pathbreaking book *The Freudian Subject*, helped shape my approach in the present book. Recently, however, Borch-Jacobsen's thought has taken a skeptical turn in a series of texts that treat hypnosis and hypnotic cures as a kind of simulated game or performance. While acknowledging the force and seriousness of his arguments, I also suggested that they are marked by the same mimetic-antimimetic tensions that have structured discussions of trauma from the start.

In chapter 6, I examined the work of the British psychiatrist William Sargant to raise questions about the use of drug-induced catharsis to treat soldiers suffering from combat fatigue or war neurosis in World War II. Again, my goal was to demonstrate how catharsis was at one and the same time conceptualized antimimetically as an impartial means for bringing traumatic memories into consciousness and narration *and* as a technique to mimetically induce actings-out that might faithfully represent the origin but might just as well take the form of fictive-suggestive performances. Typically, the tension between mimesis and antimimesis is palpable within the work of a single author. For example, at times Sargant appears to have accepted the widely held view that cure depended on getting the traumatized patient to remember memories of the past. Yet he also challenged that view by arguing that the cure depended on an intense, mimetically or hypnotically induced discharge of emotional scenes that *might* be authentic copies of the original but were frequently fictional in character. Indeed, at the limit Sargant conceived of catharsis as a means not for remembering but for liquidating the past along lines close to those adopted by Janet. I remarked that in recent years therapists have again revived the use of hypnosis to produce recollection and integration of the traumatic truth, even as their hypnotic techniques have led to a suspicion of mimetic confabulation that they are keen to allay.

Finally, in chapters 7 and 8 I examined the work of two post-Holocaust, post-Vietnam theorists of trauma, Bessel van der Kolk and Cathy Caruth, in whose writings mimetic and antimimetic conceptions of trauma are once more intertwined—but with a difference. What is new in their work, to begin with, is a fascination with, almost a relishing of, the currently modish idea that the domain of trauma is the unspeakable and unrepresentable. This allows Caruth, a literary critic, to assimilate trauma theory to a version of the "deconstructive" views on language and meaning of the literary critic and theorist Paul de Man. For van der Kolk, a physician, the same idea of the unspeakable and unrepresentable is expressed in the notion—shared by Caruth, who finds support in van der Kolk's work just as he does in hers—that the victim of trauma has been radically affected by an external event that has somehow imprinted itself literally (a key term) on or into the subject's mind and brain in such a way as to make the event inherently unsymbolizable and unrepresentable. At the same time, for both authors the traumatic memory haunts the victim in the form of a replay or "flashback" that exactly repeats the past event in all its literalness and immediacy.

Because traumatic memory is understood by van der Kolk and Caruth as above all visual, a scene, it may seem the antimimetic idea *par excellence*. But because it is imagined as other than a representation—this is the force of the notions of literalness and replay—it would also seem to be aligned with the mimetic pole of our basic opposition. However, as I argued, there is no warrant in the mimetic theory for their insistence—I am tempted to say, their tendentious claim—that the traumatic experience stands outside or beyond representation as such. Similarly, both authors may seem to embrace a version of mimesis when they emphasize the tendency of trauma to infect or contagiously influence others. In van der Kolk's work this takes the modest form of proposing that the therapist may be so affected by his patient's suffering that he or she comes to be traumatized in turn. In Caruth's work the topos of infection takes the more dramatic form of proposing that the trauma experienced by one generation can be contagiously or mimetically transmitted to ensuing generations, with the result that each of us can be imagined as receiving a trauma that we never directly experienced. But their respective accounts of how this infection operates—visually, specularly, with the impact of an incomprehensible external event—seems more closely aligned with the antimimetic idea.

So once again we find an unresolved tension, if not quite between mimesis and antimimesis, at any rate between sets of assumptions that bear a certain relation to those poles. One indication of that tension is the status of hypnosis in Caruth's and van der Kolk's work. Caruth simply avoids the question of hypnosis, suggested confabulation, and simulation. Van der Kolk confronts the issue of hypnosis directly by denying that false memories can be suggested; he even advocates the use of barbiturates and hypnosis to discover the truth of the traumatic past. Contradictorily, however, he also concedes that traumatized patients tend to be so suggestible that their memories may not be so literal after all.

The reader who has come this far will not need to be told that my discussion of Caruth and van der Kolk has a critical edge that is absent from my analysis of earlier figures. In the first place, I am dismayed by the low quality of van der Kolk's scientific work. Again and again, as I tried to show, there are slippages and inconsistencies in his arguments about the literal nature of traumatic memory, arguments that are inadequately supported by the empirical evidence he adduces. It is all the more necessary to insist on this because humanists tend to take his claims to scientific accuracy at face value. As for Caruth, I feel a similar impatience with the sloppiness of her theoretical arguments; in the name of close reading she produces interpretations that are so arbitrary, willful, and tendentious as to forfeit all claim to believability. Finally, I am unsympathetic to the way in which she tends to dilute and generalize the notion of trauma: in her account the experience (or nonexperience) of trauma is characterized as something that can be shared by victims and nonvictims alike, and the unbearable sufferings of the survivor as a pathos that can and must be appropriated by others.

What, then, are the implications of my book? I shall close by mentioning two. The first is that current debates over trauma are fated to end in an impasse, for the simple reason that they are the inescapable outcome of the mimetic-antimimetic oscillation that has determined the field of trauma studies throughout this century. The understandable but misplaced desire to *resolve* the mimetic-antimimetic oscillation means that discussions of trauma are characteristically polarized between competing positions each of which can be maintained in its exclusiveness only at the price of falling into contradiction or incoherence.

Elaine Showalter's entertaining but shallow *Hystories: Hysterical Epidemics and Modern Media* (1997) typifies this state of affairs: purporting to throw critical light on current debates about trauma and hysteria, it is better understood as a work that simply replicates the tensions inherent in the mimetic-antimimetic conflict. Thus Showalter accepts the skeptical condemnations of Freud's seduction theory by Frederick Crews, Richard Webster, and others, who regard the claims of Freud's patients that they had been sexually traumatized as a product of Freud's imperious suggestions. She also criticizes the credulity of those feminists, such as Herman, who are unable to recognize the suggested nature of many recovered memories. (As she observes, with regard to recovered memories and charges of ritual abuse the tide seems to have turned, as courts are reversing decisions and releasing from jail persons who have been convicted of child abuse; this is largely because of the suggested-tainted nature of the testimony involved.) More broadly, Showalter regards multiple personality disorder, Gulf War syndrome, chronic fatigue syndrome, and the belief in recovered memory, alien abduction, and satanic ritual abuse as modern psychological epidemics all of which have been produced by the interaction between American political paranoia, religious fundamentalism, the suggested-contagious effects of hypnotic therapies, and a rumor-mongering mass media.

At the same time, Showalter wants to defend Freud's insights about the unconscious against a reigning skepticism, maintaining that hysteria is not merely an effect of suggestion but a projection of the subject's repressed sexual fantasies and desires. What she fails to do, however, is take seriously the relationship *between* suggestion and desire, although the question of their interconnection haunts Freud's entire oeuvre in ways that made it impossible for him ever to separate the two. It is as if Showalter wishes antimimetically to purge the world in general, and psychoanalysis in particular, of the entire problem of mimesis and suggestion, so that hysterical symptoms might appear "purely," or say naturally, rather than having been contagiously appropriated by paranoids and the media. But mimesis returns to disquiet her argument when she silently makes a distinction between *two kinds* of suggestion, the bad suggestion of the media that causes hysterical plagues, and the good suggestion of scholars like herself who ought to use the same media in order to *counter*suggest the pernicious, mimetic-infectious chatter. In these and other respects, Showalter reproduces rather than critically analyzes the basic problems posed by trauma throughout this century.

By now it should not need stating that as a historian or genealogist of trauma my project has been to reveal and investigate the tensions inherent in the mimetic-antimimetic structure, without for a moment attempting to settle those tensions or—except in the most general way—to take sides. (I cannot pretend, for example, not to have expressed a respect for Freud, or for that matter a sympathy for a thinker like Ferenczi, both of whom grappled tenaciously with intractable material.) By the same token, the argument of my book does not yield a meta-position from which to assess the messy and intrinsically painful conundrums of the field. One example of a painful conundrum: in the first legal test of the movement to make rape an indictable war crime, lawyers for the accused were able to exploit the fact the Muslim victim of repeated rape was diagnosed as suffering from PTSD by demanding a halt to the trial on the grounds that the traumatized woman's suggestibility made her testimony and memories suspect. In this instance, the three-judge panel of the United Nations tribunal charged with trying the case reopened the lawsuit so that the victim could be cross-examined, and eventually found against the accused.[1]

A second implication of my book is simply this: if it is true that the entire discourse on trauma in the West has been structured by an unresolvable tension or conflict between mimesis and antimimesis, then it would be a mistake for therapists to think that treatment for the victims of trauma should follow theory in some direct way, because that theory will continue to be subject to the alternations and contradictions inherent in the mimetic-antimimetic structure. Another way of putting this is to say that to the extent that my account of the genealogy of trauma is found persuasive, it would seem to follow that the soundest basis for a therapeutic practice would be an intelligent, humane, and resourceful pragmatism. The best practitioners have always been pragmatic in this sense, making use of whatever psychotherapeutic, medical, and other methods are available to help their patients (including in certain cases the attempt to verify by documentation or other independent means the reality of past traumas) without worrying too much about the exact fit between practice and theory. My book suggests that one cannot ask for more.

1. See Marlise Simons, "Landmark Bosnia Rape Trial: A Legal Morass," *New York Times*, 29 July 1998; "Rape in Bosnia Is War Crime, UN Court Rules," *International Herald Tribune*, 11 December 1998.

INDEX

A

Abraham, Nicolas, 92n. 19, 286n

Abraham, Karl, 142

abreaction: artificial state, 209; and combat neuroses, 202, 207, 303; compared to hypnotic catharsis, 200, 213–15; effectiveness of, 216–28; Pavlovian approach, 208–11; Sargant's discovery of, 191–92; veridical memories, 205; vicissitudes of, 216–28. *See also* catharsis; hypnotic catharsis

"Abreaction Re-Evaluated" (van der Hart and Brown), 227

accident neurosis, 271, 277. *See also* railway disasters

acquired characters, inheritance of, 284, 285n. 21, 286

acting (stage), and hypnosis, 50–52

acting out, 26, 100, 165, 166, 168–69, 202

adolescence: in American girls, 63; homosexuality in, 61–62

Adolescence (Hall), 68

"Aetiology of Hysteria" (Freud), 20

affect, 96–100. *See also* emotion

age regression, 98, 146

aggressor: identification with, 8, 9, 32, 125, 135–36, 138, 158, 174n, 175, 194; and the maternal, 69–71. *See also* identification

alexithymia, 248

alters (multiple personalities), 46

American Psychiatric Association, 2, 230, 231–32, 244

amnesia, 9, 107, 113, 118, 119; and drug therapy, 52, 198, 212–13, 212n. 50, 215; hypnotic, 104, 166n, 172, 180; in multiple personalities, 133; and pathological reflexes, 209; as source of worry, 197; and traumatic shock, 299

amygdala, 256, 260

analytic honesty, importance of, 121–23

Anna O, 102, 103, 167, 168

anxiety, 27–28

Appel, John, 219

Archer, William, 50, 169n. 23

assimilation, of traumatic memory, 114–16, 117

autoerotic stage, 69

autognosis, 90, 198

automatic writing, 53, 57, 58

autonomic arousal, 2

autosymbolism, 148n

B

Babinski, Joseph, 4, 91, 173

barbiturates, 14, 198, 214, 215, 221; as truth serum, 206.

Bashkirtseff, Marie, 62

behaviorism, influence of, 165

Bellamy, Edward, *Dr. Heidenhoff's Process*, 107–8

Bennett, Robert S., 18

Bergson, Henri, 94, 98n. 28

Bernheim, Hippolyte, 49, 164

Beyond the Pleasure Principle (Freud), 23, 29, 143, 230, 272–74, 292–93